W9-AFK-112

Theory of Arithmetic

Theory of Arithmetic

THIRD EDITION

John A. Peterson
University of Montana

Joseph Hashisaki
Western Washington State College

John Wiley & Sons, Inc.
New York · London · Sydney · Toronto

Library of Congress Catalogue Card Number: 70-132855

ISBN 0-471-68320-5

Printed in the United States of America

10 9 8 7 6 5 4 3 2 1

Preface

The fine work of both the Mathematical Association of America and the National Council of Teachers of Mathematics through support from the National Science Foundation is beginning to be evidenced in the improved mathematics background of the students entering the colleges and universities. Because of this they are more receptive to mathematical concepts new to them, more critical and more open-minded. This revision is written for this new generation of prospective teachers with their greater sophistication and background.

The third edition of *Theory of Arithmetic* includes minor changes in the first three chapters. They are the introduction of new concepts, the deletion of less important material, the improvement of problem sets, and the rearrangement of material.

A major change has been made in Chapters 4 and 5. They now are combined to form a complete treatment of the whole numbers, including number systems in bases other than ten. Algorithms are introduced in bases other than ten to add interest and to provide motivation for their study.

Major changes have been made in the last three chapters. The material dealing with the rational number and real number systems has been reorganized and rewritten to provide better continuity and teachability. Topics from geometry have been revised and expanded to include topological concepts of the kind that are being taught in elementary schools in several countries of Europe today.

Other changes that were suggested by many interested users also have been made. As with the previous editions, the third edition appeals to plausible reason rather than to rigor.

The primary objective of this textbook remains unchanged. It is still the desire of the authors to provide the elementary teacher with the background necessary to teach mathematics in the elementary school. The curriculum in the elementary school is not restricted to arithmetic but includes other topics in mathematics as well. Accurate knowledge of the subject matter is still considered to be an essential prerequisite to good teaching.

Beginning with a short historical development which includes systems of numeration, the material covers an introduction to the language of sets and the fundamental concept of relations. The material on relations was extremely new to the elementary teachers at the time of the first edition. The stronger background of the present students permits a more formal development. The language of sets and the concept of relations are then used to develop both the algebraic and order properties of the systems of whole numbers, integers, rationals, and reals. The latter portion of the book is devoted to a treatment of geometry designed as background material for the elementary teacher made particularly appropriate by the inclusion of topics of motivational and historical interest.

By formulating precise definitions and by simple, but accurate, presentation of fundamental ideas, *Theory of Arithmetic* makes sophisticated concepts of modern mathematics accessible to the practicing and prospective teacher.

The problems and examples are a particularly significant feature of the revision. An adequate number of carefully selected problems have been thoroughly classroom tested and found to be challenging, informative, and helpful to the students. The problems have been purposely distributed throughout each chapter to encourage the students to check their comprehension before proceeding on to new concepts. This format is also well suited for homework assignments and for testing. Problem arrangement and selection permit the book to be used in a wide variety of classroom situations ranging from small controlled sections to large lecture sections. Particular attention has been given to problems which stress basic understanding of fundamental concepts as well as problems to test proficiency in computational techniques.

The material is presented in sufficient detail so that it can be taught to students who have had the usual college preparatory mathematics. Chapters 2 through 7 include the key ideas in the development of this material and should serve as the core of any course for which this book is used as a text. Under the quarter system the first eight chapters, minus a few selected topics, have proven adequate for a five-hour one-quarter course. Time allotted to the topics including time for review and examination, has been essentially as follows:

Hours 1–4. Chapters 1 and 2, with most of the time spent on the language of sets.

Hours 5–10. Chapter 3, with a careful treatment of relations and their properties.

Hours 11–23. Chapter 4, introduction to operations and the first development of a number system with emphasis on the development of a complete understanding of the place-value system of numeration.

Hours 24–33. Chapter 5, extension of the concepts of the system of whole numbers to the set of integers with careful consideration of prime factorization, the division algorithm, the Euclidean algorithm, and related topics.

Hours 34–42. Chapter 6, with emphasis on understanding the rational number as an element of the rational number system, rational numbers as equivalence classes, the concept of denseness, and the usual interpretations of rational numbers.

Hours 43–48. Chapter 7, with emphasis on decimal approximations, the real numbers as infinite decimals, completeness, the real line, and the approximation of square roots.

By careful consideration of all topics of the first seven chapters and inclusion of Chapter 8 this could be extended to a three-hour, two-quarter sequence. Under the semester system, the material is adequate for a four- or five-hour, one-semester course.

In regard to the arrangement of the material in the text, each chapter consists of several sections and possibly subsections. These are identified by a number followed by a period, then a number or a combination of numbers and a letter. The first number identifies the chapter and the one following the period identifies the section or subsection. For example, the symbol 5.7 is used to identify Chapter 5, Section 7; the symbol 5.7a identifies the first subsection of that section. Definitions and exercises are numbered in a similar manner for easy reference. Definition 5.5a is the first definition of Section 5, Chapter 5. References to other material mentioned briefly in the context are listed completely in alphabetical order at the end of each chapter. Answers to selected exercises appear at the end of the book.

We wish to acknowledge the constructive criticism and suggestions received from the many mathematicians and mathematics educators who have used and/or reviewed the book. In particular, we wish to express our appreciation to Professor Roy Dubisch, University of Washington, for his critical examination and careful editing. Special acknowledgment should be given to the Committee on Educational Media Writing Group of the Mathematical Association of America and the National Council of Teachers of Mathematics Writing Group, both of which made use of the first edition of *Theory of Arithmetic* as a reference and source book. The experience gained from participating in these writing groups is no doubt reflected in this revision. We also wish to express our appreciation to the publishers for their helpful assistance and guidance and to the many students who over the years have assisted in the development of this material.

We have made every effort to incorporate the helpful suggestions of users in this revision and feel that the result is a much improved textbook. We will certainly appreciate suggestions and criticisms, as with the first edition, and hope they will be forthcoming as they were in the past.

JOHN A. PETERSON
JOSEPH HASHISAKI

January 1971

Contents

CHAPTER 1

The Origin of Numerals and Systems of Numeration

1.1 INTRODUCTION

This book treats *arithmetic* as a unification of several distinct concepts. In the beginning, the recognition and realization that there are *systems of numeration* for naming numbers, on the one hand, and *algebraic systems* that provide *structures* within which we operate, on the other hand, will prove helpful in understanding this way of presenting the theory of arithmetic. More important, it hopefully will help the prospective teacher to identify difficulties that the elementary student is having and to offer a clue as to how to help the students overcome their problems.

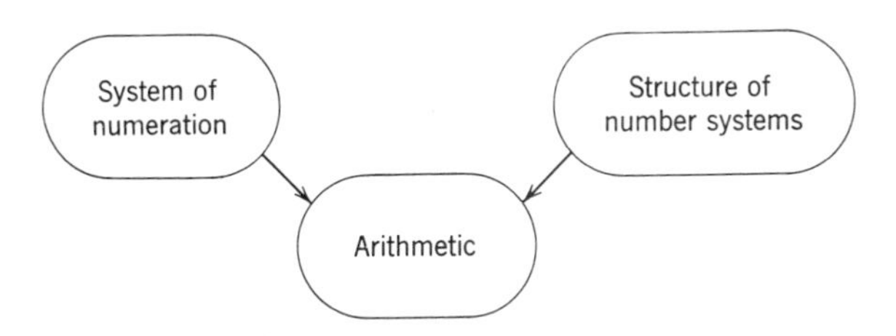

By a *system of numeration* we mean a *set of symbols that is used according to some established scheme for assigning numerals or number symbols to numbers.* In order to appreciate the significance and role of the system of numeration in arithmetic, try to multiply two numbers written in Roman numerals without converting to the decimal system.

MMMDCXCVII
× MCCCLXXIX

Indeed, what we do in a given computation does depend on what a particular set of number symbols means. On the other hand, what we are permitted to do in arithmetic is something like what we are permitted to do and are not permitted to do in such games as monopoly or baseball. Most games of this kind are played according to a set of rules, and it is the rules that determine the structure of the particular game.

Slow-pitch ball and baseball are different, but they are more alike than baseball and football. They have different structures and yet they have structures that are more alike than when compared to football. We shall discuss structure in terms of a *set of objects, things to do,* and *rules.*

1.2 SYSTEMS OF NUMERATION

Our study begins with a brief summary of the history of numerals and systems of numeration. These are like the alphabet for the arithmetician, and some knowledge of their origin leads to a better understanding of their use. We are also interested in other systems of numeration because the algorithms—the "how to do" of arithmetic computations—depend very much on the system used for naming numbers.

Statements about the origin of the concept of numbers must necessarily be conjectural. It seems logical to assume that man always had some intuitive notion of "more than" and "less than." In the development of civilization, the quantitative aspects of the environment dictated the development of some means of answering the question, "How many?" This can be done without numerals. Some sort of tally system is all that is needed to answer this question. The tally system might involve pebbles in a bag, sticks tied in a bundle, notches cut in a stick, knots tied in a rope, or marks in the sand. Whatever the type of device used to form the *reference set*, the principle involved is the same, namely, a matching between the objects being counted and the reference set. The set consisting of the fingers of the hands was, and for some purposes still is, the most convenient reference set. From this, man developed a set of words to use as a more convenient reference set for keeping track of "how many." The next step was from the oral words to written words, then to symbols and the development of *systems of numeration.*

A *system of numeration* is a set of symbols and a scheme for assigning numerals, or number symbols, to numbers. In order to understand better and appreciate our own system of numeration, we shall investigate some other systems that have been, or are now, used. We shall be interested not so much in the particular symbolism of each system as in the principles and concepts involved.

All systems of numeration have certain characteristics in common. The number of basic symbols is *finite*, varying from as few as two in some systems to thirty or more in others.

Example 1

The basic symbols in the Roman system of numeration are I, V, X, L, C, D, and M. The basic symbols in the system of numeration we use are 0, 1, 2, 3, 4, 5, 6, 7, 8, and 9.

Since the number of basic symbols must necessarily be finite and the number of numbers to be symbolized is infinite, it is necessary at times to use the same symbol more than once in the representation of a number. This is true with every system of numeration but, in some systems, the symbol may have a different meaning.

Every system possesses a symbol for the number "one." In some systems, subsequent numbers are written by repeated use of this symbol. The idea of *grouping* objects is essential to a systematic way of keeping track of things, and special symbols for these groups evolved into the idea of a *base*. This puts a bound on the number of times the one-symbol had to be used. In other systems distinct single-character symbols are used to represent subsequent numbers up to a certain number. Then a symbol of two or more characters is introduced.

In addition, there are underlying principles that play varying roles in different systems and certain concepts that occur in some and not in others. The concepts and principles are important and, once understood, allow us to invent new symbols and to construct other systems of numeration. These principles form our basis for classifying the systems of numeration. We consider several systems with these thoughts in mind.

Students have often asked if they are expected to memorize the symbols and their meanings in the systems of numeration used as illustrations in the following sections. This is not at all necessary. The numeration systems are chosen to illustrate underlying principles and concepts.

1.3 ADDITIVE SYSTEMS OF NUMERATION

Additive systems of numeration are characterized as those systems that rely primarily on the additive principle to determine the number represented by a given set of symbols. They have symbols for the number one, for the base, and powers of the base, and sometimes for multiples of powers of the base. The *repetitive principle*, that is, repeated use of the same symbol, is used in representing numbers between powers of the base. The number represented by a particular set of symbols is simply the sum of the numbers each symbol of the set represents. This is the *additive principle*.

To clarify these underlying principles, let us examine some examples.

1.3a. The Egyptian Hieroglyphic System

The Egyptian hieroglyphic system dates back as early as 3000 B.C. and was used for some 2000 years. Table 1 is a partial list of the symbols used, with the decimal equivalents and a word description.

Table 1 Egyptian Hieroglyphics

Hindu-Arabic or Decimal	Egyptian	Description
1	𓏺	A staff (vertical stroke)
10	𓎆	A heel bone (arch)
100	𓍢	A scroll (coiled rope)
1000	𓆼	A lotus flower
10,000	𓂭	A pointing finger
100,000	𓆐	A bourbot fish (tadpole)
1,000,000	𓁨	A man in astonishment

These symbols are used to represent a number in the same way that coins and bills are used to make up a given sum of money. The one-symbol is used repeatedly to represent the numbers one through nine. Essentially this amounts to constructing a representative set for a matching between the objects being counted and the repeated one-symbols. The other symbols represent a compounding of the previous symbols, and any number is expressed by using these symbols additively. Consider the following examples:

Example 1

𓆼𓆼𓍢𓎆𓎆𓏺𓏺𓏺𓏺 means $1000+1000+100+10+10+1+1+1+1$, which is 2124 as a decimal numeral.

Example 2

𓁨𓏺𓏺𓏺𓏺𓏺𓏺𓍢𓆼𓎆 means $1{,}000{,}000+1+1+1+1+1+1+100+1000+10$, which is 1,001,116 as a decimal numeral.

Example 3

𓎆𓎆𓎆𓎆𓎆𓎆𓎆𓎆𓏺𓏺𓏺𓏺𓏺𓏺𓏺 means $10+10+10+10+10+10+10+10+1+1+1+1+1+1$ $+1$, which is 87 as a decimal numeral.

Note that in a simply additive system of numeration such as this, the order in which the symbols appear is immaterial.

Addition and subtraction of numbers written in Egyptian numerals are as easy as making change. Multiplication and division must have been somewhat difficult with their cumbersome system, especially without pencil and paper. The Egyptians had methods for performing these operations. Also, fractions were understood and used by Egyptians (see Eves, pp. 39–40). We do not present the details of computation in the various systems here, for our interest is primarily in the properties of the systems of numeration in comparison with our own system.

To summarize, the Egyptian hieroglyphic system possesses the following properties. The symbols are hieroglyphs and as many as 45 of six different symbols are required to represent numbers to and including 100,000. The repetitive principle is used in representing numbers between one and the base, which is ten, and between powers of the base. The additive principle applies to any set of symbols to determine the number represented.

Exercise 1.3a

1. Express the following in Egyptian numerals:

(a) 77 (b) 629 (c) 90,909
(d) 2,507,916 (e) 2124 (f) 1,001,116
(g) 808

2. Write the decimal equivalent of the following Egyptian numerals:

(a) (c)

(b) (d)

3. Write in Egyptian numerals:

(a) the sum of (a) and (b) of problem 2.
(b) the sum of (a) and (c) of problem 2.
(c) the sum of (b) and (d) of problem 2.

4. Which of the numbers represented by the numerals of problem 2 is the largest and which the smallest?

1.3b The Roman System

The Roman numerals and system of numeration date from the time of the ancient Romans. These numerals were commonly used in bookkeeping in European countries until the eighteenth century, although the decimal numerals were generally known as early as 1000. The introduction of the printing press saw a rapid change in the use of the decimal numerals, although the Roman numerals continued to be used in some schools until about 1600, and still are to a limited extent. Examples are numerals on a clock face, chapter numbers in a book, and section numbers in an outline.

Much has been written about Roman notation. It is not our purpose to examine the system in detail but simply to look at some of the general characteristics (see Newman, pp. 447–449).

As now used, the symbols for Roman numerals are summarized in Table 2.

Table 2 Roman Numerals

Decimal	1	5	10	50	100	500	1000
Roman	I	V	X	L	C	D	M

The Roman system is also essentially an additive system of numeration in that a number designated by a set of symbols is simply the sum of the numbers represented by each of the symbols in the set.

Unlike the Egyptian system where the arrangement of the symbols had no special meaning, the Roman system uses the concept of order in its scheme. As with the decimal system Roman numerals are written with the symbol for the larger number to the left in any set of symbols which represents a number. Exceptions will be noted later.

A symbol with a bar above it indicates the number represented by that symbol multiplied by 1000. A double bar means multiplication by one thousand thousand or 1,000,000. This is an example of the *multiplicative principle.*

Example 1

MDCCCLXII = 1000 + 500 + 100 + 100 + 100 + 50 + 10 + 1 + 1.
= 1862 in the decimal system.

With so few symbols in an additive system, a great deal of repetition is necessary to express some large numbers. For example, the decimal numeral 387 is written CCCLXXXVII. How many symbols would be necessary to designate the number that in decimal numerals is written as 8888?

Present-day use of the Roman system also involves the *subtractive principle.* This principle states that if a symbol of a smaller number precedes a symbol of a larger number, the two are considered as a pair. The number represented by the pair is the larger number minus the smaller number. Today this principle is restricted to the numerals for four and nine, forty and ninety, four hundred and nine hundred, etc. Table 3 summarizes this property.

Table 3 Roman Numeral Subtractive Principle

IV = IIII	XL = XXXX	CD = CCCC
IX = VIIII	XC = LXXXX	CM = DCCCC

Any extension of its use to more than pairs of symbols could lead to ambiguous results; for example, IXC could be interpreted as $100-9$, or as $100-10-1$, resulting in a designation for 91 or 89, depending on the interpretation. Note that the subtractive principle depends on the *order* of the symbols.

In summary, the Roman system possesses the following properties. The symbols are, at present, letters of the alphabet and as many as 20 of the different symbols along with the multiplicative principle are required to represent numbers to and including 100,000. The repetitive principle is used in representing numbers between those for which distinct symbols are available. This system is what might be called a modified base ten system in that intermediate symbols are introduced for five, fifty, and five hundred. The additive principle applies, as does the subtractive principle. The multiplicative principle is also used in representing large numbers. The ordering of the symbols is an essential part of the scheme.

Exercise 1.3b

1. Express the following in Roman numerals:

(a) 26 (b) 39 (c) 49
(d) 342 (e) 431 (f) 449
(g) 1551 (h) 1961 (i) 2409

2. Express each of the following in decimal numerals:

(a) XXXVII (b) XLIX (c) XCIV
(d) CCCLXII (e) CDLVII (f) DCXLIV
(g) MCLI (h) MCMXLV (i) MMCMXCIX

3. State some advantages of the Roman system over the Egyptian system.

4. For numbers less than 1000, the addition facts for the Roman system are:

IIIII = V VV = X
XXXXX = L LL = C
CCCCC = D DD = M

(a) Add: MDCCCLXII + CXLIV
(*Note:* Here it is helpful to write CXLIV as CXXXXIIII.)
(b) Add: MDXII + DCVII
(c) Subtract: MCCVI − DCLXIII

5. (a) What follows MCMXLIX? (b) What follows MCM?

6. Which number is larger, MMCMXXIX or MMDCCXXIX?

7. Double each number in Exercise 2.

1.3c The Ionic Greek System

The Ionic Greek numeral system was also a system of the additive type but with a more complicated scheme involving many more symbols. It consisted of the 24 letters of the Greek alphabet plus three additional symbols for the obsolete digamma, sampi, and koppa. Initially capital letters were used, later the small letters. The system necessitated memorizing the set of symbols given in Table 4.

Table 4 Ionic Green Numerals

1 α alpha	10 ι iota	100 ρ rho
2 β beta	20 κ kappa	200 σ sigma
3 γ gamma	30 λ lambda	300 τ tau
4 δ delta	40 μ mu	400 υ upsilon
5 ϵ epsilon	50 ν nu	500 ϕ phi
6 obsolete digamma	60 ξ xi	600 χ chi
7 ζ zeta	70 o omicron	700 ψ psi
8 η eta	80 π pi	800 ω omega
9 θ theta	90 obsolete koppa	900 obsolete sampi

With this system numbers could be written in a more compact form, although still of the additive type, for example,

$$\lambda\gamma = 33 \qquad \chi\xi\epsilon = 665 \qquad \pi\eta = 88$$

For the multiples of 1000 the first nine symbols were used with a "prime," thus: $\alpha' = 1000$, $\beta' = 2000$, etc. For 10,000 M was used, and the multiplicative principle was applied for larger numbers; for example, $\beta M = 20{,}000$; $\zeta M\beta'\upsilon\nu\beta = 72{,}452$.

Note that there are advantages and disadvantages in the Ionic Greek system over the other systems we have discussed. The principal advantage is the economy of symbols. For numbers up to 1000 the number of symbols required to express a number is the same as in the decimal system of numeration. The principle disadvantages are that there are so many symbols to memorize and that they are letters of the alphabet and may be confused with words.

Characterization of this system includes the following facts. The symbols are letters of an early Greek alphabet, 27 in number, and as many as 6 of these along with the multiplicative principle are required to represent numbers up to and including 100,000. This is a base ten system, and the additive principle applies. It is noteworthy that the repetitive use of the symbols is not as marked in the Greek system of numeration compared to others, but this particular feature comes at the cost of having so many different symbols.

Exercise 1.3c

In the following, use *d* for digamma, *k* for koppa, and *s* for sampi.

1. Express the following decimal numerals in the Ionic Greek system:

(a) 36 (b) 39 (c) 49
(d) 342 (e) 431 (f) 449
(g) 1551 (h) 1961 (i) 2409

2. Express the following Ionic Greek numerals in decimal notation:

(a) $\mu\delta$ (b) $\pi\zeta$ (c) $\chi\nu\gamma$
(d) $\sigma\delta$ (e) $\rho o\beta$ (f) $\rho\iota\gamma$
(g) $d'\upsilon\lambda\epsilon$ (h) $\theta'\psi\kappa\epsilon$ (i) $\epsilon M\delta'\phi\xi\zeta$

3. State some advantages and disadvantages, in addition to those previously mentioned, of the Ionic Greek system as compared to the Roman and Egyptian systems of numeration.

4. What number follows (a) $\omega\pi\gamma$? (b) $\phi\mu\beta$?

5. Which is the largest and which the smallest in the following set?

$\{\chi\xi\delta, \quad \chi\pi\eta, \quad \chi\nu\delta\}$

6. Double each number in Exercise 2, (a), (d), (g) and write the result in Greek numerals.

1.4 MULTIPLICATIVE SYSTEMS OF NUMERATION

In these systems symbols are chosen for one, two, three, etc., up to the base, and another set chosen to represent powers of the base. These symbols are then used with the multiplicative and additive principles to represent any number.

1.4a The Chinese-Japanese System

The traditional Chinese-Japanese numeral system is of this type. A partial list of symbols with an example appears in Table 5.

Table 5 Chinese-Japanese Numerals

1	一	ichi	10	十	ju	2,345	
2	二	ni	100	百	hyaku	二	2000
3	三	san	1000	千	sen	千	
4	四	shi				三	300
5	五	go				百	
6	六	roku				四	40
7	七	shichi				十	
8	八	hachi				五	5
9	九	ku					

Exercise 1.4a

1. Express the following decimal numerals in the Japanese system:

(a) 42	(b) 54	(c) 36
(d) 125	(e) 246	(f) 782
(g) 2146	(h) 1984	(i) 5469

2. Express the following Japanese numerals as decimal numerals:

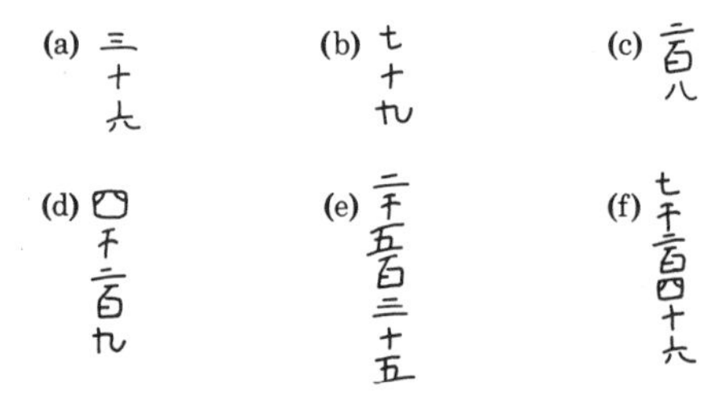

3. List some of the advantages and disadvantages of the Japanese system as compared to the systems previously mentioned.

4. Construct the tables of elementary facts for addition and multiplication for the Japanese system of numeration.

1.5 PLACE-VALUE SYSTEMS OF NUMERATION

This is the type of system of numeration with which we are most familiar, for "our" system, the decimal system with Hindu-Arabic symbols, is a place-value system. In this type of system, symbols are chosen for zero, one, two, etc., up to, but not including, the base. In the decimal system these symbols are referred to as *digits*. For a system of base "*b*" there are "*b*" such symbols. Then any number can be expressed uniquely as a sum of terms, each of which is one of the basic symbols times a *power* of the base. The power of the base by which each of the basic symbols is to be multiplied is determined by its placement in relation to a reference point. In the decimal system this is called the *decimal point*.

1.5a The Hindu-Arabic Symbols

The symbols we use in our present-day arithmetic are referred to as Hindu-Arabic—Hindu since they were probably originated by the Hindus, and Arabic because they came to Europe in the Arabic language. The earliest preserved examples of our present numerals are found on some stone columns in India dating from about 250 B.C. Other early examples are found among records cut about 100 B.C. on the walls of a cave in a hill near Poona, India, and in some inscriptions of about A.D. 200 carved in the caves at Nasik, India. These early examples contain no zero and do not employ place value. Place value, however, and also zero, must have been introduced before A.D. 800, for the Persian mathematician al-Khorârizmî describes such a completed system in a book dated

A.D. 825. Just how the new numerals were transmitted to Europe is not historically clear. Probably they were brought by traders and travelers. The Arabs invaded the Iberian Peninsula in A.D. 711 and, no doubt, introduced the new symbols to the Spaniards.

In a book written in Spain, dated A.D. 976, the following set of symbols was recorded:

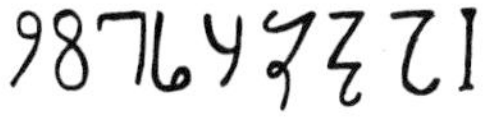

Changes in the shape of the symbols from the earliest known form to those in use today can be attributed primarily to the scribes who did the copy work. With the advent of the printing press at about the middle of the fifteenth century the symbols were fairly well standardized, only slight changes in form having occurred since then.

Consider the individual symbols: it is quite clear that the symbol for "one," 1, was a natural outgrowth of such things as one tally, one stick, etc. The symbol for "two" probably began as || or $=$. The former followed a rather natural evolution from || to N to ٢, which is the symbol for two used in the present-day Arabic language. Similarly, the latter symbol may have changed from $=$ to Z, and then to 2, which is the symbol we use. The symbol for "three" possibly changed from ||| to W to ٣ to μ, which is again the present-day symbol used in the Arabic language, or from $\equiv$ to ≷ to 3, our present-day symbol. To conjecture the origin of the symbol we use for "four" is much more difficult. The Arabic symbol for four, ٤, is the only symbol involving four joined straight lines for the number four. Little is known or conjectured about the origin of the other symbols. The table on page 454 of Newman, Vol. I, gives an interesting comparison of the shapes of the Hindu-Arabic symbols from the twelfth century to the advent of the printing press in the fifteenth.

1.5b Exponents

Since it is easier and more convenient to use exponential notation in discussing place-value systems, we review exponents briefly at this time.

You may recall that the idea of exponent is a convention in notation that we adopted by definition. Just as it is easier to write $5n$ instead of $n+n+n+n+n$, so we also agreed to write 10^3 instead of $10 \cdot 10 \cdot 10$. The numeral "3" is called the *exponent* and the numeral "10" is called the *base*. In general, b^n means $b \cdot b \cdot b \cdot b \ldots$ to n factors. The superscript "n" is called the *exponent* and the "b" is called the *base*. The whole symbol, "b^n," is called a *power* of the base, b.

For convenience we review some simple consequences of our convention of writing b^n instead of $b \cdot b \cdot b \cdot b \ldots$ to n factors. Rather than prove the results, which requires the use of mathematical induction and the associative law for multiplication, we simply state the results without proof and cite a few examples to make the general statements seem plausible. (In the following $a \neq 0$.)

1. $a^m \cdot a^n = a^{m+n}$.
2. $(a^m)^n = a^{m \cdot n}$.

If we define

$$a^0 = 1,$$

and

$$a^{-n} = \frac{1}{a^n},$$

then

$$3. \ \frac{a^m}{a^n} = a^{m-n}.$$

Example 1

$3^2 \cdot 3^3 = (3 \cdot 3)(3 \cdot 3 \cdot 3) = (3 \cdot 3 \cdot 3 \cdot 3 \cdot 3) = 3^5.$

Example 2

$(2^3)^2 = (2^3)(2^3) = (2 \cdot 2 \cdot 2)(2 \cdot 2 \cdot 2) = (2 \cdot 2 \cdot 2 \cdot 2 \cdot 2 \cdot 2) = 2^6.$

Example 3

Case 1. If m is greater than n,

$$\frac{a^5}{a^3} = \frac{a \cdot a \cdot a \cdot a \cdot a}{a \cdot a \cdot a} = a \cdot a = a^2 = a^{5-3}.$$

Case 2. If $m = n$,

$$\frac{a^3}{a^3} = \frac{a \cdot a \cdot a}{a \cdot a \cdot a} = 1 = a^{3-3} = a^0.$$

Case 3. If m is less than n,

$$\frac{a^2}{a^3} = \frac{a \cdot a}{a \cdot a \cdot a} = \frac{1}{a} = a^{2-3} = a^{-1}.$$

Cases 2 and 3 of Example 3 illustrate the desirability of the definitions that for any nonzero a, $a^0 = 1$ and $a^{-n} = 1/a^n$. For a discussion of the case where a is permitted to be 0 in a^0, see the article by Herbert E. Vaughan in *The Mathematics Teacher*, cited in the references at the end of this chapter.

Example 4

What is 10^{-1}? $\quad 10^{-1} = \frac{1}{10} = 0.1$

What is 4^{-2}? $\quad 4^{-2} = \frac{1}{4^2} = \frac{1}{16}$

Exercise 1.5b

1. Simplify each of the following:

(a) $10^3 \cdot 10^4$ (b) $(a^2)^3$
(c) $2^3 \cdot 2^2$ (d) $3^0 \cdot 3^2$

2. Find the value of:

(a) $10^3 \cdot 10^2$ (b) 2^{-2}
(c) 3^{-1} (d) 4^{-2}
(e) $2^{-3} \cdot 8$ (f) $10^x \cdot 10^2$

3. Find the value of:

(a) $(\frac{1}{2})^3$ (b) $(3^3)^3$
(c) $3^{(3^3)}$ (d) $(2^2)^2$
(e) $2^{(2^2)}$

1.5c The Decimal System

The decimal system uses the ten Hindu-Arabic symbols, 0, 1, 2, 3, 4, 5, 6, 7, 8, 9, and this includes a symbol for zero. It has base ten and is a place-value system. Any number may be expressed as a sequence of symbols and is interpreted as a sum of terms made up of these symbols times the appropriate power of ten. The power of ten is determined by the *position* or *place* the symbol occupies with reference to the decimal point. If the decimal point is omitted, as is usually the case with whole numbers, it is understood that the reference point is immediately to the right of the sequence of digits; for example, the symbols "241" and "241." represent the same number.

Table 6 is an abbreviated table of place values for numbers in the decimal system.

When we see the numeral 386 we recognize that the 6 is in the units position and the place it occupies becomes the reference position for determining the place values associated with the other digits. Another way of indicating the reference point or reference position is to put a dot

Table 6 Place Values

10^9	1,000,000,000	billions
10^8	100,000,000	hundred millions
10^7	10,000,000	ten millions
10^6	1,000,000	millions
10^5	100,000	hundred thousands
10^4	10,000	ten thousands
10^3	1000	thousands
10^2	100	hundreds
10^1	10	tens
10^0	1	units
10^{-1}	0.1	tenths
10^{-2}	0.01	hundredths
10^{-3}	0.001	thousandths
etc.		

immediately after the 6, that is, 386., which serves to identify the units position. The numerals "386" and "386." are names for the same number. The dot actually serves two purposes: the first as mentioned previously, is to indicate the units position, and the other will be discussed in Section 6.13. This dot is called the *decimal point.* It is used in the United States and England, but the English write it higher in the line of print than we do. In other European countries a comma is used instead.

A symbol such as 2145.67 is interpreted as

$$2145.67 = 2 \cdot 10^3 + 1 \cdot 10^2 + 4 \cdot 10^1 + 5 \cdot 10^0 + 6 \cdot 10^{-1} + 7 \cdot 10^{-2}.$$

The sum of multiples of powers of the base is called the *expanded form,* or *polynomial form*, of the numeral.

The chief advantage of the decimal system of numeration over the systems we have discussed previously is its economy of symbols and its adaptability to computation.

Exercise 1.5c

1. Write 1,020,304 in expanded form.

2. Write 12 in expanded form.

3. Write 10 in expanded form.

4. Write 10,000 in expanded form.

5. What is the meaning attached to the digit 3 in the numeral

345? 435? 453?

6. Simplify each of the following:

(a) $2^3 \cdot 2^6$ (b) $5^p \cdot 5^q$
(c) $(a^2)^4$ (d) $(x^3)^2 \cdot (x^4)^3$
(e) $\frac{m^7}{m^5}$, $m \neq 0$ (f) $b^{3x} \div b^x$, $b \neq 0$
(g) $x^2 \div x^6$, $x \neq 0$

7. Find the value of

(a) $10^3 \cdot 10^2$ (b) $3^2 \cdot 2^3$
(c) $2^3 - 2^0$ (d) $3^2 \cdot 6^{-2}$
(e) $(5^{-4})^{-1}$ (f) $5^{2x} \cdot 5^{-2x}$
(g) $8^{-1} \cdot 2^0$ (h) $a^{-3} \div a^0$, $a \neq 0$

1.5d Expanding and Reading Large Numbers

Example 1

$$56{,}146{,}929 = 5 \cdot 10^7 + 6 \cdot 10^6 + 1 \cdot 10^5 + 4 \cdot 10^4 + 6 \cdot 10^3 + 9 \cdot 10^2 + 2 \cdot 10^1 + 9 \cdot 10^0.$$

$$3{,}050{,}060{,}992 = 3 \cdot 10^9 + 0 \cdot 10^8 + 5 \cdot 10^7 + 0 \cdot 10^6 + 0 \cdot 10^5 + 6 \cdot 10^4 + 0 \cdot 10^3 + 9 \cdot 10^2 + 9 \cdot 10^1 + 2 \cdot 10^0.$$

The first number in Example 1 is read "fifty-six million, one hundred forty-six thousand, nine hundred twenty-nine." The commas are used to mark off groups of three digits. These groups of three digits are called *periods*. This grouping facilitates reading the numbers. Starting with the first group on the right and reading toward the left, we have hundreds, thousands, millions, billions, trillions, quadrillions, quintillions, and so forth.

Very large and very small numbers are usually written in what is called the *scientific notation*. In this notation the numbers are expressed as some number greater than or equal to one but less than ten, times the appropriate power of ten. For example, 1,000,000 could be written as $1 \cdot 10^6$; 23,000,000,000 is simply $2.3 \cdot 10^{10}$. The speed of light is approximately $3 \cdot 10^{10}$ cm/sec. This would be read "three times ten to the tenth centimeters per second."

The following are given as examples of physical constants presented in scientific notation:

Velocity of light	$2.99776 \cdot 10^{10}$ cm/sec
Avogadro number	$6.0228 \cdot 10^{23}$/mole
Velocity of sound	$3.3 \cdot 10^4$ cm/sec (approx.)
An Angstrom unit	10^{-8} cm
Constant of gravitation	$6.673 \cdot 10^{-8}$ dyne
Electronic charge	$4.803 \cdot 10^{-10}$ esu
Mass of electron	$9.107 \cdot 10^{-28}$ grams
Mass of hydrogen atom	$1.673 \cdot 10^{-24}$ grams

Exercise 1.5d

1. Write the following numbers in expanded form:

(a) 12 (b) 121
(c) 302 (d) 10,504
(e) 10,000 (f) 9090
(g) 11 (h) 1,001,001

2. Write the following in two different forms:

Example: $10^1 \cdot 10^2 = 10^{1+2} = 10^3$ or $10^1 \cdot 10^2 = 10 \cdot 100 = 1000$.

(a) $10^2 \cdot 10^3$ (b) $10^0 \cdot 10^0$
(c) $10^3 \cdot 10^7$ (d) $10^1 \cdot 10^1$
(e) $2^5 \cdot 2^2$ (f) $2^2 \cdot 2^2$
(g) $2^0 \cdot 2^1$ (h) $2^{10} \cdot 2^{10}$

3. Write the numbers given in problem 1 in

(a) the Roman system of numeration.
(b) the Egyptian system of numeration.
(c) the Ionic Greek system of numeration.

4. The sum of the digits of a two-digit number is 10. If the digits are interchanged, the number thus formed is 54 less than the original number. Find the number.

5. In a two-digit number, the units digit is three times the tens digit. If the digits are interchanged, the number is increased by 18. Find the number.

6. Carry out the following calculations in scientific notation:

(a) $(3 \cdot 10^5)(2 \cdot 10^7) = ?$
(b) $(8 \cdot 10^8) \div (2 \cdot 10^6) = ?$
(c) $(6 \cdot 10^{23})(3 \cdot 10^{-18}) = ?$
(d) Using the speed of light as approximately $3 \cdot 10^{10}$ cm/sec and one mile as approximately $1.6 \cdot 10^5$ cm, determine the speed of light in miles per second.

7. Write the following numbers using scientific notation:

(a) 0.0000065 (b) 0.000000087
(c) eight billionths (d) 5,700,000,000,000
(e) 6,000,000,000,000 (f) twenty-one trillionths

8. Write the following numbers in standard notation:

(a) $8.7 \cdot 10^9$ (b) $6.23 \cdot 10^{-6}$
(c) $8.7 \cdot 10^{-8}$ (d) $6.02 \cdot 10^{23}$
(e) $5 \cdot 10^{-9}$ (f) $1.08 \cdot 10^{-7}$

9. Multiply the following, expressing the product in scientific notation and in standard form:

(a) $(10^{-3})(10^5)$ (b) $(10^{-4})(10^{-3})$
(c) $(10^{-6})(10^6)$ (d) $(3.75 \cdot 10^{-5})(2.24 \cdot 10^6)$
(e) $(7.25 \cdot 10^5)(2.16 \cdot 10^{-8})$ (f) $(6.75 \cdot 10^8)(2.42 \cdot 10^{-5})$

1.6 THE COUNTING BOARD

Initially man knew but one use for numbers, namely, for counting objects. The development of addition, subtraction, and multiplication was gradual and, since the symbols were awkward to work with, special devices were invented to aid in computation.

The Romans used a counting table, or counting board, on which lines were drawn, each line representing units, tens, hundreds, and so on—with the space between lines representing fives, fifties, five hundreds, and so on. They tallied with small round discs. By placing these on the lines and between the lines they were able to register any number they pleased, and, by adding additional discs and simplifying, they were able to carry out addition problems. A schematic diagram of their counting board would appear somewhat like Figure 1. The number represented in the schematic diagram would be 2837 in decimal notation.

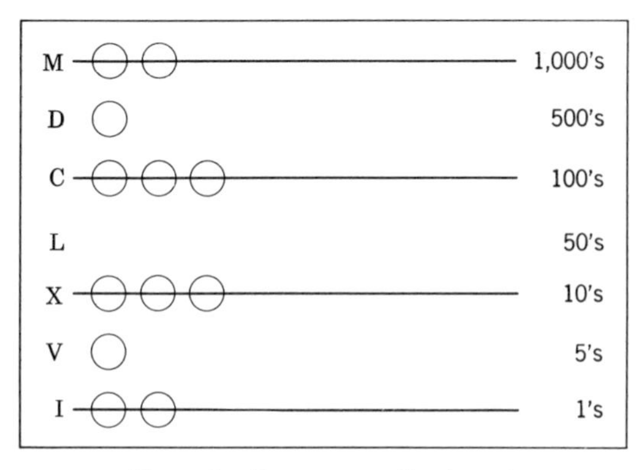

Figure 1. Roman counting board.

The Roman merchants displayed and sold their goods over these counting tables. The words *counting table* were shortened to *counter*. This is the origin of the word we employ for the fixtures in stores that are used to display goods.

It is interesting to note that in the use of their counting board the Romans essentially had a place-value system of numeration, but they did not recognize it as such.

1.7 THE ABACUS

The devices developed by various peoples of the world to aid in computation differed in appearance and, to some extent, in design. The most popular, however, appear to have the same underlying scheme and are classified as some form of the abacus. Basically the scheme was that lines, grooves, or rods were used to represent units and the powers of the base. Counters or beads were then placed on the lines, grooves, or rods to designate how many of each of the units and powers of the base were to be used in representing a number. These counters were not removed from the device. Instead their position indicated whether they were to be counted or not in the representation of a number.

The Roman abacus was a bronze or wood board. Grooves were cut in the board and small round pebbles placed in these grooves to represent numbers. These pebbles were called calculi, which is the plural of calculus. This tells us the origin of the word *calculate* and of the word *calculus* when it designates a mathematical discipline.

A schematic diagram of the Roman abacus is given in Figure 2. In this diagram a symbol with a bar above it indicates the number represented by that symbol multiplied by 1000. The double bar indicates multiplication by 1,000,000. Beads placed at the bottom of the groove, toward the operator, were in the neutral position and were not to be counted. Beads placed at the top of the groove were in the "active" position and were to be counted. The number, in our system of numeration, indicated by

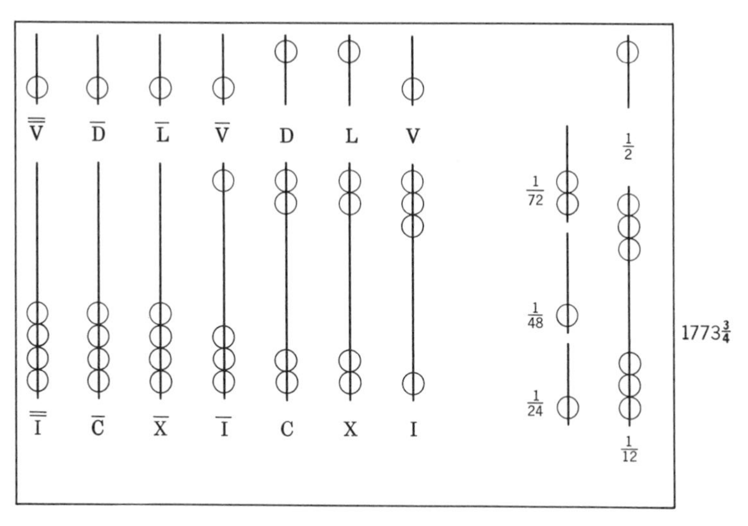

Figure 2. Roman abacus.

the position of the beads in Figure 2 would be

$$1000 + 500 + 200 + 50 + 20 + 3 + \tfrac{1}{2} + \tfrac{3}{12} = 1773\tfrac{3}{4}.$$

The abacus developed by the Chinese, called the *suan pan*, was of the rod and bead variety. A dividing bar separated sets of two beads and five beads on each bar. Each bead above the bar had associated with it a value five times that of a bead below the bar on the same rod. The active position for the beads was toward the dividing bar. Beads toward the outside were in the passive, or neutral, position. Figure 3 shows a typical

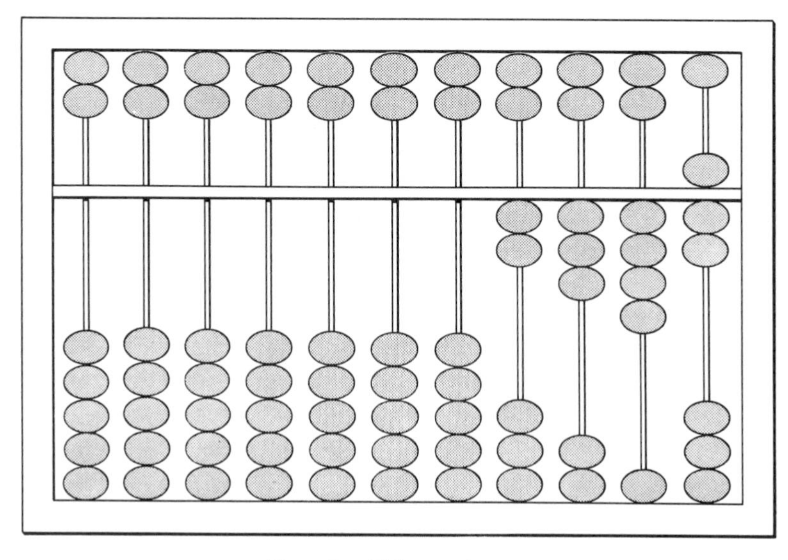

Figure 3. Chinese abacus.

arrangement for the *suan pan* with the number 2347 designated by the beads. It is interesting to note that the *suan pan* is still in general use by the Chinese.

The abacus developed by the Japanese, called the *soroban*, is very similar to that of the Chinese except that instead of the five-two bead arrangement on each rod, the Japanese model is generally of the four-one or five-one bead arrangement. Figure 4 shows a typical arrangement with the number 27,483 indicated.

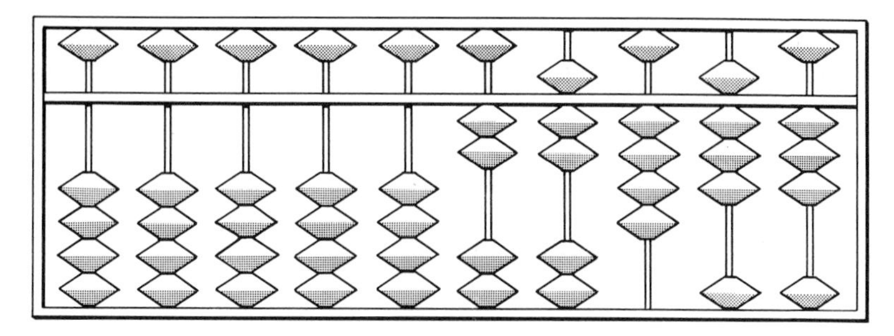

Figure 4. Japanese abacus.

Most of us, when we think of the abacus, associate it primarily with the Orient. Actually the abacus, in one form or another, was used throughout Europe and Asia until people became acquainted with and accepted the decimal system of numeration and associated methods of computation. This change did not take place rapidly. There were those who favored the use of other systems of numeration and the use of the abacus in computation. They were called the *abacists*. Opposed to these were the advocates of the decimal system with its algorithms, or procedures, for computation. They were called the *algorists*. It took approximately 500 years for the algorists to gain general acceptance of their techniques in computation. But by the year 1600 they had achieved their goal and had established the arithmetic techniques which have remained in general use up to the present time.

The abacus was then placed in a semiretired status, but this ancient device is finding its way back into the classrooms of today. Teachers are finding it helpful in the teaching of place value, addition, and subtraction. With the advancements made in recent years in the field of computers, it may be interesting to note that the abacus may be classified as one of the first "digital computers".

Exercise 1.7

1. Write 385 and 583 in

(a) Egyptian hieroglyphics (b) Traditional Japanese
(c) Ionic Greek

2. How many different symbols must one memorize in order to write numbers less than 1000 in

(a) Ionic Greek? (b) Roman?

3. Systems of numeration can be characterized to a certain extent by whether or not they possess the following properties:

(a) Additive (b) Symbol for zero
(c) Subtractive (d) Multiplicative
(e) Place value (f) Repetition

They may be further characterized by

(a) Number of symbols (b) Type of symbols
(c) Base

Give the characteristics of the systems studied in this chapter.

4. (a) In a place value system, base two, let 0 denote zero and 1 denote one. Write the first 25 numbers in this system.
(b) What are the decimal numerals of the numbers written as follows in the base two system?

10101, 10001000, 1110011, 1001101

5. What reason(s) would you give to the parents of a youngster to justify the teaching of number systems other than our own?

(a) They help in the understanding of our own system of numeration.
(b) A youngster should know how to compute in more than one system of numeration because it is required by state law.
(c) They help a youngster develop an appreciation of our own system of numeration.
(d) They show the youngster the advantages of our system of numeration.

6. Why are we studying the Egyptian and Greek systems of numeration?

7. Simplify:

(a) $(2^5 \cdot 3^4)(2^3 \cdot 3^2) =$
(b) $(2^3 \cdot 3^{-2})(2^{-4} \cdot 3^5) =$

8. What is the double of (a) 5000? (b) 2^{20}? (c) 10^6?

9. What is one half of (a) 2^{10}? (b) 10^{12}?

10. What is meant by *a system of numeration?*

11. All systems of numeration have certain characteristics in common. List a few.

12. Use the symbols I, □, △, and ⊡ to represent the decimal numbers 1, 4, 16, and 64 respectively, and let the symbols O, I, L, and F represent the decimal numbers 0, 1, 2, and 3. Use whatever symbols are necessary to write the decimal numbers 25, 100, and 197 in

(a) an additive system.
(b) a multiplicative system.
(c) a place-value system.

13. (a) In an additive system, base five, let 1, 5, 5^2, and 5^3 be represented by I, L, F, and E. Express the numbers 360, 252, 78, and 33 in this system.
(b) In a place-value system, base five, let 0, 1, 2, 3, and 4 be represented by O, I, L, F, and E. Express the numbers 360, 252, 78, and 33 in this system.

14. The great pyramid of Gizeh was erected about 2900 B.C. In order to appreciate some of the engineering and mathematical problems which had to be solved using the hieroglyphic numerals, write a report on the pyramids and, in particular, include some statistics on the largest of them.

REFERENCES

Eves, Howard, *An Introduction to the History of Mathematics*, revised edition, Holt, Rinehart, and Winston, New York, 1964.
Newman, James R., *The World of Mathematics*, Simon and Schuster, New York, 1956.
Vaughan, Herbert E. "The Case of 0°", *The Mathematics Teacher*, Vol. LXIII, No. 2, February 1970.

CHAPTER 2

Sets

2.1 INTRODUCTION

There is much in arithmetic that is clear and concise, simply and easy to understand, interesting and intriguing. At the same time, as it is usually taught, there is much that is not clearly understood by the teachers and difficult for them to explain and, unfortunately, widely accepted as being difficult to master. Many people would be reluctant to confess that they did poorly in English or history but readily admit that they never could do arithmetic. This attitude is due primarily to vague and incorrect presentation of fundamental concepts. We hope to promote understanding and interest by being reasonably precise in their presentation and by replacing vagueness with clarity. We begin with a fundamental notion in mathematics, the idea of a set.

2.2 SETS

In everyday life we use such words as collection, class, group, set:

a collection of stamps or coins
a set of dishes
a group of boys
the class of '67

These words are used intuitively and freely without any thought of defining them. They are used synonymously, and the same word may be used in a variety of situations. It is in this spirit that the mathematician uses the word "set." The word *set* will be used to denote a collection of objects that we shall call *elements* of the set.

Example 1

In a set of crayons, each crayon is an *element* of the *set* of crayons. In a set of dishes, each dish is an *element* of the *set* of dishes. In a set of positive integers, each integer is an *element* of the *set* of positive integers. In a set of ideas, each idea is an *element* of the *set* of ideas.

The notion of a set of elements is a creation of the mind (an idea). The mind unconsciously organizes objects into sets. The process starts at a very early age. A child is shown a picture of a horse or sees a horse, and when the animal has been given the name *horse*, all similar animals are readily distinguished and identified as horses. The word *horses* then can be thought of as the name of a very large set. Each *element* of this *set* is a particular horse. Different sets with different elements will be brought to mind by such words as *cows, cars, people, schools, students*, etc. As soon as the words are spoken or read, *particular elements* of each *set* come to mind. One can speak of a rancher in Montana, a professional baseball player, a teacher. Each is an arbitrary element of a set and it is the generic element that is the object of interest. In other situations one might speak of *the* people in Montana, *the* professional baseball players, *the* school teachers. In each of these, the *set* is the object of interest.

The decision as to membership, that is, whether or not a particular element belongs to a set, is quite easy to make for some sets. For others it is much more difficult. As an example, suppose we were discussing the set of even natural numbers (consisting of 2, 4, 6, 8, and so on). Given a particular number, it is easy to decide whether or not it is an element of the set of even natural numbers. Thus 16 is an element of the set and 17 is not; 52 is an element of the set and 113 is not. The set of even natural numbers is one of the type for which there is an easily tested criterion for belonging and for which there is a definite procedure for determining whether or not a given object is in the set.

As an example of the other type, consider the set of all people in Chicago who are ill. Even if we knew all the people in Chicago and were qualified as doctors of medicine we might have difficulty in making decisions as to whether a certain resident of Chicago belonged to the set under discussion or not.

In mathematics we can usually be quite precise in establishing the criterion for belonging when we speak of sets. It is well to bear in mind, however, that there are sets for which deciding membership is difficult. If you are discussing a set with someone else, be sure that you both have the same criterion for making your decisions as to membership.

2.2a Set Notation

Let us adopt the convention of using capital letters to represent sets and small leters to represents elements of a set. Thus we may refer to the set X consisting of the elements which we might label a, b, c, etc. The symbol

$$a \in X$$

shall mean "the object a is a member of the set X," or "the object a belongs to the set X," or simply "a is in X." We shall also read this as "a is an element of the set X." The symbol

$$a \notin X$$

represents the negation "a is not an element of X."

A set might be specified by listing its elements. Thus the pictured set A (see Figure 1) consists of an apple, an orange, a pencil, and a table. The pictured set B consists of four books. This method of specifying a set is expressed by the notation used in the following examples:

$$C = \{2, 4, 6, 8\}.$$

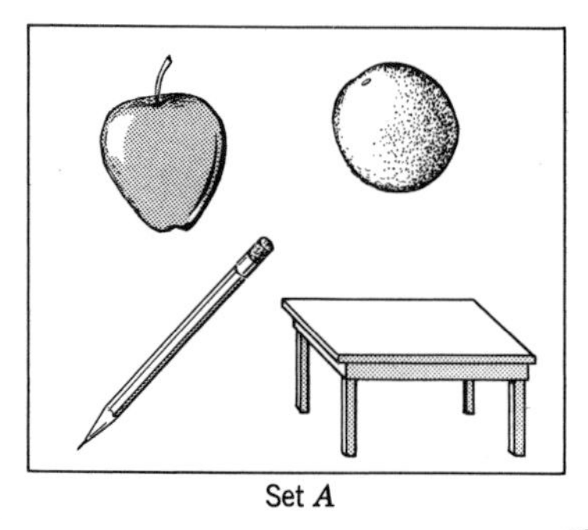

Set A

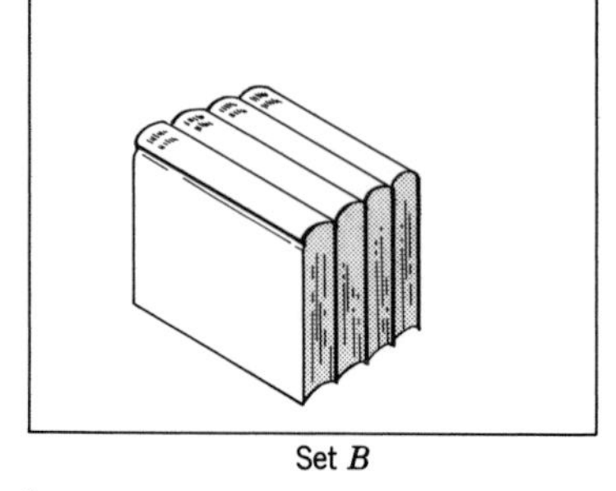

Set B

Figure 1

This is interpreted, "C is a name for the set consisting of the elements 2, 4, 6, 8." This is also read "the set C is composed of the elements 2, 4, 6, and 8." (Note that the symbol $=$ is used in the sense of "is a name for.")

$$D = \{x, y, z\}.$$

D is a name for the set consisting of the elements x, y, and z.

$$E = \{1, 2, 3, \ldots, 20\}.$$

E is a name for the set consisting of the first 20 natural numbers. The dots in the listing of the set E mean that the sequence of natural numbers is to continue to, and include, 20.

A set might also be specified by giving a criterion for belonging to the set. This is usually done by describing a common property of the elements, for example: the set X consists of the set of all red-haired people; the set

Y consists of all red-haired men; the set Z consists of all red-haired men in Montana. Notice that increasing the number of conditions or properties that an element must have in order to be a member of a set tends to decrease the size of the set.

We also use the following "set builder" notation to indicate sets:

$$A = \{x \mid x \text{ has a certain property}\}.$$

We read this, "A is the set of *all* x such that x has a certain property." (Here we have shortened "is a name for" to "is.")

The symbol x is called a *variable*. Some writers refer to it as a *pronumeral* because it is a symbol related to a numeral in the same way that a pronoun is related to a noun. It is used to denote any element of a particular set. The set is called the *domain* of the *variable*. We will specify the domain unless the set is easily understood from the context. We will use generally letters from the last part of the alphabet as variables. The statement "x has a certain property" is a *condition* on the variable x.

Exercise 2.2

The set of whole numbers is $\{0, 1, 2, 3, 4, 5, 6, 7, \ldots\}$.

1. What is the difference between 32 and $\{32\}$?

2. Classify each set as to whether it is easy or difficult to establish membership, and tabulate the elements of the set if possible.

(a) All months of the year that have exactly 30 days.
(b) All months of the year that have exactly 29 days.
(c) All even integers greater than 43.
(d) All integers that are perfect squares and less than 61.
(e) All numbers whose squares are zero.
(f) The five students not attending the University of Montana who learned the most in high school.
(g) All good boys.
(h) All fractions between zero and one.
(i) All healthy men in Chicago.

3. (a) List ten elements in the following set:

$$X = \{m \mid m = 2k \text{ and } k \text{ is a whole number}\}.$$

(b) Is it easy to determine membership in this set?
(c) What are elements of this set called?
(d) What is the domain of the variable k?
(e) What is the condition on the variable m?

4. (a) List ten elements in the following set:

$$Y = \{n \mid n = 2k + 1 \text{ and } k \text{ is a whole number}\}.$$

(b) What are the elements of this set called?
(c) What is the domain of the variable k?
(d) What is the condition on the variable n?

2.3 SUBSETS

We often find elements of one set which are also elements of another. If all the elements of one are also elements of another we have a special useful relationship which we define as follows.

> ***Definition 2.3a.*** The set A is a *subset* of the set B if every element of A is an element of B.

We use the symbol $A \subseteq B$ to denote this. This is usually read "A is a subset of B." Notice that the possibility that A and B are different names for the same set is not excluded.

Example 1

Let $A = \{1, 2, 3, 4, 5, 6\}$ and $B = \{1, 2, 3, 4, 5, 6, 7, 8, 9\}$; then A is a subset of B, for every element of A is an element of B.

$$\overbrace{\underbrace{1, 2, 3, 4, 5, 6}_{A}, 7, 8, 9}^{B}$$

Example 2

If $A = \{4, 3, 5\}, B = \{3, 5, 4\}$, then $A \subseteq B$.

Example 3

Let A be the set consisting of all people in the United States and let B be the set consisting of all the students in the first grade in the United States. B is a subset of A. If we let C be the set consisting of the girls in the first grade in the United States, then C is a subset of B. Notice also that C is a subset of A.

According to our definition,

1. $A \subseteq A$ and
2. If $C \subseteq B$ and $B \subseteq A$, then $C \subseteq A$.

> ***Definition 2.3b.*** The set A is a *proper subset* of B if every element of A is an element of B, but at least one element of B is not an element of A.

We use the symbol $A \subset B$ to denote this. This is read, "A is properly contained in B." This can also be written $B \supset A$ and is read "B properly contains A."

Example 4

Let $A = \{a, b\}$. The sets $\{a\}$ and $\{b\}$ are proper subsets of A. Notice that they are also subsets of A. On the other hand, the set $\{a, b\}$ is a subset but not a proper subset.

The word *animals* refers to a set of living objects. The word *horses* refers to another set of living objects. The statement that every horse is an animal is the same as the statement that the set of horses is a subset of the set of animals. Figure 2 will help you visualize these sets.

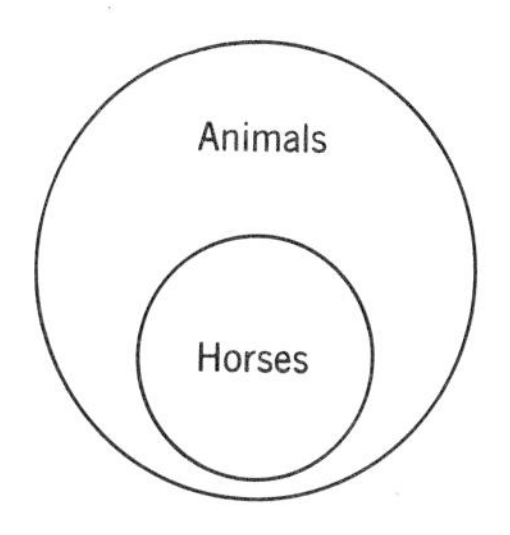

Figure 2

Notice that this is a *pairing* of the terms "animals" and "horses" in a particular way, that is, the set of horses is related to the set of animals. The set of horses is "included" in the set of animals. We say that the set of horses is related to the set of animals by inclusion, and we use the notation

$$\{\text{horses}\} \subset \{\text{animals}\}.$$

Inclusion is a relation between sets in the same way that "is a relative of" is a relation between people.

Example 5

$F = \{n | n$ is a whole number and is divisible by $4\}$.
$R = \{y | y$ is a whole number and $y = 32\}$.
$S = \{x | x$ is a whole number, x is even, and x is between 30 and $34\}$.

Note that the domain of the variable n in the set F is the set of whole numbers. There are two conditions on the variable in F. What are they? What are the conditions on the variable x in the set S? Notice that R and S are both sets which consist of the same element 32. (This is sometimes read as "singleton 32.")

We have indicated two ways of specifying sets. It may happen that the same set may be specified in several different ways. We need some criterion for determining whether a set specified in one way is the same, or distinct from, a set specified in another way.

Definition 2.3c. Two sets, A and B, are the same if every element of A is an element of B and every element of B is an element of A. We denote this $A = B$.

From the definition it follows that $A = B$ if $A \subseteq B$ and $B \subseteq A$. According to this criterion, the sets R and S of Example 5 are the same: $R = S$.

Example 6

Let $A = \{a, b, 3, 7\}$ and $B = \{3, b, a, 7\}$. Then $A = B$. It is also true that $A \subseteq B$ and that $B \subseteq A$.

Example 7

Let $C = \{a, b, a, a\}$ and $D = \{a, b\}$. Then $C = D$. It is also true that $C \subseteq D$ and that $D \subseteq C$.

Notice in Example 7 that every element in the set C is in the set D and every element in the set D is in the set C; therefore $C = D$. It will be our practice to treat the same letter or same number or same object which appears more than once in any listing of the elements of a set as only one letter or one number or one object. For instance, the letter a appears in the listing of the elements of the set C three times. For our purposes, the set C has only two distinct elements. In some of the textbooks for kindergarten and primary, this practice seemingly does not prevail. To simplify the art work, a set of three rabbits or four triangles will have drawings identical in appearance representing the three rabbits or the four triangles. However, the practice is to treat the rabbits, the triangles, or other objects as being distinct. This does not lead to learning wrong concepts since at that level the objects are used for counting, matching, etc., and not to teach inclusion relationships between sets. (For an excellent discussion of *set equality* see the article of that name by Prof. Roy Dubisch in the May 1966 issue of *The Arithmetic Teacher*.)

2.3a The Empty Set

It will be useful to introduce a new set called the *empty set* (or void set) which we denote by $\emptyset$ or $\{\ \}$. The empty set $\emptyset$ is the set that has no elements. Its occurrence seems natural and convenient if we consider the following examples.

Example 1

Let $A = \{s | s \text{ is a student under 21 years of age}\}$
and $B = \{s | s \text{ is a student over 21 years of age}\}$.

The set of elements common to both A and B is empty; that is, there is no student who is both under 21 years of age and over 21 years of age.

Example 2

Let $A = \{a, b, c\}$ and $B = \{1, 2, 3\}$.

The set of elements common to both A and B is empty.

Example 3

Let $A = \{0, 1, 2, 3\}$ and $B = \{0, 4, 5, 6\}$.

The set of elements common to both A and B is $\{0\}$. It is not the empty set. It is the set containing the number zero.

Two sets, X and Y, which have no elements in common are said to be *disjoint* (see Definition 2.4c).

2.3b The Universal Set

In discussions of sets we will have in mind some fixed class of objects to which the discussion is limited, and the sets we mention will be sets of elements from this fixed class. The fixed class is referred to as the *universal set* or the *universe*. The universal set is not the same for all problems. When discussing sets consisting of natural numbers, the universal set could be the set of all natural numbers. In other problems the universal set may be the set of all people, the set of all students, the set of all real numbers, etc. We designate the universal set by the symbol U.

2.3c Counting Subsets

Counting is not always as easy as it may appear. Some problems involving counting may be difficult because of the extremely large numbers involved, whereas other problems involving counting may be difficult because the "things" being counted may be hard to distinguish.

The man who, in return for a favor to the king, modestly asked for one grain of wheat on the first square of a checker board, two grains on the second, four on the third, each time doubling the amount on the previous square for the 64 squares on the checker board, posed a problem of the first type.

Counting subsets of a given set could be a problem of the second type without some helpful hints. It is particularly useful to observe the following principle.

If an event can occur in M ways and, after it has occurred in any one of these ways, a second event can occur in N ways, then the two successive events can occur in $M \cdot N$ ways.

Example 1

If there are two routes from city A to city B and three routes from city B to city C, then there are six routes from A to C. The six routes identified by number and letter are $(1, a)$, $(1, b)$, $(1, c)$, $(2, a)$, $(2, b)$, $(2, c)$ (see Figure 3).

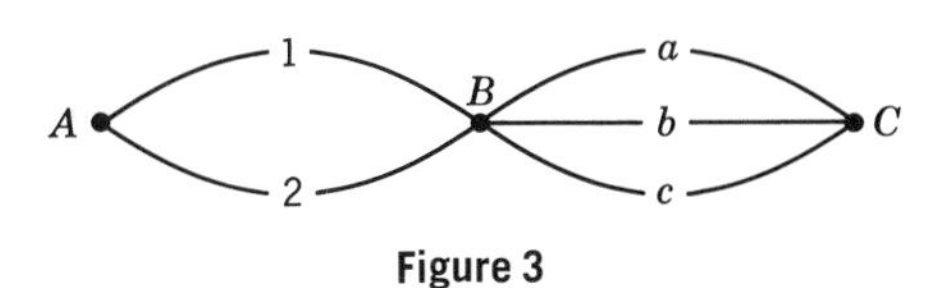

Figure 3

In counting subsets of a set with n elements, an "event" will be "placing an element." This event can occur in two ways: either place the element in a subset or do not place an element in a subset. After the first element has been disposed of, the second element is *placed.* This can also be done in two ways. Since there are n elements, there are

$$\underbrace{2 \cdot 2 \cdots 2}_{n \text{ factors}} = 2^n$$

ways of forming subsets of a set with n elements. If the choice "do not place" is made for each of the elements of the set, the resulting set is the empty set. If the choice "place" is made for each of the elements of the set, the resulting set is the set itself. Therefore, when we say there are 2^n subsets of a set of n elements, this includes the empty set and the set itself.

Exercise 2.3

1. Let $P = \{2, 3, 5, 7\}$; $S = \{2\}$; $Q = \{3, 5, 7\}$; $\emptyset = \{\ \}$.

(a) Which of the sets are subsets of the set P?
(b) Which are proper subsets?
(c) Which sets are subsets but not proper subsets?
(d) How many subsets has the set P?

2. Is $S \in P$? Explain.

3. Is $S \subseteq P$? Explain.

4. Is $S \in S$? Explain.

5. Is $S \subseteq S$? Explain.

6. Consider the set $S = \{0, 1, 2, 3, 4, 5, 6, 7, 8\}$.

(a) Select a subset of S so that each number in the subset is even; odd.
(b) Select a subset of S that contains all the numbers in S that are multiples of 3; of 1.
(c) Select a subset of S that contains all the numbers in S that added to 3 give 5; that added to 3 give 2; that added to 3 give a number in S; that added to 0 give a number in S; that added to 4 give a number not in S.
(d) Select a subset of S so that twice each number in the subset is not in S; three times each number in the subset is in S.
(e) Select a subset of S so that five more than twice each number in the subset is in S.

7. Let S be the set of all even numbers and T the set of all odd numbers. Use the set-builder notation to specify the sets S and T.

8. How many committees can you form if you have three people to choose from?

9. How many committees can you form if you have four people to choose from?

10. (a) Is $0 \in \emptyset$?
(b) Is $\emptyset \in \emptyset$?
(c) If A is a set, is $\emptyset$ a subset of A?
(d) Is $\emptyset$ a subset of $\emptyset$?

11. (a) Specify the singleton set {George Washington} in three different ways.
(b) Specify the set $\{2, 4, 6, 8\}$ in two different ways.
(c) Specify the set $\{4, 2, 8, 6\}$ in two different ways.
(d) What can you say about the set consisting of the girls who are graduates of West Point Military Academy?
(e) Is the set, whose only element is the empty set, empty? That is, is $\{\emptyset\} = \emptyset$?

12. Read the article, "Set Equality", by Dubisch. (See references.)

2.4 NEW SETS FROM OLD

Because we can do more than just discuss them, sets are of mathematical interest. We shall see how to construct new sets in terms of given sets, thereby defining *operations* involving sets.

2.4a Union of Sets

Definition 2.4a. The *union* of two sets, A and B, is the set of all elements that are in A, *or* in B, *or* in both.

We denote this $A \cup B$. (Note that this is the *inclusive* use of "or.") An element will belong to $A \cup B$ if the element belongs to A, or B, or both. In our set-builder notation

$$A \cup B = \{x|x \text{ is an element of } A \textit{ or } B \textit{ or } \text{both}\}.$$

Example 1

Let $A = \{a, b, c\}$ and $B = \{1, 5, a, x, y\}$; then $A \cup B = \{a, b, c, 1, 5, x, y\}$.

Example 2

Let $A = \{x|x \text{ is a student in the freshman class}\}$, and
$B = \{x|x \text{ is a freshman boy}\}$; then
$A \cup B = \{x|x \text{ is a student in the freshman class}\} = A$.

2.4b Intersection of Sets

Definition 2.4b. The *intersection* of two sets, A and B, is the set of all elements that are in both A and B.

We denote this $A \cap B$. An element will belong to $A \cap B$ if it belongs to both A and B. In our set-builder notation

$$A \cap B = \{x|x \text{ is an element of } A \textit{ and } B\}.$$

In Example 1, Section 2.4a, $A \cap B = \{a\}$. In Example 2, $A \cap B = B$, which denotes that the intersection of the set which consists of all the freshman boys with the set consisting of all the freshmen is the set of all freshman boys.

Definition 2.4c. Two sets X and Y, whose intersection is the empty set are said to be *disjoint*, that is, the sets X and Y are disjoint if $X \cap Y = \emptyset$.

Example 1

Let $A = \{1, 2, 3, 4, 5\}$, $B = \{3, 4, 5, 6, 7, 8, 9\}$, and $C = \{11, 12\}$; then $A \cap B = \{3, 4, 5\}$, $A \cap C = \emptyset$, and $B \cap C = \emptyset$. A and C are disjoint. B and C are disjoint.

Example 2

In the diagrams in Figure 4 let A be the set of all points inside and on the circle marked A; let B be the set of all points inside and on the circle marked B; and let C be the set of all points inside and on the circle marked C.

These diagrams are called *Venn diagrams.* The sets corresponding to the intersection and union of the various sets are shaded and marked.

Exercise 2.4

Draw Venn diagrams as in Figure 4 and shade the regions corresponding to the following unions and intersections. Label each diagram. When you have finished, decide which of the symbols in problems 1 through 12 represent the same sets.

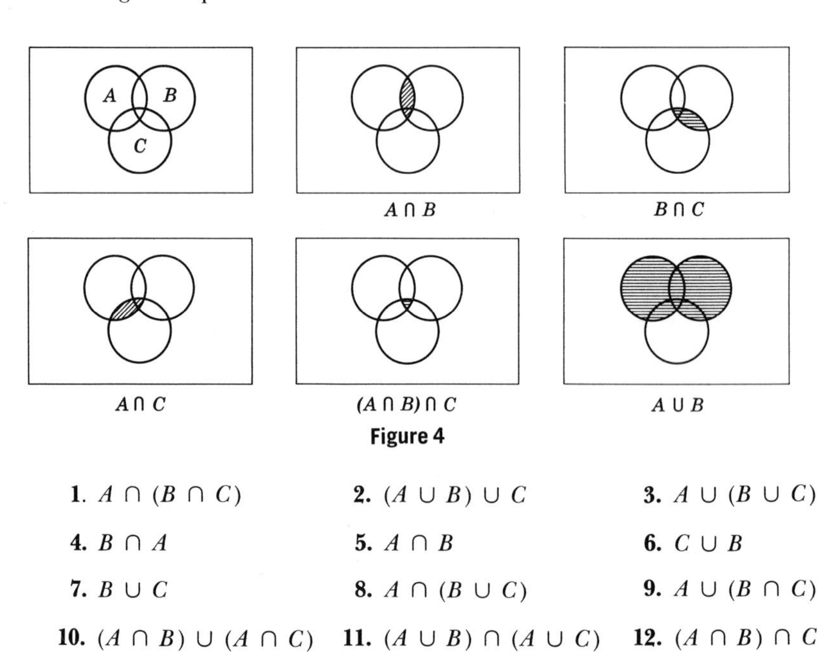

Figure 4

1. $A \cap (B \cap C)$ **2.** $(A \cup B) \cup C$ **3.** $A \cup (B \cup C)$

4. $B \cap A$ **5.** $A \cap B$ **6.** $C \cup B$

7. $B \cup C$ **8.** $A \cap (B \cup C)$ **9.** $A \cup (B \cap C)$

10. $(A \cap B) \cup (A \cap C)$ **11.** $(A \cup B) \cap (A \cup C)$ **12.** $(A \cap B) \cap C$

13. In each of the following diagrams, give the symbolic description of the set indicated by the shaded region. Make it as simple as possible.

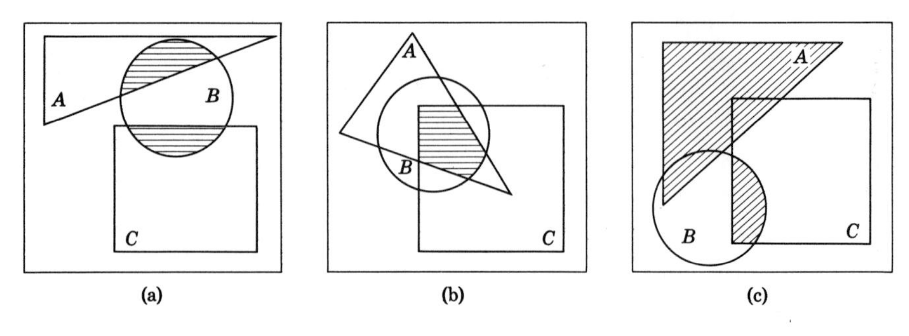

14. In planning for a birthday party for her daughter, Mrs. Jones made a list of the children according to their preferences in refreshments. It was as follows:

Ice Cream		*Cake*		*Apples*	
Tom	Ted	Ted	Tobe	Sue	Jack
Jim	Jack	Tim	Jane	Ted	John
Joan	June	John	Sono	Jill	June
Sam	Tim	Jill	Jan	Sara	Tim

There were no two children with the same name.

(a) List the names of the children who like both ice cream and cake.
(b) List the names of the children who like both ice cream and apples.
(c) List the names of the children who like both cake and apples.
(d) List the names of the children who like all three.
(e) Illustrate these sets with Venn diagrams.

2.5 THE COMPLEMENT OF A SET

Definition 2.5. If A is a subset of the universal set U, the set of all elements of U that are not in A is called the *complement* of A.

We use the symbol A' to signify the complement of A.

Example 1

If U is the set of students in a university and M is the subset consisting of the men students, then M' is the set of women students.

Example 2

Let $U = \{a, b, c, d, e\}$, and
$A = \{a, e, c\}$; then
$A' = \{b, d\}$.

In general, if

$A \cup B = U$, and
$A \cap B = \emptyset$,

then A and B are *complementary sets*, and

$A' = B$, and
$B' = A$.

In particular,

$U' = \emptyset$, and
$\emptyset' = U$.

2.6 MEMBERSHIP TABLES

Earlier we described set membership. It is convenient for the purpose of identifying equality of compound statements involving sets to introduce the concept of membership table.

Let x be an arbitrary member of the universal set U; and let A, B, C, and so on, be subsets of U. Then the element x in U can be related to a particular subset of U in exactly one of two ways: it either belongs or it does not belong to that set. That is, if $x \in U$ and $A \subseteq U$, then $x \in A$ or $x \notin A$, but not both. Thus, according to our way of counting established in section 2.3c, if we have two subsets of U there are 2^2 or 4 possibilities. Three subsets would give us 2^3 or 8 possibilities. The possibilities for membership in two subsets of the universe along with the membership table (Table 1) for the basic set operations appears in Figure 4.

Table 1

Possibilities for membership for $x \in U$		Union of A and B	Intersection of A and B	Complement of A
$x \in A$	$x \in B$	$x \in (A \cup B)$	$x \in (A \cap B)$	$x \in A'$
T	T	T	T	F
T	F	T	F	F
F	T	T	F	T
F	F	F	F	T

Each symbol in the columns tells whether the statement above it is true or false. For example, if an arbitrary element $x \in U$ is a member of the set A, the letter T appears; if it is not, the letter F appears. Reading across the first row tells us that if an arbitrary element of the universe is an element of A and an element of B, then it is an element of $A \cup B$, but it is not an element of the complement of A.

Example 1

Construct the membership table for $A \cap (B \cup C)$ and for $(A \cap B) \cup (A \cap C)$ and draw a conclusion about these sets (Table 2).

The first three columns of Table 2 indicate the eight possibilities for membership in the three subsets of U. The other columns were constructed in order from left to right to indicate membership in the combinations of sets we are interested in. We note that the membership table for $A \cap (B \cup C)$ is identical with that for $(A \cap B) \cup (A \cap C)$. We conclude that $A \cap (B \cup C) = (A \cap B) \cup (A \cap C)$.

Table 2

$x \in A$	$x \in B$	$x \in C$	$x \in (B \cup C)$	$x \in (A \cap (B \cup C))$	$x \in (A \cap B)$	$x \in (A \cap C)$	$x \in ((A \cap B) \cup (A \cap C))$
T	T	T	T	T	T	T	T
T	T	F	T	T	T	F	T
T	F	T	T	T	F	T	T
T	F	F	F	F	F	F	F
F	T	T	T	F	F	F	F
F	T	F	T	F	F	F	F
F	F	T	T	F	F	F	F
F	F	F	F	F	F	F	F

Exercise 2.6

1. Translate the following into verbal statements:

(a) $A \cup A = A$, $A \cup \emptyset = A$, and $A \cup U = U$.

(b) $A \cup B \supseteq A$ and $A \cup B \supseteq B$.

(c) If $C \subset A$ and $C \subset B$, then $C \subset A \cup B$.

(d) $A \cup B = A$ if and only if $A \supseteq B$.

(e) $A \cap A = A$, $A \cap U = A$, and $A \cap \emptyset = \emptyset$.

(f) $A \cap B \subseteq A$ and $A \cap B \subseteq B$.

(g) If $C \subset A$ and $C \subset B$, then $C \subset A \cap B$.

(h) $A \cap B = A$ if, and only if, $A \subseteq B$.

2. Find the union and the intersection for each of the following pairs of sets:

(a) $A = \{1, 3, 5\}$, $B = \{2, 4\}$

(b) $A = \{1, 3, 5\}$, $B = \{1, 3, 5\}$

(c) $A = \{1, 3, 5\}$, $B = \{1, 2, 3\}$

(d) $A = \{2, 3, 4\}$, $B = \{2, 4\}$

(e) $A = \{2, 3\}$, $B = \{1, 2, 3\}$

(f) $A = \{2, 4, 6\}$, $B = \{2, 3, 5\}$

3. Let U be the set of natural numbers $\{1, 2, 3, 4, \ldots\}$, E the set of even natural numbers, A the set of odd natural numbers, and B the set $\{1, 2, 3, 4, 5, 6\}$. Perform the following set calculations giving each result in two ways: by *tabulation* (perhaps incomplete) and by *description*.

(a) $E \cup U$

(b) $E \cup A$

(c) $A \cup B$

(d) $E \cup B$

(e) $E \cap U$

(f) $E \cap A$

(g) $A \cap B$

(h) $E \cap B$

(i) $(A \cup B) \cap E$

(j) $A \cup (B \cap E)$

(k) $(E \cap A) \cup B$

(l) $E \cap (A \cup B)$

4. What is $A \cap B$ if

(a) A and B are disjoint?

(b) A and B are identical?

(c) $A \subset B$?

(d) $A \supset B$?

(e) A and B overlap?

5. What is $A \cup B$ if

(a) A and B are disjoint?

(b) A and B are identical?

(c) $A \subset B$?

(d) $A \supset B$?

(e) A and B overlap?

6. Let $U = \{1, 2, 3, 4, 5\}$, $C = \{1, 3\}$, and A and B be nonempty sets. Find A in each of the following:

(a) $A \cup B = U$, $A \cap B = \emptyset$, and $B = \{1\}$.

(b) $A \subset B$, $A \cup B = \{1, 5\}$, and $A \cap C = \emptyset$.

(c) $A \cap B = \{3\}$, $A \cup B = \{2, 3, 4\}$, and $B \cup C = \{1, 2, 3\}$.

(d) A and B are disjoint, B and C are disjoint, and the union of A and B is the set $\{1, 2\}$.

7. In the following diagram, let U be the set of all points inside and on the rectangle and let A, B, and C denote the points inside and on the circles as marked. Use slanted lines to indicate the set A'; use lines slanted in another direction to indicate the set B'. Now draw a similar figure and shade the set $(A \cup B)'$. Verify that

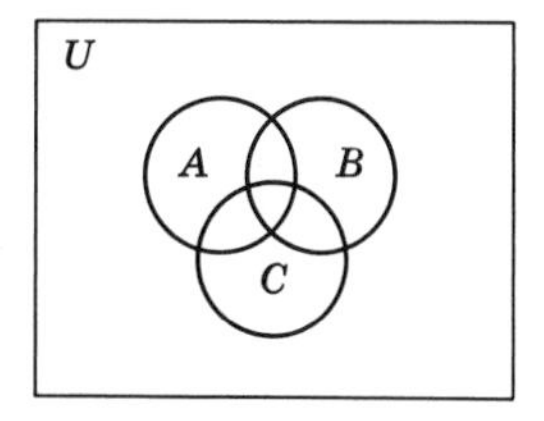

(a) $(A \cup B)' = A' \cap B'$ (b) $(A \cup C')' = A' \cap C$

(c) Verify these equalities using membership tables.

8. Follow the instructions of problem 7 and diagram $(A \cap B)'$, A', and B'.

(a) How are $(A \cap B)'$, A', and B' related?

(b) What is $(A \cup B \cup C)'$?

(d) Test your conclusions with membership tables.

9. How many subsets can a set with five elements have? How many proper subsets?

10. An oversimplified human blood classification may be considered as accomplished in three tests, the A-test, the B-test, and the Rh-test, to *each* of which a person's blood either *reacts* or *does not react*. Let us designate these tests A, B, and Rh, respectively. Let **A** be the set of *people* whose blood reacts to the A-test, **B** the set of *people* whose blood reacts to the B-test, and **RH** the set of *people* whose blood reacts to the Rh-test.

(a) What is meant by $\mathbf{A} \cap \mathbf{B} \cap \mathbf{RH}'$?

(b) What is meant by $(\mathbf{A}') \cap \mathbf{B} \cap \mathbf{RH}'$?

Under this oversimplified classification there are eight mutually exclusive classes into which people fall by blood groups as indicated in the following table:

Sets of People Classified by Blood Group	International Designation of Blood Group	Sets of Tests to which People React	Approximate Percent of Population*
$\mathbf{A}' \cap \mathbf{B}' \cap \mathbf{RH}$	O Positive	$\{Rh\}$	38
$\mathbf{A}' \cap \mathbf{B}' \cap \mathbf{RH}'$	O Negative	$\{\ \}$	7
$\mathbf{A} \cap \mathbf{B}' \cap \mathbf{RH}$	A Positive	$\{A, Rh\}$	31
$\mathbf{A} \cap \mathbf{B}' \cap \mathbf{RH}'$	A Negative	$\{A\}$	6
$\mathbf{A}' \cap \mathbf{B} \cap \mathbf{RH}$	B Positive	$\{B, Rh\}$	11
$\mathbf{A}' \cap \mathbf{B} \cap \mathbf{RH}'$	B Negative	$\{B\}$	2
$\mathbf{A} \cap \mathbf{B} \cap \mathbf{RH}$	AB Positive	$\{A, B, Rh\}$	4
$\mathbf{A} \cap \mathbf{B} \cap \mathbf{RH}'$	AB Negative	$\{A, B\}$	1

**Hyland Reference Manual of Immunohematology.*

Let D be the set of *tests* to which a prospective donor's blood reacts and let R be the set of *tests* to which a prospective recipient's blood reacts. One of the conditions for a safe transfusion expressed in set language is $D \subseteq R$.

(c) Who is a potential donor for a person of the set $\mathbf{A} \cap \mathbf{B} \cap \mathbf{RH}'$?
(d) Who is a potential donor for a person of the set $\mathbf{A}' \cap \mathbf{B} \cap \mathbf{RH}'$?

Although "universal donor" and "universal recipient" are terms which are not completely accepted in present-day hematology, under our oversimplified classification the following questions are meaningful:

(e) A "universal donor" belongs to which set of people?
(f) A "universal donor's" blood reacts to which set of tests?
(g) A "universal recipient" belongs to which set of people?
(h) A "universal recipient's" blood reacts to which set of tests?

2.7 THE CARTESIAN PRODUCT OF SETS

The Cartesian product of two sets A and B is quite unlike the sets $A \cup B$ and $A \cap B$. The elements of the Cartesian product are not elements from A nor elements from B but rather what we call *ordered pairs*.

Ordered pairs occur quite naturally. We encounter ordered pairs when locating places on maps. Highway maps usually have a sequence of numerals spaced equally along one border of the map, say the bottom border, and a sequence of letters equally spaced along one of the vertical borders. A list of the towns will have after each name a number and a letter. For instance, on a map of Montana we find Missoula—4-D. Following along the lower border until we find the numeral 4, we then move along a vertical line until we are opposite the letter D on the vertical border. Missoula is located in this way by the pair $(4, D)$. The 4 is called the first component of the ordered pair, and D is called the second component. On road maps, letters and numerals are used to strengthen the concept of the ordered pair. The labels on the borders of military maps are usually numerals. Thus a hill may be designated by the ordered pair $(705, 600)$. It is understood that one reads the first component along the horizontal edge and the second component along the vertical edge. In this way the ordered pair locates a single point. If the pair $(705, 600)$ were not understood to be an ordered pair, we would have to check the pair $(600, 705)$ as a possibility for locating the hill.

Definition 2.7a. The *Cartesian product* of the sets A and B is the set of all ordered pairs (a, b) where the element in the first place, a, is an element of A and the element in the second place, b, is an element of B.

In order to distinguish the elements of the Cartesian product from one another and to avoid duplication, we must establish a criterion for "sameness." When are two elements the same?

Definition 2.7b. Ordered pairs of the Cartesian product are the same, and we write $(a, b) = (c, d)$ if and only if $a = c$ and $b = d$.

We denote the Cartesian product of sets A and B by the symbol $A \times B$, and we read this as "the Cartesian product of A and B" or simply as "A cross B". In our notation

$$A \times B = \{(a, b) \mid a \text{ is in } A \text{ and } b \text{ is in } B\}.$$

Example 1

Let $A = \{a, b\}$ and let $B = \{0, 1, 2\}$. Some elements of the set $A \times B$ are $(b, 2)$, $(a, 1)$, and $(b, 0)$. Is the ordered pair $(1, a)$ an element of $A \times B$? It is not because the first place element 1 is not in A nor is the second place element a in B. Does (a, a) belong to $A \times B$? No, because the second place member of the ordered pair does not belong to B. Write a complete list of the elements of $A \times B$.

Example 2

Let A be the set consisting of the whole numbers 1 through 9. What are some elements of the Cartesian product $A \times A$? Is $(1, 1)$ an element of $A \times A$? Yes; in fact,

$$A \times A = \{(n, m) \mid n \text{ can be any of } 1, 2, 3, 4, 5, 6, 7, 8, 9, \text{ and } m \text{ can be any of } 1, 2, 3, 4, 5, 6, 7, 8, 9\}.$$

Example 3

The following type of question appears on tests. Draw a line from each word in the column labeled A to a word with the opposite meaning in the column labeled B.

A	B
good	small
large	light
dark	bad
	up
	smooth

In this question, one is simply specifying certain elements of the Cartesian product of A and B such as (good, bad), (large, small), (dark, light).

The Cartesian product of any two sets may be represented pictorially using perpendicular lines and the same scheme used by map makers. Along the horizontal axis we label equally spaced points with the elements of the first set and equally spaced points along the vertical line with elements of the second set. Following the usual convention, we read along the horizontal axis first and then along the vertical axis. The pictorial representation of the Cartesian product of the sets $A = \{\text{a}, b, c\}$ and $B =$

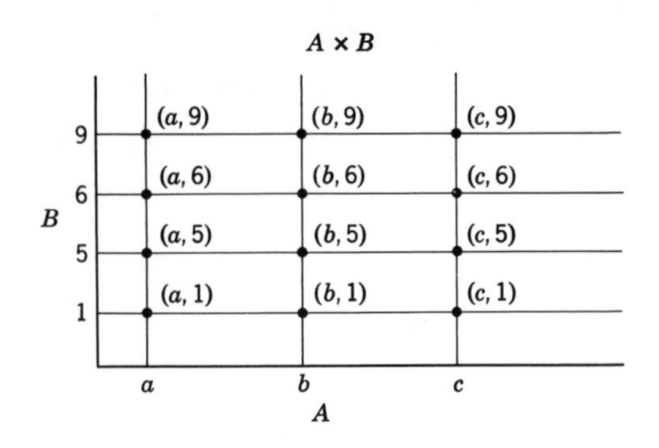

$\langle 1, 5, 6, 9\}$ would be the points whose addresses or coordinates are the ordered pairs of $A \times B$.

Exercise 2.7

1. Let $A = \{\text{blue, green, gray}\}$ and let $B = \{31, 43, 47, 59\}$. Tabulate the set $A \times B$.

2. (a) List the subsets of $E = \{0, 1\}$.
(b) What is $E \times \emptyset$?
(c) What is $E \times \{2\}$?

3. Specify each of the following sets in another way:

(a) $X = \{n | n$ is a whole number and has remainder 0 when divided by 2$\}$.
(b) $Y = \{m | m$ is a whole number and is not a multiple of 2$\}$.
(c) $Z = \{k | k$ is a whole number and has remainder 1 when divided by 2$\}$.
(d) $W = \{p | p = 2n$ and n is a whole number$\}$.

4. Let the sets A, B, and C be the points inside and on the circle, as in Example 2, Section 2.4b. Draw Venn diagrams of the following sets:

(a) $A \cap B$ (b) $A \cap B \cap C$
(c) $A \cup (B \cap C)$ (d) $(A \cap B) \cup (A \cap C)$

5. Let $A = \{a, b, c, d\}$. How many subsets has $A \times A$?

6. Let $A = \{1, 2, 3\}$, $B = \{2, 3, 4\}$, $C = \{3, 4, 5\}$, and $D = \{4, 5, 6\}$. Tabulate each of the following sets:

(a) $A \cap B$ (b) $B \cap C$
(c) $A \cap C$ (d) $A \cap B \cap C$
(e) $(A \cap B) \cap (C \cap D)$ (f) $(A \times A) \cap (B \times B)$
(g) $(A \cap B) \times (A \cap B)$ (h) $(C \times C) \cap (D \times D)$
(i) $(C \cap D) \times (C \cap D)$ (j) What can you say about (f) and (g)?
(k) What can you say about (h) and (i)?

7. What does $A \cap (B \cup C)$ equal if

(a) A and B are disjoint?
(b) B is identical to C?
(c) $A \subset C$?
(d) $A \supset B$? ($A \cap C \neq \emptyset$.)
(e) C is the empty set?

8. Let $X = \{a, b, c\}$, $A = \{a\}$, $B = \{b\}$, $C = \{c\}$, $D = \{a, b\}$, $E = \{a, c\}$, $F = \{b, c\}$. Then $S = \{\emptyset, A, B, C, D, E, F, X\}$ is the family of all subsets of X, sometimes called the *power set* of X. Insert the correct element of S in each square of the following tables:

$\cup$	$\emptyset$	A	B	C	D	E	F	X
$\emptyset$								
A								
B								
C								
D								
E								
F								
X								

$\cap$	$\emptyset$	A	B	C	D	E	F	X
$\emptyset$								
A								
B								
C								
D								
E								
F								
X								

9. Illustrate that $A \cap (B \cup C) = (A \cap B) \cup (A \cap C)$ by the use of Venn diagrams, such as in Figure 4.

10. In problem 14, Exercise 2.4, let I denote the children who like ice cream, C denote the children who like cake, and A denote the children who like apples. Describe each of the following sets by listing the names in each.

(a) $I \cap A$
(b) $C \cap A$
(c) $C \cap I$
(d) $I \cup A$
(e) $I \cap A \cap C$
(f) $C \cup A$
(g) $I \cup A \cup C$

How many children attended the party?

11. You, as a representative of a company selling soft drinks, are interested in putting soft-drink dispensers in the Student Union. The company is interested in finding out how many people like orange soda, grape soda, and strawberry soda. You hire a boy for \$50 to poll 1000 students. He is to count only those who indicate a liking for at least one of the drinks. You observe your helper and see him drinking coffee in the lounge most of the time. Later he comes to you with the results of his poll as follows:

Orange	596
Grape	710
Strawberry	427
Strawberry and Orange	274
Orange and Grape	430
Grape and Strawberry	309
All three	212

You have serious doubts about how he obtained these figures, but you are willing to pay him the $50 if the figures "add up." Would you pay him? Use a Venn diagram to help you solve this problem.

REFERENCES

Dubisch, Roy, "Set Equality," *The Arithmetic Teacher*, Vol. 13, No. 5, May 1966.

Hyland Reference Manual of Immunohematology, 3rd Edition, Hyland Laboratories, Los Angeles, California, March 1955.

CHAPTER 3

Relations and Their Properties

3.1 INTRODUCTION

In mathematics, the concept of relation is extremely powerful and useful. Its power lies in its simplicity and its usefulness lies in its generality. A formal treatment of this concept as a mathematical object is of recent origin, even though mathematical relations have been subject to intensive investigation for a long time. In particular, the treatment of relations as sets of ordered pairs has not been universal and certainly not yet widely adopted. However, the realization of just how basic and fundamental this concept is in arithmetic and in mathematics is recognized and unchallenged. If anything is to be labeled "new math" in the elementary curriculum, the concept of relation must be included. The study of inequalities and the introduction to the concept of functions are "new" and both are special cases of the broader concept of relation.

The concept of relation is introduced early because special instances of relations, such as equals and equivalence relations, are helpful to the clear understanding of the abstraction called number, of fractions as rational numbers, as well as other aspects of mathematics. Precise knowledge of other relations, such as "order" and "inequalities," is essential to the orderly development and correct mathematical treatment of the numbers we use. Finally, the modern approach of arithmetic emphasizes the algebraic structure, the patterns and properties associated with the algebraic operations. The importance and the role of the order properties of the numbers and the notions which derive from them will be given adequate treatment in this book.

3.2 RELATIONS

Objects and ideas are seldom thought of or spoken of alone. They occur in thought and in speech with other objects and other ideas. Consciously or subconsciously we associate, we compare, we classify, we evaluate, and in so doing, we think or speak of objects in the light of relations they bear to other objects. That is, much of human activity is concerned with pairing elements of sets according to some condition, some rule, or some formula. This pairing of elements of sets according to some criterion is what is meant by a *relation*.

When we compare, we are pairing things according to some specific condition or criterion.

John is taller than Susan.
It is colder in West Yellowstone than at Denver.
The Dow-Jones industrial average was lower today.
Set A is a subset of set B.

When we classify, we are pairing in such a way as to form nonoverlapping, identifiable sets.

Roses are red and violets are blue.
That is a rhododendron, not a camellia.
That car is like this car. They are both 1925 Model T Fords.

When we evaluate, we are pairing objects and things to a particular ordered set.

Helen is a B student.
The Yankees are in the second division.
He charged 47 dollars for his labor.

Different as these examples cited may appear to be, each is a specific instance of a *relation, a pairing of objects according to some criterion.* We are interested in the relations as well as the things which are related.

3.3 PROPERTIES OF RELATIONS

In order to recognize and identify the properties of relations we wish to consider enough examples so that we can observe that some relations have these particular properties and others do not (Table 1). For some of the relations we appeal to the reader's intuition for the conditions or criteria for the relations to hold true. For others we shall be more precise because of their importance in mathematics.

3.3a The Reflexive Property

The pairing of objects can be described in many ways, and we have used terminology that suggests the relation. Using this language we note that some relations pair an object with itself. For example, grass "is the same color as" grass; the set A "is a subset of" the set A. For other relations this is not true. The relation "is a daughter of" does not have this

Table 1

Relation	Set on which Defined	Symbol	Criteria or Condition
"is a daughter of"	People	None	Understood
"is the same color as"	Objects	None	Understood
"is a subset of"	Sets	$\subseteq$	See Definition 2.3a
"is parallel to"	Straight lines in	$\parallel$	No points in common or all points in common
"is less than"	Whole numbers	$<$	$a < b$ if there is a nonzero whole number k such that $a + k = b$
"divides"	Whole numbers	$\mid$	$a \mid b$ if there is a unique whole number k such that $b = ak$
"divides"	Natural numbers	$\mid$	$a \mid b$ if there is a unique natural number k such that $b = ak$
"is equal to"	Sets	$=$	See Definition 2.3c
"is a classmate of"	Children in a graded elementary school	None	Understood
"is taller than"	People	None	Understood

property. Neither does the relation "is taller than." The relations that pair an object with itself are said to have the *reflexive* property.

> ***Definition 3.3a.*** We say that the relation ® is *reflexive* if it is true that each element x of the set S in which the relation is defined is related to itself; that is, ® is reflexive if x ® x for all x in S.

Example 1

Let us define a relation Ⓗ in the set consisting of the students in a class in the following way. A student in a class, whom we may call x, will bear this relation to a student, y, if they are the same height. Here we might encounter ambiguity in interpretation, that is, what do we mean by "same height?" To avoid such complications let us say that heights shall be expressed to the nearest half inch. We may symbolize our relation, x Ⓗ y. We can readily see from the definition of this relation that for any student x, x Ⓗ x. The relation as defined is reflexive, or the relation Ⓗ possesses the reflexive property.

We appeal to the reader's tolerance to permit us to say that such statements as "a boy 'is a classmate of' himself" and "an orange 'is the same color as' itself" are true.

Does the relation "is parallel to" have the reflexive property?

Does the relation "divides" in the natural numbers have the reflexive property?

Does the relation "divides" in the whole numbers have the reflexive property?

Which of the other relations in Table 1 have this property?

3.3b The Transitive Property of Relations

The statements "quantities equal to the same quantity are equal to each other" and "lines parallel to the same line are parallel to each other" are familiar to the reader. More generally, the statement "things related to the same thing are related to each other" exemplifies the *transitive property* of relations. A less common but more useful way of expressing this property is to say, for elements a, b, and c in a set in which the relation is defined, that if a "is related to" b and b "is related to" c, then a "is related to" c.

> ***Definition 3.3b.*** A relation ® is said to be *transitive* if the hypothesis that a is related to b and b is related to c leads to the conclusion that a is related to c; that is, ® is transitive if a ® b and b ® c implies a ® c.

Example 1

The relation "is taller than" has this property. If we know that Roy "is taller than" John and John "is taller than" Bill, we can conclude that Roy "is taller than" Bill.

Does the relation "is the daughter of" have the transitive property?
Does the relation "is the same color as" have the transitive property?
Does the relation "is a subset of" have the transitive property?
Which of the other relations in Table 1 have the transitive property?

3.3c The Symmetric Property of Relations

Refer back to Table 1, and note that it is not true that if Mary "is the daughter of" John, then John "is the daughter of" Mary. In fact it is not only untrue but it even sounds ridiculous! On the other hand, it is true that if Mary's eyes "are the same color as" John's, then John's eyes "are the same color as" Mary's. The relation "is the same color as" has a property that "is a daughter of" does not. The relation "is a classmate of" also has this property. The relation "is a subset of" does not. This property is called the *symmetric property*.

> ***Definition 3.3c.*** A relation ® defined on a set S has the *symmetric* property if for any elements a and b in S, whenever a ® b, then b ® a.

Does the relation "is parallel to" have the symmetric property?
Does the relation "is less than" have the symmetric property?
Which of the other relations in Table 1 have the symmetric property?

Exercise 3.3c

Make a list of the relations in Table 1 and indicate which properties each relation possesses.

3.4 EQUIVALENCE RELATIONS

The discussion of the relations in the previous section shows that relations can be quite distinct and yet can share certain properties. The *properties* that we have so far attributed to relations are the *reflexive property*, the *symmetric property*, and the *transitive property*. If we use the symbol ® to represent an arbitrary relation, and a, b, and c to represent any elements of a set S, then

1. ® is reflexive if a ® a for all a in S.
2. ® is symmetric if whenever a ® b, then b ® a.
3. ® is transitive if whenever a ® b and b ® c, then a ® c.

The relation "is the same color as" is reflexive, symmetric, and transitive. The ordinary equals relation that you have been using in mathematics has all three of these properties.

In general, there are many other relations that exhibit these three properties. Because of the role such relations play in mathematics, they have been given a special name.

Definition 3.4a. Any relation ® that is reflexive, symmetric, and transitive is called an *equivalence relation.*

Example 1

Let us consider the set S consisting of all animals and define a relation in this set in the following way:

An animal x is related to an animal y and we shall use the language x "is the same species as" y if they are of the same species. We shall assume that "a dog is the same species as a dog" is a true statement.

The relation "is the same species as" is an equivalence relation. It is reflexive since an animal "is the same species as" himself. It is symmetric because if a first animal is the same specie as a second animal, then the second animal is the same species as the first. It is transitive because if a first animal is related to a second and the second animal is related to a third, then they are all of the same species and the first animal is related to the third.

This relation partitions the animals into classes which, in this case, are called species.

Any equivalence relation ® defined in a set S has the effect of partitioning the set into disjoint subsets, or classes, which are called *equivalence classes*. In the foregoing example the equivalence classes are the species.

Definition 3.4b. If ® is an equivalence relation in a set S and a is in S, the *equivalence class* of a is defined as the set of all elements of S which are related to a. That is, $[a] = \{x \in S | x \text{ ® } a\}$.

Also, $[a] = \{x \text{ in } S \mid a$ Ⓡ $x\}$ since an equivalence relation is symmetric.

Notice that a is in $[a]$, since a Ⓡ a because an equivalence relation is reflexive.

Notice also that if a is in $[b]$, then b is in $[a]$ because an equivalence relation is symmetric.

Finally, note that if c belongs to both $[a]$ and $[b]$ then, since a Ⓡ c and c Ⓡ b, it would follow that a Ⓡ b by the transitivity of an equivalence relation. This implies that $[a] = [b]$.

Thus, if $[a]$ and $[b]$ are equivalence classes, either $[a] = [b]$ or $[a]$ and $[b]$ are disjoint, that is, $[a] \cap [b] = \emptyset$.

Exercise 3.4

1. Let the relation Ⓣ be defined on the set of students in a particular school. Student x is related to student y by the relation Ⓣ if, and only if, x is taller than y. Jim Ⓣ Tom if and only if Jim is taller than Tom. Which of the "R, S, T" properties does the relation Ⓣ have?

2. Let the relation Ⓟ be defined on the set of natural numbers as follows: The number m is related to the number n if either both are even or both are odd. We write m Ⓟ n if, and only if, m and n are both even or m and n are both odd. Which of the "R, S, T" properties does the relation Ⓟ have?

3. Let the relation © be defined on the set of people living in a particular city. Let person x be related to person y if x is a son or daughter of y. We write x © y if, and only if, x is a child of y. Which of the "R, S, T" properties does © have?

4. Let the relation $\equiv$ be defined on the set consisting of the natural numbers and 0 as follows: $m \equiv n$ if m and n have the same remainder when divided by 7. List the "R, S, T" properties of the relation $\equiv$.

5. What can you say about the numbers related to 0 in the last problem?

6. Let the relation $\equiv$ be defined on the set consisting of the natural numbers and 0 as follows: $m \equiv n$ if they have the same remainder when divided by 2. What can you say about the numbers related to 0? What can you say about the numbers related to 1?

Example 3

The relation in problem 6, Exercise 3.4, is an equivalence relation. It is the same relation as the one defined in problem 2 of the same section. The class to which 5 belongs is [5]. The class to which 4 belongs is [4]. Notice that 5 is also in the class [1], that is, [1] = [5]. There are really only two classes, the even numbers [0] and the odd numbers [1].

Example 4

The relation in problem 4, Exercise 3.4, ($n \equiv m$ if m and n have the same remainder when divided by 7) is an equivalence relation. We use this example to illustrate the effect of an equivalence relation defined on a set. We write the natural numbers and zero in an array as follows:

$0,\ 7, 14, 21, 28, \ldots$
$1,\ 8, 15, 22, 29, \ldots$
$2,\ 9, 16, 23, 30, \ldots$
$3, 10, 17, 24, 31, \ldots$
$4, 11, 18, 25, 32, \ldots$
$5, 12, 19, 26, 33, \ldots$
$6, 13, 20, 27, 34, \ldots$

Notice that all the numbers in each row are related to each other, that is, each row is an equivalence class. There are exactly seven classes, for there are only seven possible remainders when a natural number is divided by 7. Every number is accounted for so every number belongs to some class. Notice also that the classes are disjoint in pairs. This is what we mean by a *partition* of a set.

Hence we see, as stated previously, that an equivalence relation has the effect of partitioning a set into disjoint subsets.* The converse is also true, namely, that any partition of a set into disjoint subsets determines a relation that is an equivalence relation. This fact will not be exploited at this time.

3.5 ONE-TO-ONE CORRESPONDENCE

Definition 3.5a. Two sets, A and B, are said to be in *one-to-one correspondence* if each element of A can be "paired" to an element of B and each element of B can be "paired" to an element of A in such a way that distinct elements of A are paired to distinct elements of B and distinct elements of B are paired to distinct elements of A, or if $A = B = \emptyset$.

If A and B can be put into one-to-one correspondence in one way, they can be put into one-to-one correspondence in at least one other way unless A has fewer than two elements.

Example 1

Let $A = \{a, b, c, d, e\}$, and let $B = \{1, 5, 4, 2, 6\}$. One possible one-to-one correspondence would be: $a \leftrightarrow 1$, $b \leftrightarrow 6$, $c \leftrightarrow 2$, $d \leftrightarrow 5$, and $e \leftrightarrow 4$. Another would be: $a \leftrightarrow 5, b \leftrightarrow 2, c \leftrightarrow 1, d \leftrightarrow 4$, and $e \leftrightarrow 6$.

Experiments have shown that the concept of one-to-one correspondence cannot be taken too much for granted. In several tests with children it was observed that when two sets in which the elements were arranged

*The formal definition of a partition includes the requirement of disjointness. We mention the requirement specifically here for clarity only.

in some orderly fashion were shown, the children responded correctly to questions about one-to-one correspondences. When the sets were shown with the elements in a disorderly array, many students responded incorrectly; also, when the number of elements in the sets were increased substantially, similar responses were observed.

Exercise 3.5a

1. Indicate how at least two other one-to-one correspondences could be established between the sets A and B of Example 1, Section 3.5.

2. How many one-to-one correspondences are possible between sets A and B of Example 1, Section 3.5? (*Hint*: In how many ways can a be "matched" to an element of B? After a has been matched, in how many ways can b be matched to the remaining elements of B?)

3. Using the definition of "divides," show that

(a) $7|35$ (b) $8|72$ (c) $3|51$ (d) $9|9$

4. Give a counterexample different from the one in Section 3.3c to show that the relation "divides" is not symmetric.

5. Let $N = \{1, 2, 3, 4, 5, \ldots, n, \ldots\}$ and $E = \{2, 4, 6, 8, \ldots 2n, \ldots\}$. Recall that the dots are to indicate the sequence is to go on and on. Define a one-to-one correspondence for these two sets.

6. Let the set A consist of the letters of the word "marbles." Describe a one-to-one correspondence of the set A with itself so that the resultant set again constitutes a word.

7. How many one-to-one correspondences are possible between the set in problem 6 and itself? (Each such correspondence is called a *permutation.*)

8. If we allow ten seconds to write down a permutation different from all previously written permutations of the first 15 letters of the alphabet, how long would it take to write down all possible permutations of the 15 letters? (Try to estimate the time required before you carry out the computation.)

Definition 3.5b. Two sets A and B are "matched,"* or satisfy the matching relation, if they can be placed in one-to-one correspondence.

We shall use the symbol "1–1" to stand for "one-to-one correspondence."

This matching process is of fundamental importance in mathematics and is introduced in the first year of a child's formal education. You may

*Some authors use the term "equivalent" where we use the term "matched." We believe that the term "matched" is more appropriate at this time.

recall from your own experience pictures containing, for example, three rabbits and three children. The child is asked to show how to distribute the rabbits so that each child has just one rabbit by drawing lines from each rabbit to a child. This is a 1–1 process. The matching relation is reflexive, symmetric, and transitive. The student is asked to verify this.

Exercise 3.5b

1. Verify that the matching relation is an equivalence relation.

2. Let S be the students in the Lincoln Grade School. Let the relation be "same age as," and, again, it is to be understood that a person is the same age as himself. Also, let us agree to specify ages to the nearest year. Verify that this is an equivalence relation and describe the equivalence classes.

3. Consider the set of problem 2 and let the relation be "same sex as." Verify that this is also an equivalence relation and describe the equivalence classes.

4. Let $J = \{\ldots, -3, -2, -1, 0, 1, 2, 3, \ldots\}$, where the dots indicate that the numbers continue indefinitely in both directions. On this set let us define the relation "is a multiple of"; as follows: For m and n in J, m "is a multiple of" n if there is a unique k in J such that $m = n \cdot k$. Assume that $0 = n \cdot 0$ for any n, and $-n = n \cdot (-1)$ for any n. Is this relation an equivalence relation?

5. On the set J of problem 4 let us define a relation as follows. If m and n are elements of J, m is related to n if their difference is a multiple of 4 (see problem 4). For example, 9 is related to 5 because $9 - 5 = 4$, and 4 is a multiple of 4. Also 25 is related to 41 because their difference is 16, and 16 is a multiple of 4. Is this relation an equivalence relation? If so, describe the equivalence classes.

6. What do you mean when you say that $1+1+1+1 = 4$? $2+2 = 4$? $1+3 = 4$? $1+3 = 3+1$? $\frac{8}{2} = 4$? $5-1 = 4$?

3.6 THE CARDINAL NUMBER OF A SET

The one-to-one correspondence or matching relation is an equivalence relation. It partitions the sets of all objects into equivalence classes. The singleton sets have the common property of "oneness." The sets with two objects have the common property of "twoness." The sets with three objects have the common property of "threeness." Thus every set has a name associated with it. This is not a profound notion. In fact, it is the idea used to teach children to count. The teacher points to pictures of sets of 3 rabbits, 3 children, 3 hats, and so forth, each time saying the word "three," and the idea of "threeness" is impressed on the child at an early age. It does not take long for a child to learn to say the words one, two, three, four, five, six, seven, eight, nine, and ten. It takes more time for a child to learn the meaning of the words.

Counting probably had its origin in keeping track of possessions, but this was possible without numerals. Pebbles in a bag, notches on a stick, and knots in a rope are some of the devices that have been used in "keeping tally" (see Newman). The essential idea was the one-to-one correspondence between the set being counted and the objects in the representative set or counting set. Man's recognition of what 2 pebbles, 2 shells, 2 notches, 2 knots, etc., have in common marked the real beginning of arithmetic. Certain words in our language seem to suggest that at one time man did not recognize, and give a single name to, the common property of matched sets. The words brace (of pheasants), yoke (of oxen), pair (of dice), couple, and others that convey the idea of "twoness" are examples.

When the abstraction of "twoness," "threeness," etc., was realized, that is, that the same identifying word or symbol could be used for all matched sets, the next step was to give names to these abstractions. The abstractions themselves are called *numbers*, and the names or symbols that we give them are called *numerals*. Thus when one counts a set of objects by calling off the number words, "one, two, three, . . . ," he is practically doing what the primitive man did when he put pebbles in the skin bag. A one-to-one correspondence is established between the counted objects and the number words.

One difference should be noted. It was immaterial which pebble was put in or taken out of the bag first, second, third, and so on, but the number words must be used in a prescribed order.

Definition 3.6. A nonempty set S will be said to have cardinal n, or cardinal number n, if and only if there exists a one-to-one correspondence between the elements in the set S and the set $\{1, 2, 3, 4, \ldots, n\}$. The empty set $\emptyset$ is said to have cardinal 0.

If S is a set, then $n(S)$ denotes the cardinal number of S, or the cardinal of S. It is read "n of S."

Example 1

If $S = \{1, 2, 3, 4, 5, 6\}$, then $n(S) = 6$.
If $A = \{a, b, c, d, e, f, g, h, i\}$, then $n(A) = 9$.
If $B = \{a, b, c, d\}$ and $C = \{c, d, e, f\}$, then $n(B \cup C) = 6$.

We can define finiteness using these notions. We say that a set is *finite* if there is a natural number n such that the set can be put in a one-to-one correspondence with the set $\{1, 2, 3, 4, \ldots, n\}$. Otherwise we say the set is *infinite*.

The ideas presented in the last few paragraphs are discussed more fully in Chapter 4.

Exercise 3.6

In the following, let A be the set of dots inside the oval marked A, let B be the set of dots inside the oval marked B, let C be the set of dots inside the oval marked C, and let D be the set of dots inside the oval marked D.

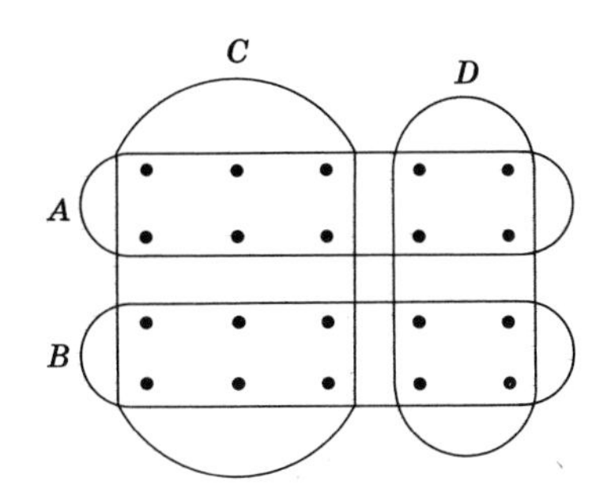

1. $n(A) = ?$

2. $n(B) = ?$

3. $n(C) = ?$

4. $n(D) = ?$

5. $n(A \cup B) = ?$

6. $n(A \cup C) = ?$

7. $n(A \cap B) = ?$

8. $n(B \cap D) = ?$

9. Verify that $n(A \cup C) = n(A) + n(C) - n(A \cap C)$.

10. Verify that $n(A \cup B) = n(A) + n(B)$.

12. Find $n(A \times A)$. Compare with $n(A) \cdot n(A)$.

11. Find $n(A \times B)$. Compare with $n(A) \cdot n(B)$.

13. A group of people were interviewed and it was found that

25 like candy
37 like movies
12 like television
9 like candy and movies
4 like movies and television
7 like candy and television
2 like candy, television, and movies

How many were interviewed?

14. If a traveler can go from Seattle to Chicago by three distinct routes and from Chicago to New York by four distinct routes, how many distinct routes are there between Seattle and New York?

15. Use the idea in problem 14 to count the number of 1-1 correspondences from the set $\{a, b, c, d\}$ to the set $\{x, y, z, w\}$.

16. Let $S = \{a, b, c\}$

(a) $n(S) = ?$ (b) $n(S \times \emptyset) = ?$
(c) $n(S \times \{0\}) = ?$

3.7 MORE ON RELATIONS IN GENERAL

In our treatment of relations in the previous sections, we did not attempt to define *relation* formally. Instead, we adopted an intuitive approach to the concept of relation by citing many examples of relations, each of which was a *pairing* of elements according to some specific criterion. It was noted that some relations had certain properties, other relations did not. We intentionally emphasized those properties that characterize equivalence relations. We were motivated by two main purposes:

1. The concept of an equivalence class will be especially useful in our discussion of the system of rational numbers.
2. The word *equals* is often used incorrectly in many different contexts and with different meanings. We hoped to achieve a better understanding and appreciation of the equals relation as indicating identity as contrasted with equivalence which, as we have seen, need not mean equality (but includes equality as a special case).

Exercise 3.7

1. State the criterion for the equals relation to hold in each of the following situations:

(a) A and B are sets. $A = B$.
(b) m and n are numbers. $m = n$.
(c) (a, b) and (c, d) are ordered pairs. $(a, b) = (c, d)$.

2. Explain the meaning of the word "equal" in the statement, "All men are born free and equal."

3. The statement, "A half plus a half is equal to a whole," is not necessarily true. Would you be willing to accept two halves of an automobile tire for a whole tire? Give other examples where the truth of the statement depends on the interpretation of the word "equals."

4. Suppose two people, both of whom like a particular cake, want to divide the cake *equally*. Cutting a cake into two *equal* parts is a difficult task. Knowing this, the two people agree that a reasonable criterion for dividing the cake equally would be that each should be satisfied with the piece he receives. How should they divide the cake with just a single cut?

5. Refer to problem 4, and decide how three people would cut the cake.

3.8 RELATIONS AS SETS

The emphasis on equivalence relations and the reflexive, symmetric, and transitive properties may be somewhat misleading. To dispel any notions that these are the only relations of interest or that there are no other properties of relations that are useful, we include a brief formal treatment of relations. This approach leads naturally to other useful concepts, in particular to the concept of *function*.

Before we become formal, let us investigate some relation-related ideas in a programmed learning exercise.

This exercise is in the form of a simple game. The effectiveness of this approach to learning will be self-evident provided you follow instructions carefully and complete the exercise. Before we begin, recall that two ordered pairs (a, b) and (x, y) are the same if and only if $a = x$ and $b = y$.

The elements of a particular set are given a few at a time. As soon as you think of an element which belongs to this particular set, write it down. As the game proceeds continue to write elements which you believe belong to the particular set.

(*a*) The elements in the first particular set are all ordered pairs. Here are a few elements: {(Martha, George), (Mary, Abe), . . .}. Before reading further, try to name an element of this set.

Here are some additional elements of the same set:

(Mary, Joseph)
(Blondie, Dagwood)

By now you should have an element which you think belongs to this set. Try to fill in the *criterion for belonging* in the following description of this set:

$$\{(x, y) | \ldots\ldots\ldots\ldots\}.$$

Fill in the blank in (Maggie, ———). How do you determine whether a particular ordered pair belongs to this particular set or not? We might say that the person whose name is the first component of the ordered pair *is the wife of* the person whose name is the second component. If we use (x, y) to denote an arbitrary element of this set as we have indicated previously, x would stand for a woman's name and y would stand for a man's name and (x, y) would belong to the set if and only if x "is the wife of" y.

Let us try a different set.

(*b*) The elements in the second particular set are also ordered pairs.

Here are some elements: {(Austin, Texas), (Atlanta, Georgia) . . .}. Before you read further, write down what you think is another element of this set.

Additional elements are:

(Albany, New York)
(Helena, Montana)

By now you should be able to name an element you think belongs to this set. To be certain, fill in the blank of each of the following:

(———, Maine), (Sacramento, ———), (Springfield, ———).

In the set-builder notation we would write

$$\{(x, y) | \ldots ? \ldots ? \ldots ? \ldots\}.$$

In this instance, x is the name of a city, y is the name of a state, and (x, y) belongs to this particular set if x "is the capital of" y.

If we want to talk about the set in (*a*), above, we might describe the set as the set of all ordered pairs (x, y), where x is the name of a woman, y is the name of a man, and (x, y) belongs to the set if, and only if, x "is the wife of" y. This description is adequate but long-winded and even somewhat redundant. It did not take too many ordered pairs in this set to know how x is related to y. On the other hand, if we know the *relation* of x to y, we would know the ordered pairs in this set. This being so, we might introduce a simple symbol to stand for "is the wife of" such as Ⓦ, and instead of writing x "is the wife of" y, simply write x Ⓦ y. We could then write this set as

$$\{(x, y) | x \text{ Ⓦ } y\}.$$

We might even ask, "What is this set?" Frequently, the set itself is called the relation and we might write Ⓦ $= \{(x, y) | x$ Ⓦ $y\}$. Here Ⓦ is used to denote the set defining the relation as well as the relation itself. This may result in some confusion, but is consistent with mathematical convention.

The set in (*b*) above is the set of ordered pairs whose first component is the name of a city and its second component is the name of a state. A particular ordered pair is in this set if and only if the first-place member is the capital of the second-place member. If we use the symbol © in place of "is the capital of" we can specify this set as

$$\{(x, y) | x \text{ © } y\}.$$

Following the convention of the previous paragraph, we could call this set the relation "is the capital of" and denote it by ©. If we do this, it should be noted that we can either write (Atlanta, Georgia) $\in$ © or Atlanta © Georgia. Both would be read, "Atlanta is the Capital of Georgia." It is worth noting again that if the relation between the first-place member and the second-place member of the ordered pairs is known, then we would know the ordered pairs and, conversely, if we know the ordered pairs, we would know the relation. This fact has led to the definition of a relation as a set of ordered pairs.

Definition 3.8a. A *relation* is a set of ordered pairs.

In the relation "is the wife of," it is understood that the names in the first component of the ordered pair (x, y) are the names of married women and the names in the second component are the names of married men. The set of married women is called the *domain* of the relation and the set of married men is called the *range* of the relation. In the relation in (*b*), the domain is the set of capital cities and the range is the set of states in the United States.

In earlier discussions we used special symbols to designate relations that occur often enough to have become conventional. We used $<$ for *less than*, > for *greater than*, | for *divides*, and ≤ for *less than or equal to*. We also used the symbol ® to denote an arbitrary relation and wrote a ® b to indicate that *a is ®-related to b*. As we have just indicated, we can also consider a relation ® as a set, each member of which is an ordered pair. Note that we write, interchangeably, x ® y or $(x, y) \in$ ®.

Definition 3.8b. The set of all first components of elements of a relation is called the *domain* of the relation.

Definition 3.8c. The set of all second components of elements of a relation is called the *range* of the relation.

In general, the domain and the range of a relation are distinct sets. If the domain and the range of a relation are in the same set, the relation is said to be *defined in the set*. A relation with domain A and range B is a subset of $A \times B$. A relation in a set A is a subset of $A \times A$.

Example 1

Consider the set $A = \{1, 2, 3, 4\}$. Then the set $A \times A$ is given by the following table:

4	(1, 4)	(2, 4)	(3, 4)	(4, 4)
3	(1, 3)	(2, 3)	(3, 3)	(4, 3)
2	(1, 2)	(2, 2)	(3, 3)	(4, 2)
1	(1, 1)	(2, 1)	(3, 1)	(4, 1)
	1	2	3	4

From the set $A \times A$ consider the subset consisting of

$\{(1, 1), (2, 2), (3, 3), (4, 4)\}$.

4	(1, 4)	(2, 4)	(3, 4)	**(4, 4)**
3	(1, 3)	(2, 3)	**(3, 3)**	(4, 3)
2	(1, 2)	**(2, 2)**	(3, 2)	(4, 2)
1	**(1, 1)**	(2, 1)	(3, 1)	(4, 1)
	1	2	3	4

Suppose you were told that these pairs were selected because the first component bears a special relation to the second and that only these ordered pairs of the set $A \times A$ have this relation. Certainly you would conclude that the relation described is the *equals* relation.

Example 2

Consider the subset $\{(1, 4), (1, 3), (1, 2), (2, 4), (2, 3), (3, 4)\}$.

4	**(1, 4)**	**(2, 4)**	**(3, 4)**	(4, 4)
3	**(1, 3)**	**(2, 3)**	(3, 3)	(4, 3)
2	**(1, 2)**	(2, 2)	(3, 2)	(4, 2)
1	(1, 1)	(2, 1)	(3, 1)	(4, 1)
	1	2	3	4

This subset also defines a relation. After examining the ordered pairs in the set, you should see they describe the relation *is less than*.

Example 3

Consider the subset {(1, 1), (1, 2), (1, 3), (1, 4), (2, 2), (2, 4), (3, 3), (4, 4)}.

4	**(1, 4)**	**(2, 4)**	(3, 4)	**(4, 4)**
3	**(1, 3)**	(2, 3)	**(3, 3)**	(4, 3)
2	**(1, 2)**	**(2, 2)**	(3, 2)	(4,2)
1	**(1, 1)**	(2, 1)	(3, 1)	(4, 1)
	1	2	3	4

This subset defines the relation *divides* in the set A. The relation *divides* is a subset of the Cartesian product $A \times A$.

Notice that if the subset of $A \times A$ contains the diagonal elements, that is, (1, 1), (2, 2), (3, 3), (4, 4), whatever the remaining elements of the subset may be, we must conclude that the relation defined by the subset is *reflexive*. Can similar statements be made with reference to the symmetric and transitive properties of relations?

We repeat: A relation with domain A and range B is quite frequently defined as a subset of the Cartesian product, $A \times B$. A relation in a set A is a subset of $A \times A$.

A continuation of our programmed learning exercise leads to part (*c*).

(*c*) The elements in this particular set are all ordered pairs of numbers. Here are some elements in this set: (7, 9), (1, 3), and (10, 12). With so few ordered pairs, the relation being suggested may not be evident. Here are some more elements in the set: (2, 4), (3, 5), and (4, 6). Can you fill in the blank in (11, ______)? Can you fill in the blank in (x, ______)?

Describe the set in set-builder notation as

$$\{(x, y) | \ldots\}.$$

An ordered pair of numbers belongs to this set if and only if the second-place number is larger by 2 than the first-place number. Thus, if the first-place number is x, the second-place number is $x+2$. We write

$\{(x, y) | \ y \text{ is 2 greater than } x\}$ or

$\{(x, x+2) | \ x \text{ is a number}\}$.

Since these two sets are the same,

$$(x, y) = (x, x+2) \quad \text{for all } x.*$$

*Of course we *might* have been considering $y = x + 2 + (x-7)(x-1)(x-10)x^2$ since when $x = 7$, this expression also yields 9 for y and similarly gives $y = 3$ when $x = 1$ and $y = 12$ when $x = 10$. But throughout this discussion we are assuming the "natural" and "simplest" relations possible.

But this is so if and only if $x = x$ and $y = x + 2$. Since x is always equal to x, we might say that this relation is determined by the equation

$$y = x + 2.$$

(*d*) What relation does the following set suggest?

$$\{(1, 1), (2, 4), (3, 9), (4, 16), (5, 25), \ldots\}.$$

Think of this as $\{(x, y) | \text{(criterion for pairing } x \text{ and } y)\}$. This set can also be thought of as $\{(x, x^2) | x \text{ is a number}\}$. Since these two sets are the same.

$$(x, y) = (x, x^2) \quad \text{for all } x.$$

Thus this relation is determined by

$$y = x^2.$$

The last two relations (*c*) and (*d*) are examples of a special kind of relation called a function.

Relations may be classified in terms of the manner in which they pair the elements. This pairing is sometimes called a correspondence. There are four kinds of correspondences. They are:

1. One-to-one correspondences.
2. Many-to-one correspondences.
3. One-to-many correspondences.
4. Many-to-many correspondences.

These correspondences occur quite naturally and are illustrated in the following diagrams.

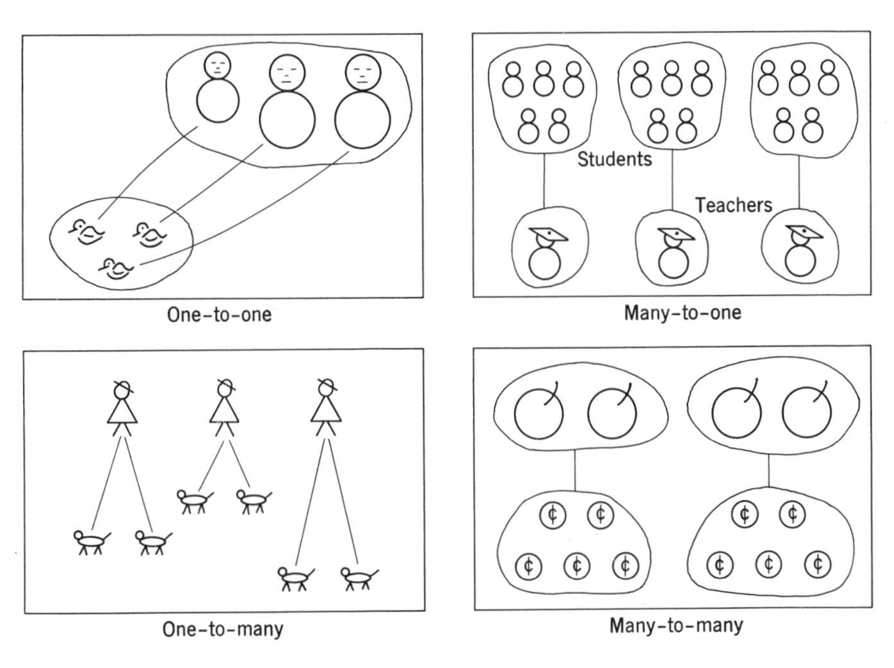

The one-to-one and many-to-one relations are of special interest and are called *functions*.

> ***Definition 3.8d.*** A relation is *single valued* if it is a one-to-one or many-to-one correspondence. This is often expressed as, x Ⓡ y and x Ⓡ z cannot happen unless $y = z$.

In set notation $(x, y) \in$ Ⓡ and $(x, z) \in$ Ⓡ cannot happen unless $y = z$.

> ***Definition 3.8e.*** Single-valued relations are called *functions*.

Functions play a central role in mathematics. For relations that are functions, single letters are used. Such relations are usually referred to as the function f, the function g, etc.

Since functions are relations, a function is a set of ordered pairs. If the first-place member is denoted x, the second-place member is frequently denoted by $f(x)$. Thus a function can be thought of as: $f = \{(x, f(x)) \mid x$ is in the domain of the function and $f(x)$ is paired to x according to a specific condition or criterion$\}$. Thus

$$(x, y) = (x, f(x)) \text{ if and only if } y = f(x)$$

and x is in the domain of f. Thus the function is determined by the equation

$$y = f(x).$$

Note that $y = x + 2$ and $y = x^2$ are both instances of this more general situation.

Every relation has an inverse.

> ***Definition 3.8f.*** The *inverse* of a relation Ⓡ is the set of all ordered pairs (y, x) for which (x, y) is in Ⓡ. We use the notation Ⓡ$^{-1}$ to denote the inverse relation.

Example 4

If Ⓡ is the relation $<$, then Ⓡ$^{-1}$ is the relation $>$.

$3 < 5$	$5 > 3$
$(3, 5) \in$ Ⓡ	$(5, 3) \in$ Ⓡ$^{-1}$

Example 5

If Ⓡ is the relation "is a factor of" in the set of natural numbers, then Ⓡ$^{-1}$ is the relation "is a multiple of."

2 is a factor of 6, $(2, 6) \in$ Ⓡ

6 is a multiple of 2, $(6, 2) \in$ Ⓡ$^{-1}$

Example 6

An alternate way of depicting the relation of example 2, Section 3.8, is by using the convention described in Section 2.7, but deleting the ordered pairs.

Here the dots in the diagram represent the ordered pairs. Those that are circled are the pairs defining the relation $\{(1, 2), (1, 3), (1, 4), (2, 3), (2, 4), (3, 4)\}$. The set of points which depicts the relation is called the *graph* of the relation. The elements of the *domain* are plotted along the horizontal axis. The elements of the *range* are plotted along the vertical axis.

Example 7

The graph of the relation discussed in (c) of this section is shown below.

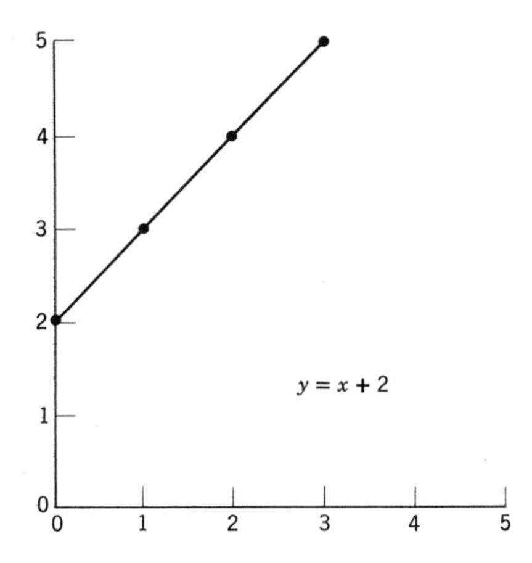

Example 8

The graph of the relation discussed in (d) of this section is shown below.

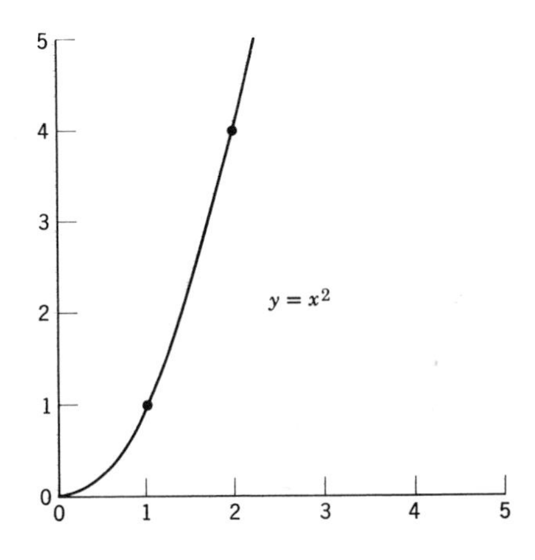

Exercise 3.8

1. Let $A = \{$Boise, Helena, Olympia, Salem$\}$.
$B = \{$Idaho, Montana, Washington, Oregon$\}$.

(a) Construct $A \times B$.
(b) Indicate the subset defined by the relation "is the capital of."
(c) Indicate the set defined by the inverse of the relation in (b).

2. If f is a function, the inverse of f is a relation. When is the inverse of f a function?

3. What is the difference in meaning between the symbols

(a) (a, b) and (b, a)?
(b) (a, b) and $\{a, b\}$?

4. (a) Which of the relations described by the following graphs are functions? (See Example 6, Section 3.8).
(b) State the domain and range of each relation that is a function.
(c) Which of the relations have inverses that are functions?

y	1	2	3	4
4	·	·	·	⊙
3	·	·	·	⊙
2	·	·	·	⊙
1	·	·	·	⊙

(a)

y	1	2	3	4
4	·	·	⊙	·
3	·	·	·	⊙
2	·	⊙	·	·
1	·	·	·	·

(b)

y	1	2	3	4
4	·	⊙	·	·
3	·	⊙	⊙	·
2	⊙	⊙	·	⊙
1	·	·	⊙	·

(c)

y	1	2	3	4
4	·	·	·	·
3	·	⊙	⊙	·
2	·	·	·	·
1	⊙	·	·	⊙

(d)

5. Define an order relation in the set consisting of the pupils at Garfield Elementary School.

6. Let W denote the set of whole numbers, $\{0, 1, 2, 3, \ldots\}$. Define a

® $= \{(a, b) | a \in W$ and $b \in W$ and $a - b$ is divisible by $3\}$.

Which of the following are true and which are false?

Justify each answer.

(a) $(1, 13) \in$ ® (b) $(-2, 13) \in$ ® (c) 1 ® 4
(d) 5 ® 27 (e) ® is symmetric (f) ® is transitive

REVIEW EXERCISE 1

1. Provide the word or phrase to replace the dash that makes each of the following a true statement. Supply enough words to make the statement *complete.*

(a) ——— is a name for a number.
(b) ——— is a set of symbols and some scheme for using these to give names to numbers.
(c) ——— is an example of an additive system of numeration.
(d) ——— is an example of a multiplicative system of numeration.
(e) ——— is an example of a place-value system of numeration.
(f) A unique feature of the decimal system of numeration is ———.
(g) MCMLXVII is the Roman numeral for ——— (decimal numeral).
(h) The symbols $a \in X$ means ———.
(i) For two sets, A and B, ——— if every element of A is an element of B and every element of B is an element of A.
(j) ——— is the set which has no elements.
(k) The union of two sets A and B is the set ———.
(l) The intersection of two sets A and B is the set ———.
(m) If $A \subseteq U$, then the complement of A with respect to U is the set of all elements ———.
(n) Two sets, A and B, are said to be disjoint if ———.
(o) There are ——— subsets of a set of n elements.
(p) The Cartesian product of two sets A and B is the set of all ordered pairs (a, b), such that ———.
(q) A relation ® defined on a set S is reflexive if, for all a in S, ———.
(r) A relation ® defined on a set S is symmetric if, for all a and b in S, ———.
(s) A relation ® defined on a set S is transitive if, for all a, b, and c in S. ———.
(t) An equivalence relation is one that is ———.
(u) If set A can be placed in one-to-one correspondence with the set $\{1, 2, 3, 4, 5, \ldots, n\}$, then n is said to be ——— of the set A.
(v) Relations described in terms of correspondences fall into four classes. They are ———.
(w) A relation defined by a set of ordered pairs, no two of whose first components are the same, is called ———.
(x) For two sets, A and B, A is a subset of B if ———.
(y) A relation defined on a set A is a subset of ———.
(z) A single-valued relation is called a ———.

2. In each of the diagrams on page 64, shade the region that is symbolized below each diagram.

3. In each of the diagrams on page 64 write the set of symbols below each diagram that describes the shaded region.

4. Let $T = \{m \mid m = 3k, k \text{ a whole number}\}$.
(a) Is T a subset of W, the set of whole numbers?
(b) Is T a proper subset of W?
(c) T (has fewer elements than), (has more elements than), (can be matched with) W. Choose one answer.

5. Let W denote the set of whole numbers, $\{0, 1, 2, 3, 4, \ldots\}$. Define the relation ® in W by

$$® = \{(a, b) \mid a \in W \text{ and } b \in W \text{ and } a - b \text{ is divisible by } 2\}.$$

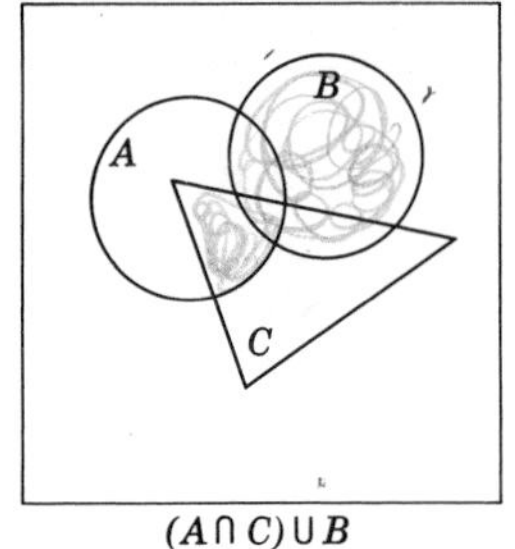

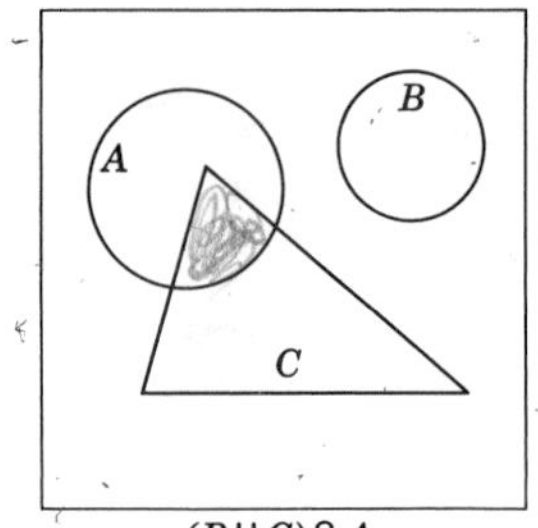

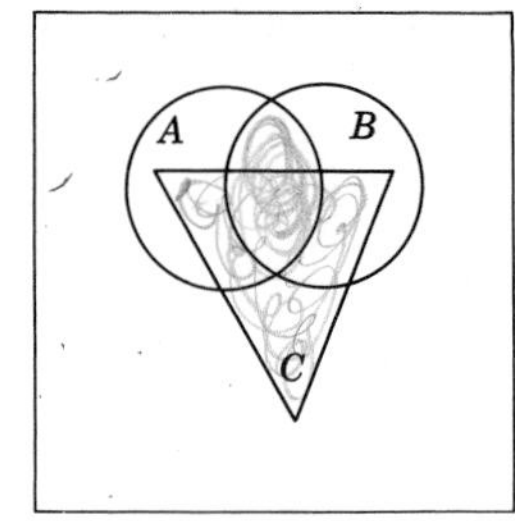

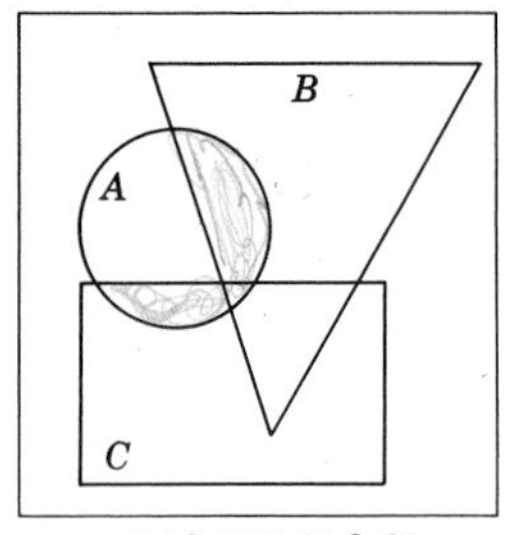

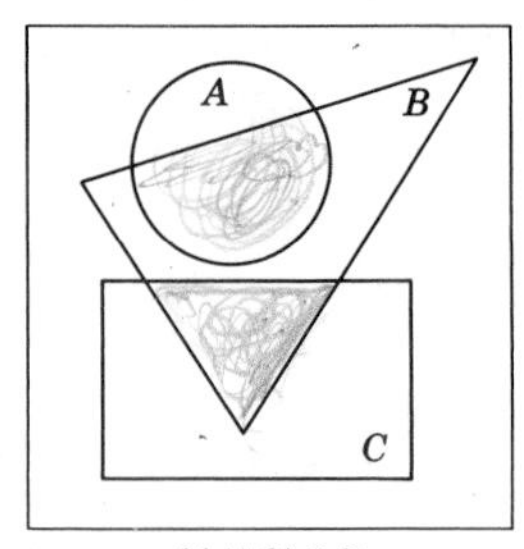

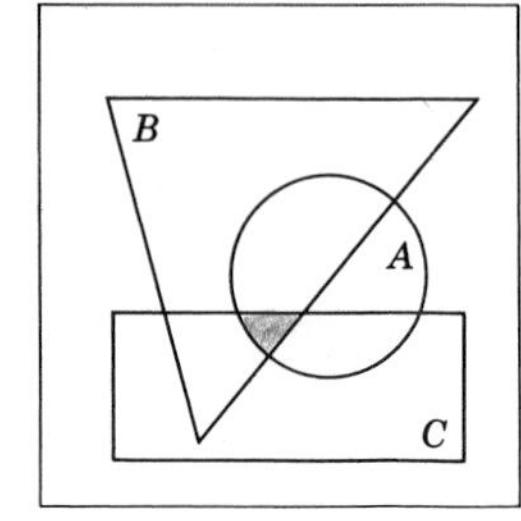

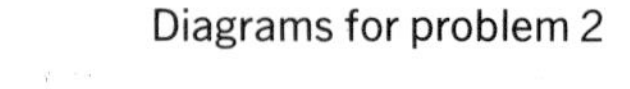

Diagrams for problem 2

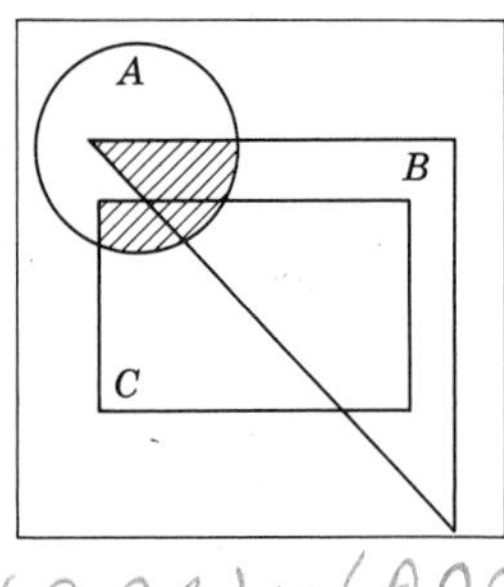

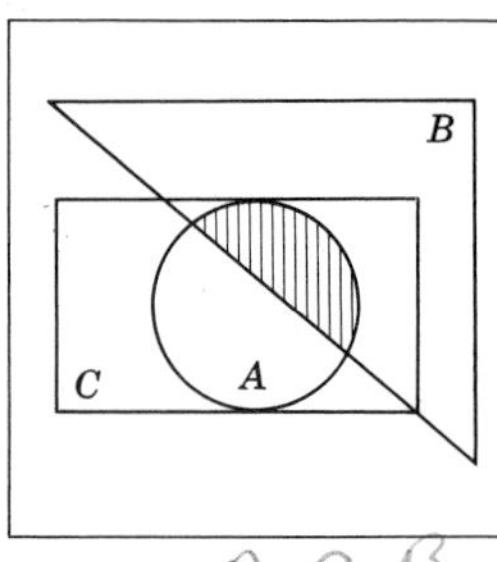

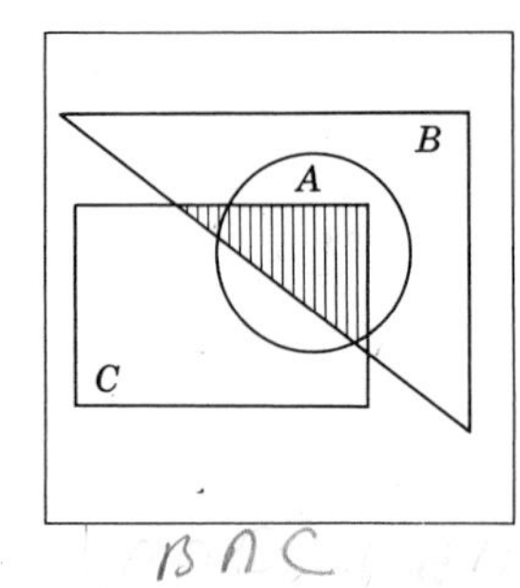

Diagrams for problem 3

State whether the following are true or false and give a reason for your answer:

(a) $(13, 1) \in$ ®. (b) $(-2, 13) \in$ ®. (c) 4 ® 1.
(d) 27 ® 5. (e) ® is symmetric. (f) ® is transitive.
(g) ® is reflexive.

6. Give the cardinal number of each of the following:

(a) $\emptyset$ (b) $\{\emptyset\}$
(c) $\{0\}$ (d) $\{(a, b)\}$

7. Simplify, but leave as powers

(a) $(10^2)^3$ (b) $\dfrac{2^7 \cdot 3^5}{(2 \cdot 3)^3}$

(c) $\frac{3^3 \cdot 5^7}{15^4}$ (d) $8 \cdot 2^{-4}$

(e) $9 \cdot 3^{-4}$

8. The following is a partial list of characteristics and distinguishing features of numeration systems:

(a) symbols for powers of the base
(b) symbol for zero
(c) order of symbols important
(d) subtraction principle
(e) repeated use of a symbol
(f) multiplicative principle
(g) additive principle
(h) place value
(i) a single distinguishable symbol for ten

Use the letters to indicate which of these are characteristics

(a) of the Egyptian system of numeration.
(b) of the Roman system of numeration.
(c) of the Chinese-Japanese system of numeration.
(d) of the decimal system of numeration.

9. Given the following correspondence of decimal numerals to symbols representing the same number:

Decimal Numeral	*Symbol*
0	O
1	I
2	L
3	F
4	E
5	B
25	B^2
125	B^3
625	B^4

(a) List the symbols that would be used in an additive system of numeration.
(b) List the symbols that would be used in a multiplicative system of numeration.
(c) List the symbols that would be used in a place-value system of numeration.
Complete the following table:

	Decimal Numeral	*Additive System*	*Multiplicative System*	*Place-Value System*
(d)	98	———	———	———
(e)	1884	$B^4B^4B^4BIIII$	———	———
(f)	1930	———	———	*FOLIO*

10. Use symbols to indicate that:

(a) x is an element of the set B.
(b) x is *not* an element of the set C.

(c) the set A is a subset of the set B.
(d) A is the set which contains no elements.
(e) C is the union of sets A and B.
(f) D is the intersection of sets A and B.
(g) sets A and B have no elements in common.

11. Thirty students go on a camping trip. Twelve students return with both sunburns and insect bites. Twenty report sunburns. How many suffered insect bites if it is known that only three students suffered neither?

12. Let $U = \{1, 2, 3, 4, 5\}$, $C = \{1, 3\}$, and A and B be nonempty sets.

(a) $A \cup B = U, A \cap B = \emptyset$, and $B = \{1\}, A = ?$
(b) $A \subseteq B$ and $A \cup B = \{4, 5\}, B = ?$
(c) $A \cap B = \{3\}, A \cup B = \{2, 3, 4\}$ and $B \cup C = \{1, 2, 3\}. A = ?$

13. Let us try to make up a system of numeration. Since the place-value system is the most efficient, we will use the symbols Δ, $-$, $\subset$, Σ, and the ideas of place value as follows:

$n(\emptyset) = \Delta.$
$n(\{\Delta\}) = -.$
$n(\{\Delta, -\}) = \subset.$
$n(\{\Delta, -, \subset\}) = \Sigma.$

(a) In this system, what is the cardinal of the set $\{c, x, w, e, r, g\}$?
(b) What number follows $\Sigma\Delta\Sigma$?
(c) What number is four times $\Sigma \subset -$?
(d) What number is one-fourth of $\Sigma \subset -$?

REVIEW EXERCISE 2

1. Let 0 represent zero, I represent one, L represent two, and a, b, c, d represent $3, 3^2, 3^3, 3^4$, respectively. Make a table as follows and, using the necessary symbols from this set, represent the given decimal numbers in (a) an additive system of numeration, (b) a multiplicative system of numeration, and (c) a place-value system of numeration, where the resepctive place values are units, threes, nines, twenty-sevens, eighty-ones, and so on.

	Additive	*Multiplicative*	*Place-Value*
9	———	———	———
25	———	———	———
109	———	———	———

2. Let I denote the set of all people.
M denote the set of all male people.
W denote the set of all female people.
R denote the set of all red-haired people.
T denote the set of all people 21 or more years old.
Describe the following subsets of I in words. (For example, $W \cap R$ is the set of all red-haired females.)

(a) $M \cap R$ (b) $R \cup T$
(c) $M \cap W$ (d) $M \cup T$

3. Let $A = \{1, 2, 3, 4\}, B = \{a, b, c\}$.

(a) Make the table of $A \times B$.
(b) Make the table of $A \times A$.
(c) Pick the subset of $A \times A$ such that the numbers of the ordered pairs are related by "less than."

4. Given the set $A = \{1, 2, 3, 4, 5, 6\}$, produce a set matched to A, that is, one that is related to A by the "matching" relation.

5. Given the set $M = \{1, 2, 3, 4,\}$ and $N = \{a, b, c, d\}$,

(a) establish a one-to-one correspondence between M and N.
(b) How many such correspondences are there?

6. Given $M = \{1, 2, 3\}$,

(a) list all subsets of M,
(b) How many subsets are there of a set of five elements?

7. Consider $U = \{1, 2, 3, 4, 5, 6, 7\}$;

$$A = \{1, 2, 3, 7\}; \qquad B = \{3, 4, 6, 7\}; \qquad C = \{4, 5\}.$$

(a) Draw a Venn diagram of the sets U, A, B, and C
(b) Describe the following sets by listing the elements:

(1) $A \cap B$	(2) $A \cup (B \cup C)$
(3) $A \cup B$	(4) $A \cap (B \cup C)$
(5) $A \cap (B \cap C)$	(6) $A \cup (B \cap C)$

8. List the properties (i.e., reflexive, symmetric, transitive) possessed by each of the following relations defined on the specified sets:

Set	*Relation*
(a) The natural numbers	"is greater than"
(b) The natural numbers	"divides"
(c) Straight lines in a plane	"is parallel to"
(d) Pupils at Paxon School	"is in the same grade as"
(e) Animals	"is the same specie as"

(f) Which of the above relations are equivalence relations?

9. Define the following:

(a) The union of two sets.
(b) The intersection of two sets.
(c) The empty set.
(d) The Cartesian product of two sets.
(e) Subset.
(f) Equivalence relation.
(g) Function.
(h) Order relation.

10. Let $A = \{1, 2, 3\}, B = \{1, 2, 3, 4\}$, and $C = \{4, 5, 6, 7\}$.

(a) Find $n(A)$, $n(C)$, $n(A \cup C)$, and $n(A \cap C)$. Write $n(A \cup C)$ in terms of $n(A)$, $n(C)$, and $n(A \cap C)$.

(b) Find $n(A)$, $n(B)$, and $n(A \times B)$. Write $n(A \times B)$ in terms of $n(A)$ and $n(B)$.
(c) Find $n(C \times B)$ and write it in terms of $n(C)$ and $n(B)$.
(d) Find $n[(A \cup C) \times B]$ and write it in terms of $n(A)$, $n(B)$, and $n(C)$.

REFERENCES

Hamilton, Norman T. and Joseph Landin, *Set Theory, The Structure of Arithmetic*, Allyn and Bacon, Boston, 1961, pp. 46–73.

Kelley, John L., *Algebra, A Modern Introduction*. D. Van Nostrand and Co., Princeton, New Jersey, 1960.

Newman, James R., *The World of Mathematics*, Simon and Schuster, New York, 1956.

CHAPTER 4

The System of Whole Numbers

4.1 INTRODUCTION

Teaching a child to count is more than teaching him to repeat the words one, two, three, etc. Teaching a child to count is actually teaching him to recognize the common property of matched sets and properly label this abstraction, just as teaching a child the various colors involves distinguishing the colors and properly labeling them. "Number blindness" is not yet listed as a physiological disorder as is "color blindness." The matching process distinguishes sets with a precision that cannot be accomplished by the eye in distinguishing colors. The number concept should be easy to teach and interesting to learn.

In previous chapters we introduced the notion of sets, relations, and properties of relations, and we examined a few examples of relations defined on sets. Various relations were seen to have certain properties in common. Those relations that are reflexive, symmetric, and transitive occur so frequently in mathematics and are so important that they are designated by a special name, *equivalence relations* (see Section 3.4). The effect of an equivalence relation on the set on which it is defined is *to partition the set* into subsets, which we call *equivalence classes*. The matching relation defined on sets in terms of the fundamental concept of *one-to-one correspondence* is an equivalence relation, and the common property of the sets in an equivalence class of this *matching relation* is called a *cardinal number*.

The names and symbols given to these abstractions are called *numerals*.

Thus the symbol "1" represents the number which is the distinguishing feature of the equivalence class of all sets with the property of "oneness." When we write the symbol "1", we can form a set that consists of this symbol. We denote this set $\{1\}$. The symbol "2" represents the number which is the distinguishing feature of the equivalence class of all sets with the property of "twoness." A representative set for this class is the set $\{1, 2\}$. In the same way we can form the sets $\{1, 2, 3\}$, $\{1, 2, 3, 4\}, \ldots,$ $\{1, 2, 3, 4, 5, \ldots, n\}$, and $\{1, 2, 3, 4, 5, 6, 7, \ldots\}$. In the set $\{1, 2, 3, 4, 5, 6, 7, \ldots\}$, the dots indicate that the sequence of numbers continues indefinitely. This set is called the set of *natural numbers.* We will use the capital letter N to denote this set.

$$N = \{1, 2, 3, 4, 5, \ldots\}.$$

4.2 COUNTING SETS

The sets $\{1\}$, $\{1, 2\}$, $\{1, 2, 3\}, \ldots,$ are representative sets of the equivalence classes under the matching relation to which each belongs. These sets we will call *counting* sets. Notice that they are *ordered.* That is, the number 1 always comes first in each set. The number 2 always follows the number 1 in these sets.

Note: Some authors use the convention of replacing the curly brackets, or braces, with parentheses when denoting *ordered sets.* Thus they would write $(1, 2, 3, 4)$ as we have done with ordered pairs. This is sound mathematical convention, but we will not be using ordered sets, except for ordered pairs, in what follows.

To introduce the *whole numbers* with the natural order referred to previously, we modify a conventional procedure for the sake of simplicity.

4.3 THE WHOLE NUMBERS

The number zero is defined as the cardinal of the empty set.

$$0 = n(\emptyset).$$

The number 1 is then defined as the cardinal of the set which contains only the element 0 as a member.

$$1 = n(\{0\}).$$

The number 1 is called the *successor* of 0. The number 2 is defined as the cardinal of the set which consists of the elements 0 and 1.

$$2 = n(\{0, 1\}).$$

The number 2 is called the *successor* of 1. The number 3 is defined in like manner, etc.

$$3 = n(\{0, 1, 2\}).$$
$$4 = n(\{0, 1, 2, 3\}).$$

$5 = n(\{0, 1, 2, 3, 4\})$.

.

.

.

etc.

The number 3 is the successor of 2, the number 4 is the successor of 3, etc. The successor of the number k is denoted by $k+1$. Continuing indefinitely we obtain the set $\{0, 1, 2, 3, 4, 5, 6, \ldots\}$. This set, which we denote by W, is called *the set of whole numbers*. (The particular choice of name for this set is a matter of personal preference. Some authors call this set the set of natural numbers. We choose to use the term *natural* numbers for the set which begins with the number 1 and which does not include the number 0.)

The idea of the successor of a number serves to distinguish the natural order of the whole numbers as opposed to another kind of order which will be discussed in detail later. It should be pointed out that it is this natural order which children must learn when they first learn numbers.

Example 1

There is a difference in the statements "4 is the successor of 3" and "4 is greater than 3."

The counting sets start with the number 1. Counting is simply the process of establishing a one-to-one correspondence between the set being counted and the appropriate counting set with this natural order. The *last* number in the ordered counting set is the cardinal of the set being counted.

Any number in the set of whole numbers is the cardinal of some *finite* set. As we said before, a set which can be put into one-to-one correspondence with one of the ordered counting sets with a last number is called a *finite* set. The empty set has cardinal number 0. A set is *countably infinite* if there is a one-to-one correspondence between the elements of the set and the set of *all* natural numbers. Thus the set of natural numbers is itself countably infinite, as is the set $\{2, 4, 6, 8, \ldots\}$.

Infinite sets can be characterized in other ways. Thus a set may be defined as infinite if it admits a one-to-one correspondence of the whole set with a *proper* subset of itself. Thus, in terms of this definition, N is infinite since it has a one-to-one correspondence with one of its proper subsets, $\{2, 4, 6, 8, \ldots\}$.

4.4 ORDINAL AND CARDINAL USE OF NUMBERS

Although there is a technical difference between ordinal and cardinal numbers, we will not make this distinction in this book but rather emphasize the ordinal and cardinal use of the numbers.

When we use a number in answer to the question, "How many?" we are making *cardinal use* of the number. That is, when a number is used to designate the "size" of a set, it is being used *cardinally*.

On the other hand, any use of these numbers that depends on the prescribed order is the *ordinal use* of the numbers. The number represented by the numeral at the top or bottom of a page in a book is an example of the ordinal use of the number. It is the *82nd* page. When a number is used to designate a counted position, it is being used ordinally. The team is in *third* place in the league standings. When we use a number to answer the question, "Which one?", we are making ordinal use of the number.

Example 1

A person entering a bowling alley is given a slip of paper with the numeral 9 on it. It means he is the *9th* person who is waiting to bowl. This is the ordinal use of the number. The bowler would have 8 people ahead of him on the waiting list. This is the cardinal use of the number 8.

Exercise 4.4

1. In Little League baseball 19 boys turn out for a team. Of these, 11 are wearing baseball shirts and 14 are wearing baseball pants. Each boy is wearing part of a uniform and, of course, some have both. Let S denote the set of children wearing baseball shirts and let P denote the set of children wearing baseball pants.

(a) What is $n(S)$?
(b) What is $n(P)$?
(c) What is $n(S \cup P)$?
(d) What is $n(S \cap P)$?
(e) When finding the cardinal of each of the sets S and P, which boys were counted twice?
(f) How are the numbers of parts (a), (b), (c), and (d) related?

2. A group of 40 students goes on a camping trip. Of these, 12 return with both sunburns and insect bites, and 20 report sunburns. How many suffered insect bites if it is known that only 13 students suffered neither?

3. If the cardinal of set A is 132, the cardinal of set B is 97, and the cardinal of set $A \cap B$ is 43, what is the cardinal of set $A \cup B$?

4. Let $T = \{m | m = 3k, k \in W\}$.

(a) What element in T corresponds to the first element in W?
(b) What element in T corresponds to the third element in W?
(c) Is T a proper subset of W?
(d) How is the cardinal of the set T related to the cardinal of the set W?

5. Let $Q = \{n | n = 2k - 1, k \in N\}$.

(a) What element in Q corresponds to the first element in N?
(b) What is another name for the set Q?
(c) If $R = \{j | j = 2i + 1, i \in W\}$, how is R related to Q?

6. If $A = \{a, b, c, d, e\}$ and $B = \{d, e, n, m, k\}$, what is the cardinal of each of the following?

(a) $n(A \cup B) = ?$
(b) $n(A \cap B) = ?$
(c) $n(A \times B) \ = ?$

7. Let $A = \{a, a, b, c, c\}$.

(a) What is $n(A \cup \emptyset)$?
(b) What is $n(A \cap \emptyset)$?
(c) What is $n(A \times \emptyset)$?

8. (a) What does it mean to say that a set is finite?
(b) What does it mean to say that a set is countably infinite?

9. Distinguish between the cardinal use and the ordinal use of the natural numbers by citing examples involving the same numbers.

10. (a) Verify that "same color as" is an equivalence relation in the set of fruit.
(b) One of the equivalence classes under this relation contains an element which has the same name as the equivalence class. What is it?

11. We define the relation "$\equiv$" on the set $W = \{0, 1, 2, 3, \ldots\}$ as follows. For any two whole numbers m and n, $m \equiv n$ if they both give the same remainder when divided by 12, for example, $14 \equiv 38$, $17 \equiv 5$, etc. Verify that this is an equivalence relation and describe the equivalence classes by listing several representative elements from each of the equivalence classes.

Special Problem (from the *Scientific American*)
You face the problem of crossing a desert that is 800 miles across. You have a vehicle that is capable of hauling enough gasoline to travel 500 miles, including regular supply and cargo. What is the least number of trips required to cross the desert?

4.5 SYSTEMS OF NUMERATION AND NUMBER SYSTEMS

A number as a counting concept is one notion. A number as an element of a number system is quite different. The harmonious integration of these two ideas comprises the fundamentals of arithmetic. For this purpose it is helpful to reiterate what we mean by a system of numeration and at the same time to give a naive, intuitive definition of a number system.

A system of numeration is a set of symbols and a scheme for using these symbols to give names to all the numbers. We have examined a few of the many systems of numeration that man has developed to meet his particular needs. These should be recalled and the symbols and schemes reexamined in broad outline.

The system offering the greatest advantages from the standpoint of simplicity, economy of symbols in expressing numbers of any magnitude, and computation is the place-value system of numeration. Such systems with different bases will be examined in detail in Chapter 5.

By a *number system* we mean, from an intuitive standpoint, a *set* of numbers, *operations* defined on the numbers in the set, and *rules* governing these operations. Without prescribed operations and specific rules determining the behavior of the elements under these operations, the numbers would be no more interesting than a bag of marbles or a handful of checkers. It is in terms of the set, the operations, and the rules that structure of the number system has meaning. By changing the set of numbers, we change the structure of the number system. By changing the rules or adding new rules, we change the structure of the number system. It is in reference to these ideas that we will examine in some detail the following number systems:

The system of whole numbers
The system of integers
The system of rational numbers
The system of real numbers

We shall approach each system intuitively, using the background of the reader, precise definitions, and a certain amount of work on the reader's part to gain insight into the structure of each system. But first we must look further into the "equals" concept.

Exercise 4.5

1. What is meant by an additive system of numeration?

2. Discuss the Egyptian system of numeration, stressing the underlying scheme.

3. How does the Roman system differ from the Egyptian?

4. How does the Greek system of numeration differ from the Roman and Egyptian systems of numeration?

5. In a base three place-value system of numeration let

$0 = n(\emptyset)$.
$1 = n(\{0\})$.
$2 = n(\{0, 1\})$.

(a) What number follows 20_{three}?
(b) What number follows 22_{three}?
(c) What number follows 12_{three}?
(d) What number immediately precedes 100_{three}?
(e) What number immediately precedes 20_{three}?

6. Which number is larger?

(a) $2^{(2^2)}$ or $(2^2)^2$
(b) $3^{(3^3)}$ or $(3^3)^3$
(c) $10^{(10^{10})}$ or $(10^{10})^{10}$

7. (a) In what ways is football like soccer?
   (b) How do these games differ?
   (c) Do they differ in structure?

8. Distinguish between the game of checkers and the game of chess. Do they differ in structure?

4.6 THE EQUALS RELATION

One of the sources of difficulty in learning mathematics is the use of symbols. The opposite should actually be true since symbols are merely linguistic tools used to aid in communicating ideas and techniques. Most often the difficulty stems from our forgetting the meanings that we have agreed to give to the symbols. At other times, there is some confusion resulting from using the same symbol in a variety of seemingly different situations. The equals relation and the symbol for equals have in the past been an example. Most often "=" is used in the sense that if a denotes an object, concrete or conceptual, and if b also denotes an object, then $a = b$ (read "a equals b") means that the object denoted by a is identical with the one denoted by b. For the uninitiated and, in particular, in arithmetic, the difficulty arises in the criterion for *identical.*

Definition 4.6. We say that two numbers, a and b, are equal and write $a = b$ if and only if a and b are names for the same number.

Example 1

The symbols $2+3$ and 5 are different numerals for the same number; therefore we write $2+3=5$.

Note that in the foregoing definition, we have defined *equals* for *numbers*. Previously, we defined *equals* for *sets*, and *equals* for *ordered pairs*. In each instance we could have used either the usual meaning of "=" or the "names of the same object" definition. We chose to define the equals relations in the different situations because in each instance we wanted to emphasize what was necessary to determine whether the "objects" are identical or not. That is, we wanted to stress the criteria for the relation to hold.

It is immediately obvious to the reader that $2+3=5$, but it is not immediately obvious that $2483 \times 17{,}986 = 44{,}659{,}238$. The symbol "=" will be used in other situations and in each instance the "names of the same object" meaning will suffice. Nevertheless, a specific criterion will usually be included to emphasize what must be done to determine whether the relation holds or not. Regardless of the explicit sense in which the term is used, the relation "equals" (=) is an equivalence relation. That is,

1. For any a, $a = a$. (reflexive)
2. If $a = b$, then $b = a$. (symmetric)
3. If $a = b$ and $b = c$, then $a = c$. (transitive)

Where we have explicitly defined the relation, we will usually ask the student to use numerical examples to illustrate these properties. It is very easy to verify them, but the details of the proofs are not always warranted.

Exercise 4.6

1. Give the specific criterion for "=" to hold in each of the following:

(a) A and B are sets. $A = B$.
(b) (a, b) and (c, d) are ordered pairs. $(a, b) = (c, d)$.

2. Let $N = \{1, 2, 3, 4, 5, \ldots\}$ and $W = \{0, 1, 2, 3, \ldots\}$.

(a) List some elements in $W \times N$.
(b) Define "$\doteq$" for elements of $W \times N$ as follows
$(a, b) \doteq (c, d)$ if and only if $ad = bc$.

Use numerical examples to illustrate the reflexive, symmetric, and transitive property of this relation.

3. Define "$\equiv$" for elements in $W \times W$ as follows:
$(n, m) \equiv (r, s)$ if and only if $n + s = m + r$.
Use numerical examples to illustrate the reflexive, symmetric, and transitive property of this relation.

4. There are numerous situations in which one particular name for a number is more appropriate than another. For instance, when buying yard goods, $17\frac{1}{3}$ is more appropriate than 52/3. Can you think of other situations for which this is true?

4.7 BINARY OPERATIONS

Addition of numbers is a *binary operation*. Multiplication of numbers is also a *binary operation*. The operation of addition, which we denote by +, assigns to each *ordered pair* of numbers, say (2, 3), a third number, 2 + 3.

$$\underset{\text{ordered pair}}{(2, 3)} \xrightarrow[\text{operation}]{(+)} \underset{\text{the assigned number}}{2 + 3}$$

Multiplication, which we denote by ·, assigns to each ordered pair of numbers, for instance (2, 3), a third number, 2 · 3.

$$\underset{\text{ordered pair}}{(2, 3)} \xrightarrow[\text{operation}]{(\cdot)} \underset{\text{the assigned}}{2 \cdot 3}$$

The term *binary* refers to the *two* numbers in the *ordered pair*. The term *ordered pair* is used because it is not immediately obvious to a beginning student that addition assigns to (9, 4) the same number as it does to the ordered pair (4, 9). In an application, adding the number 4 to the number 9 is not always the same as adding the number 9 to the number 4. (Ask any first-grade pupil!)

Returning to the assigned number, we should note that there has been and still is the tendency to look at $29+33$ as something to do, rather than as a number. It is, in fact, a number. So also is $29 \cdot 33$. Computation is simply the process of finding other names for these numbers. (Admittedly, the symbols are numerals, but we will not overemphasize the distinction between numbers and numerals unless it is essential for clarity.) Even though the reader knows how to add, multiply, subtract, and divide, we review the basic operations of addition and multiplication in terms of sets in such a way as to make meaningful many of the ideas previously learned in arithmetic by rote. In other words, the binary operations themselves become objects of study.

> ***Definition 4.7.*** A *binary operation*, denoted by $*$, defined on a set S assigns to each *ordered pair* (m, n) of elements of S a *uniquely* determined element which we denote $m * n$.

That the element $m * n$ is uniquely determined means that the binary operation $*$ assigns to each ordered pair of elements (m, n) *one and only one* element. The element $m * n$ may have other names. One of the tasks of arithmetic is to discover systematic procedures for finding other names of the resultant when the binary operation is addition or multiplication. This is nothing more or less than arithmetic computation.

In some of the steps involved in these computations it may be more convenient to use different names for the same number; for example, we may use $3+4$ instead of 7 or $5 \cdot 1$ instead of 5. This is a property of the "equals" relation, called the *substitution property* of equals. This property is implicit in our statement that the resultant of a binary operation is uniquely determined. It is stated in general as follows:

1. A number may be substituted for its equal in any expression.

As a consequence of the *substitution property* it follows that

2. If equal numbers are added to equal numbers, their sums are equal that is, if $a = b$ and $c = d$, then $a+c = b+d$.

3. If equal numbers are multiplied by equal numbers, their products are equal, that is, if $a = b$ and $c = d$, then $ac = bd$.

We shall often use this property in what follows and refer to it by saying either "unique products or sums" or "the substitution property."

The *process* of finding other names of the resultant of the binary operation of addition is also called addition. Similarly, the *process* of finding another name of the product of the binary operation of multiplication is also called multiplication. That is, the terms addition and multiplication are used for both the process and the operation. We shall discuss the properties that addition and multiplication have as operations, and then see how they are used in making addition and multiplication meaningful as processes.

4.8 PROPERTIES OF BINARY OPERATIONS

4.8a The Closure Property

Let us use the symbol $\oplus$ to denote an arbitrary binary operation defined in the set S. Let the letters a, b, c, etc., denote elements of the set S. The specific operation may be addition, or multiplication, or some other binary operation which we may use as an illustration. The assigned element $(a \oplus b)$ may or may not be in the set S. If the assigned element $(a \oplus b)$ is a uniquely determined element in the set S for all ordered pairs of elements of S, we say that the binary operation $\oplus$ has the *closure property*. Another way of saying this is that the set S *is closed* with respect to the binary operation.

Note that the closure property depends both on the operation and the set. Also note that $(a \oplus b)$ is treated as an element rather than something to do!

Example 1

The set of odd numbers is not closed under addition, but it is closed under multiplication.

Example 2

The set of even numbers is closed with respect to both addition and multiplication.

4.8b The Commutative Property

The binary operation $\oplus$ is said to have the *commutative property* if it assigns to the ordered pairs (a, b) and (b, a) the same element in the set S. We indicate this by writing

$$a \oplus b = b \oplus a.$$

for all a and b in S.

Generally, (a, b) is different from (b, a), but if the binary operation $\oplus$ is commutative, $(a \oplus b)$ and $(b \oplus a)$ are different names for the same element in S.

To digress momentarily, recall the common meaning associated with the word "commute." A commuter is one who travels from home to work, from work to home. He "changes places." When two elements commute with respect to a binary operation, they "change places."

It is quite common to speak of adding the number 2 to the number 9. We indicated earlier that in the concrete sense this is quite different from adding the number 9 to the number 2. The commutative property of addition says that it is immaterial whether you add the number 2 to the number 9 or add the number 9 to the number 2; the sum will be the same. That is, the order of adding numbers is immaterial.

4.8c The Associative Property

The binary operation $\oplus$ is said to have the *associative property* if for any a, b, and c in the set S

$$(a \oplus b) \oplus c = a \oplus (b \oplus c).$$

Initially, the parentheses in the expression $(a \oplus b) \oplus c$ indicate that we first determine the element $(a \oplus b)$ and then find $(a \oplus b) \oplus c$.

$$\begin{aligned} &(a, b) \longrightarrow \oplus \longrightarrow (a \oplus b) \\ &((a \oplus b), c) \longrightarrow \oplus \longrightarrow (a \oplus b) \oplus c. \end{aligned}$$

In the expression $a \oplus (b \oplus c)$, the parentheses indicate that we first find the element $(b \oplus c)$ and then $a \oplus (b \oplus c)$.

$$\begin{aligned} &(b, c) \longrightarrow \oplus \longrightarrow (b \oplus c) \\ &(a, (b \oplus c)) \longrightarrow \oplus \longrightarrow a \oplus (b \oplus c). \end{aligned}$$

The associative property of the binary operation $\oplus$ permits us to group (i.e., associate) the elements as we choose. In fact, it also permits us to insert parentheses or remove parentheses:

$$a \oplus (b \oplus c) = a \oplus b \oplus c = (a \oplus b) \oplus c.$$

We are interested in the associative property because there are both associative and nonassociative systems in mathematics and associative and nonassociative situations in physics.

4.8d The Existence of an Identity

The set S is said to have an *identity* with respect to the binary operation $\oplus$ if there is an element, which we denote by i, in the set S such that for any element a in the set S, both of the following are true:

$$a \oplus i = a \quad \text{and} \quad i \oplus a = a.$$

The most familiar identities are the identity for addition and the identity for multiplication. The term unity element is also used and the identity element for multiplication is often called the unit. This is not true, however, for the identity for addition.

4.8e Inverses

An element a in the set S is said to have an *inverse* with respect to the binary operation $\oplus$ if there is an element in S, which we denote by a^{-1}, such that the binary operation assigns to the pair a and a^{-1} the identity element. That is, a^{-1} is the inverse of a with respect to the binary operation $\oplus$ if

$$a \oplus a^{-1} = i \quad \text{and} \quad a^{-1} \oplus a = i.$$

The inverse with respect to multiplication of numbers is sometimes called the *reciprocal*. The inverse with respect to addition of numbers is sometimes referred to as the *negative* and sometimes as the *opposite*.

Example 1

The inverse of 3 with respect to addition is -3 and the inverse of 3 with respect to multiplication is $\frac{1}{3}$.

Exercise 4.8

1. Let S denote the set of whole numbers which can be written in the form $2k$, where k is a whole number. That is,

$S = \{m | m = 2k, \text{ and } k \in W\}$.

(a) Does the set S have an identity element with respect to ordinary addition?
(b) Does the binary operation of addition have the closure property? Use numerical examples to help justify your answer.
(c) Does the set S have inverses with respect to the binary operation of addition? Use numerical examples to help justify your answer.
(d) Does the binary operation of addition have the commutative property? The associative property? Use numerical examples to help justify your answers.

2. Employ the set S of Exercise 1 and let the binary operation be ordinary multiplication.

(a) Does the set S have an identity element with respect to the binary operation of multiplication?
(b) Does the binary operation of multiplication have the closure property?
(c) Does the set S have inverses with respect to the binary operation of multiplication?
(d) Does the binary operation of multiplication have the commutative property? The associative property? Illustrate.

3. Let Q denote the set of whole numbers which can be written in the form $2k+1$, where k is a whole number.

$Q = \{m | m = 2k+1, \text{ and } k \in W\}$.

Answer the same questions as in problems 1 and 2 for the set Q.

4. Let the set T denote the set of whole numbers which can be written in the form $5k$, where k is a whole number. That is,

$T = \{m | m = 5k, \text{ and } k \in W\}$.

Answer the same questions as in problems 1 and 2 for the set T.

5. Let W denote the set of whole numbers. We define the binary operation $\oplus$ as follows. For m and n in W, let $\oplus$ assign to the ordered pair $m \oplus n$ be the number obtained by taking m to the power n. That is,

$m \oplus n = m^n$.

(a) What is $5 \oplus 3$?
(b) Does the binary operation $\oplus$ have the closure property?
(c) What is $(2 \oplus 3) \oplus 2$?
(d) What is $2 \oplus (3 \oplus 2)$?
(e) What is $3 \oplus (3 \oplus 3)$?
(f) What is $(3 \oplus 3) \oplus 3$?
(g) What is $10^{10^{10}}$?
(h) Is there an identity element with respect to this binary operation?
(i) Does the binary operation have the commutative property? Use a numerical example to justify your answer.
(j) Does the binary operation $\oplus$ have the associative property? Use a numerical example to justify your answer.

6. Let W denote the set of whole numbers and define the binary operation $\ominus$ as follows. For m and n in W, let $\ominus$ assign to the ordered pair (m, n) the element in the first component. That is,

$m \ominus n = m.$

(a) What is $3 \ominus 7$?
(b) Does the binary operation $\ominus$ have the closure property?
(c) Does the binary operation $\ominus$ have the commutative property?
(d) Does the binary operation $\ominus$ have the associative property? Use a numerical example to justify your answer.

7. Let W denote the set of whole numbers and define an operation as follows. For m and n in W, let $m \,.\, n$ be a whole number which divides both m and n. (The whole number a divides the whole number b if there is a unique whole number c such that $b = a \cdot c$.)

(a) Does the operation . have the closure property?
(b) Is $12 \,.\, 18$ unique?
(c) Does this operation satisfy Definition 4.7?

8. Let $U = \{a, b, c, d\}$ and let S denote the set whose elements are subsets of the set U. Recall that there are 16 such sets including the empty set $\emptyset$ and the universal set U. Let the binary operation $\cup$ defined on the set S be the operation which assigns to an ordered pair of sets the union of the two sets.

(a) Does the binary operation $\cup$ have the closure property? Use specific subsets to help justify your answer.
(b) Does the binary operation $\cup$ have the commutative property? Use specific subsets to help justify your answer.
(c) Does the binary operation $\cup$ have the associative property? Use specific subsets to help justify your answer.
(d) Does the set S have an identity element with respect to the binary operation of union?

9. Using the sets U and S of the previous problem, let the binary operation $\cap$ assign to ordered pairs of elements of S the intersection of the two sets.

(a) Does the binary operation $\cap$ have the closure property? Use specific subsets to help justify your answer.
(b) Does the binary operation $\cap$ have the commutative property? Use specific subsets to help justify your answer.
(c) Does the binary operation $\cap$ have the associative property? Use specific subsets to help justify your answer.
(d) Does the set S have an identity element with respect to the binary operation of intersection?

4.9 ADDITION AND MULTIPLICATION OF WHOLE NUMBERS

We define addition and multiplication of whole numbers so that the properties of each as binary operations are more plausible. Once established, we choose to call the properties *laws* because they guide us in what we can do and what we cannot do in arithmetic and indeed, in mathematics. We will see how these laws enable us to save work in computation, how they help us to find and understand shortcuts, as well as make meaningful many of the things we may have learned in the past by rote.

4.10 ADDITION IN W

For whole numbers, and *only for whole numbers*, we define addition as follows. The whole numbers were introduced as the cardinal of sets. Thus if a and b are whole numbers, a is the cardinal of a set A, and b is the cardinal of a set B. If the sets A and B are disjoint, the whole number $a+b$ is the cardinal of the set $A \cup B$.

Definition 4.10a. If $a=n(A)$ and $b=n(B)$ are whole numbers and $A \cap B=\emptyset$, the binary operation of addition (+) assigns to the ordered pair (a, b) the whole number $a+b$ which is the cardinal of the set $A \cup B$. We write this

if $a=n(A)$, $b=n(B)$, and $A \cap B=\emptyset$, then

$a+b=n(A \cup B)$.

Example 1

Let $A=\{a, b, c\}$ and $B=\{x, y, z, w\}$. Then $3=n(A)$, $4=n(B)$, and $A \cap B=\emptyset$. $3+4=n(A \cup B)=n(\{a, b, c, x, y, z, w\})=7$.

This is not a practical way to add, but this definition does lend itself to establishing the properties of the binary operation of addition. For example, if A is the set of automobiles licensed in the State of New York in 1971 and B is the set of automobiles licensed in the State of California in 1971 we would not normally find $a+b$ by counting the elements in the set $A \cup B$. However, it is the way we actually teach children how to add whole numbers.

Without going into unwarranted details of proof, we state simply that two finite sets that can be matched under a one-to-one correspondence have the same cardinal. In particular, two equal sets obviously have the same cardinal. Closure follows from the fact that the union of two finite sets is also finite. What is needed for uniqueness is that if, also, $a=n(C)$ and $b=n(D)$, $C \cap D=\emptyset$, then $n(A \cup B)=n(C \cup D)$. This follows not only from A–C and B–D but from the (unproved) fact that if A–C and B–D, $A \cap B=C \cap D=\emptyset$, then $A \cup B$–$C \cup D$.

For any sets A, B, and C we have

$$A \cup B=B \cup A.$$
$$(A \cup B) \cup C=A \cup (B \cup C).$$
$$A \cup \emptyset=\emptyset \cup A=A.$$

In Sections 4.10 and 4.11, a is the cardinal of the set A, b is the cardinal of the set B, c is the cardinal of the set C, and the sets A, B, and C are mutually disjoint, i.e. $A \cap B=\emptyset$, $A \cap C=\emptyset$, and $B \cap C=\emptyset$.

4.10a The Closure Law for the Addition of Whole Numbers

For any two elements a and b in the set W, the sum $a+b$ is a uniquely determined element in W.

As indicated earlier, this simply states that the sum of two whole numbers is a whole number. We will use this fact in an interesting way when we compare numbers.

4.10b The Commutative Law of Addition of Whole Numbers

For any two elements a and b in W it is always true that

$$a+b=b+a.$$

To prove this we write

$$a+b=n(A)+n(B)=n(A \cup B).$$
$$b+a=n(B)+n(A)=n(B \cup A)=n(A \cup B).$$

By the transitive property of equals.

$$a+b=b+a.$$

That is, addition of whole numbers is commutative.

4.10c The Associative Law of Addition of Whole Numbers

For any three elements a, b, and c in W

$$(a+b)+c=a+(b+c).$$

To prove this we write

$$(a+b)+c = n(A \cup B)+n(C) = n((A \cup B) \cup C)$$
$$= n(A \cup B \cup C).$$
$$a+(b+c) = n(A)+n(B \cup C) = n(A \cup (B \cup C))$$
$$= n(A \cup B \cup C).$$

By the transitive property of equals,

$$(a+b)+c = a+(b+c).$$

That is, addition of whole numbers is associative.

Repeated applications of the commutative law and the associative law permit us to rearrange and regroup terms in an indicated addition without changing the sum. That is, by the commutative law we can change the order of the numbers in an indicated addition, and by the associative law we can group the numbers in any manner which makes the addition as a *process* easier without changing the final outcome.

4.10d The Identity for Addition of Whole Numbers

There is a unique whole number 0 such that for any element in W

$$a+0 = 0+a = a.$$

To prove this we note that $A \cap \emptyset = \emptyset$ for all sets A and write

$$a+0 = n(A)+n(\emptyset) = n(A \cup \emptyset) = n(A) = a.$$

Then by the transitive property of equals, for all whole numbers a

$$a+0 = a.$$

That is, zero is the additive identity.

Historically the symbol 0 played the role of a place holder in early place-value systems of numeration. In our discussion it was introduced as the cardinal of the empty set. As an element of the system of whole numbers it is the identity element for the binary operation of addition. It is that unique element of the number system which when added to any number gives a sum that is identically equal to that number. Thus $7+0 = 7$. Everyone is familiar with this fact. However, everyone is not familiar with how to use this fact to advantage in arithmetic and in mathematics. Some instances will be presented later.

Exercise 4.10d

The following are examples of the application of either the commutative law of addition or the associative law of addition. Designate which law is being applied in each example.

1. $2+3 = 3+2$

2. $(2+3)+4 = 2+(3+4)$

3. $a+b=b+a$

4. $x+(y+z)=(x+y)+z$

5. $(a+b)+c=(b+a)+c$

6. $2+3+5+7=(2+3)+(5+7)$

7. $2+(a+3)+b=(2+a)+(3+b)$

8. Group the numbers 2, 3, 5, and 8 as a sum in several different ways; carry out the addition and compare the final sums.

9. Distinguish between the cardinal use of the number 0 and 0 as an element of a number system.

10. Use the definition of the additive identity to show that there is only one additive identity in W, namely 0.

11. Let $A=\{u, v, w\}$, $B=\{a, a, b, b, c\}$, and $C=\{e, i, o, k\}$. Verify each of the following:

(a) $A \cup B = B \cup A$
(b) $A \cup (B \cup C) = (A \cup B) \cup C$
(c) $A \cup \emptyset = \emptyset \cup A = A$

4.11 MULTIPLICATION IN W

Many of us learned multiplication of whole numbers as repeated addition. This is a valid approach, tried and tested. For instance,

$$5 \cdot 4 = 4+4+4+4+4 = 20.$$

This approach is consistent with sound mathematical convention. It is common practice to write in place of the expression $a+a+a+a$ the shorter expression $4 \cdot a$ or simply $4a$. In the special instance that a is a number, this is nothing more nor less than multiplication.

Definition 4.11a. If a and b are whole numbers, the binary operation of multiplication assigns to the ordered pair (a, b) the whole number $a \cdot b$ or simply ab which is another name for the sum $b+b+b+\cdots$ until we have a terms of b.

$$a \cdot b = \underbrace{b+b+b+\cdots+b}_{(a \text{ terms})}.$$

The problem with the repeated addition approach is that it fails when $a=0$. (To argue that 0 terms of b is 0 is sheer sophistry!)

On the other hand, in keeping with the set-theoretic approach to arithmetic, we may, as we did for addition, define multiplication of whole numbers in the language of sets.

Example 1

Let $A = \{1, 3, 9\}$ and $B = \{J, K, M, P\}$. Then

$$A \times B = \begin{Bmatrix} (1, J), (1, K), (1, M), (1, P) \\ (3, J), (3, K), (3, M), (3, P) \\ (9, J), (9, K), (9, M), (9, P) \end{Bmatrix}$$

$n(A) = 3$,
$n(B) = 4$,
$3 \cdot 4 = n(A) \cdot n(B) = n(A \times B) = 12$.

Definition 4.11b. If a and b are whole numbers, the binary operation of multiplication assigns to the ordered pair (a, b) the cardinal of the set $A \times B$, where

$a = n(A)$ and $b = n(B)$. That is, for all whole numbers a and b,

$$a \cdot b = n(A) \cdot n(B) = n(A \times B).$$

Observe that, except when $a = 0$, this definition is equivalent to the addition approach.

Neither the repeated addition nor the Cartesian product approach is universally accepted. Certainly it would not be reasonable to expect anyone to multiply 789×987 by forming a sum of 789 terms of 987 any more than it would be to expect anyone to count the elements in the set $A \times B$ where $789 = n(A)$ and $987 = n(B)$. Sound arguments can be proposed by the advocates of both procedures for defining multiplication of whole numbers. Neither approach is entirely free of severe criticism from some authoritative sources. In the interest of providing the widest possible background for the preservice teacher, we discuss both approaches in some detail. Since there is no way of knowing which approach will be stressed in a particular school curriculum, some familiarity with both is desirable and helpful.

Children can be led to discover that 3×6 is the same as 6×3, $4 \times 7 = 7 \times 4$, and so on. They can even be led to generalize. That is, they can be led to conjecture that, in general, when two numbers are multiplied, it is immaterial whether the first is multiplied by the second or the second is multiplied by the first. That is, for all whole numbers a and b,

$$a \cdot b = b \cdot a.$$

There should be no expectation that children should be able to prove this fact or even to understand a proof. This fact and others should be made reasonable and understandable. The two definitions of multiplication of whole numbers just presented offer two ways of making reasonable the facts about multiplication.

4.11a The Repeated Addition Approach

The repeated addition definition of multiplication of whole numbers has the historical advantage of being more familiar to the average teacher, which is a very important factor in their teaching effectiveness. Furthermore, this definition requires less in the way of sophistication and background. Finally, it is the method most often resorted to in learning the elementary facts for multiplication. This will be borne out when we work with arithmetic in other bases (see Chapter 5).

4.11b The Cartesian Product Approach

The Cartesian product approach to multiplication of whole numbers has the advantage that once certain facts about Cartesian products have been established, the properties of multiplication follow easily and reasonably. This is not to imply that this approach is simpler than the repeated addition approach. The logical difficulties simply occur in a different way and in a different setting.

The following facts about Cartesian products can be easily verified for finite sets A, B, and C with small numbers of elements as you will be asked to do in the exercises. They are also true for all sets A, B, and C.

1. There is a one-to-one correspondence between the elements of $A \times B$ and $B \times A$, and so

$$n(A \times B) = n(B \times A).$$

2. There is a one-to-one correspondence between the elements of $(A \times B) \times C$ and $A \times (B \times C)$, and so

$$n[(A \times B) \times C] = n[A \times (B \times C)].$$

3. There is a one-to-one correspondence between the elements of $\{e\} \times A$ and A.

$$n(\{e\} \times A) = n(A).$$

4. $A \times (B \cup C) = (A \times B) \cup (A \times C)$.

5. $\emptyset \times A = \emptyset$ for any set A.

Exercise 4.11b

Let $A = \{x, y, z\}$, $B = \{0, 3\}$, and $C = \{i, j, k\}$.

1. Exhibit a one-to-one correspondence between $A \times B$ and $B \times A$.

2. Exhibit a one-to-one correspondence between $(A \times B) \times C$ and $A \times (B \times C)$.

3. Exhibit a one-to-one correspondence between the set A and the set $\{e\} \times A$.

4. Show that the relation $A \times (B \cup C) = (A \times B) \cup (A \times C)$ holds for the sets A, B, and C.

5. Show that $\emptyset \times A = \emptyset$.

4.11c Closure Law

That the product of two whole numbers is a uniquely determined element in the set W is a consequence of the definitions and some facts about sets.

4.11d The Commutative Law of Multiplication

If a and b are whole numbers, it is always true that

$$a \cdot b = b \cdot a.$$

This is a consequence of statement 1 about Cartesian products.

$$a \cdot b = n(A) \cdot n(B) = n(A \times B) = n(B \times A) = b \cdot a.$$

4.11e The Associative Law of Multiplication

Earlier we used parentheses to denote ordered pairs and, in general, ordered sets. Parentheses are also used to denote numbers such as $(2 \cdot 4)$ and $(4+9)$. If such expressions occur with other operations or other parentheses, it is understood that the operations within the innermost parentheses or other signs of grouping such as brackets are to be executed first.

Examples

$3 \cdot (5+6) = 3 \cdot 11 = 33.$

$2 + (3 \cdot 4) = 2 + 12 = 14.$

$5[3+2(3+4)] = 5(3+2 \cdot 7) = 5(3+14) = 5 \cdot 17 = 85.$

$(2 \cdot 4) + (2 \cdot 6) = 8 + 12 = 20.$

If a, b, and c denote whole numbers, it is always true that

$$a \cdot (b \cdot c) = (a \cdot b) \cdot c.$$

The associative law of multiplication of whole numbers is a consequence of statement 2 about Cartesian products.

$$\begin{aligned} a \cdot (b \cdot c) &= n(A) \cdot n(B \times C) = n[A \times (B \times C)] \\ &= n[(A \times B) \times C] = n(A \times B) \cdot n(C) = (a \cdot b) \cdot c. \end{aligned}$$

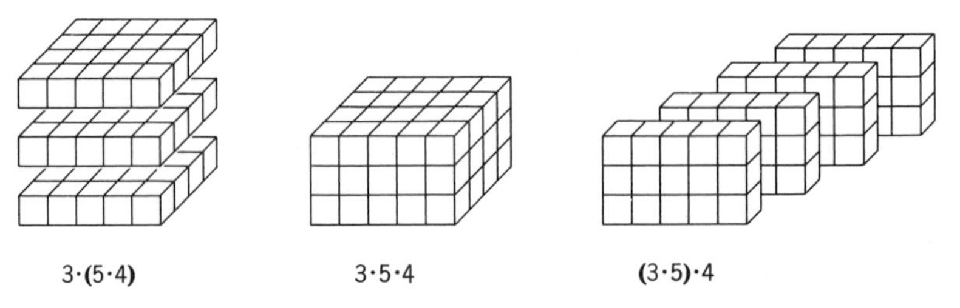

Figure 1

That multiplication of whole numbers satisfies the associative law means that in an indicated product of more than two factors, it is immaterial where the parentheses are placed. In such a product parentheses may be removed or inserted. This is simply another way of saying that it is immaterial how the factors in a product are *grouped* (i.e., associated), the result is the same.

Example 1

Let x, y, and z denote numbers. Then

$$(2x) \cdot (y \cdot 3z) = 2xy3z = (2xy) \cdot (3z) = (2x)(y3) \cdot z.$$

A combination of the associative and commutative laws for multiplication permits us to carry out the following manipulations: $(a \cdot b) \cdot c = a \cdot (b \cdot c) = a \cdot (c \cdot b) = (a \cdot c) \cdot b = (c \cdot a) \cdot b$, etc. In particular, $a \cdot b \cdot c$ is meaningful even though the operation of multiplication is a *binary* operation. It dosn't matter whether we consider $a \cdot b \cdot c$ to mean $(a \cdot b) \cdot c$ or $a \cdot (b \cdot c)$, since $(a \cdot b) \cdot c = a \cdot (b \cdot c)$.

Exercise 4.11e

1. The following are examples of the application of either the commutative law of multiplication or the associative law of multiplication. Designate which law is being exemplified by each example.

(a) $5 \cdot 4 = 4 \cdot 5$ (b) $(5 \cdot 4) \cdot 8 = 5 \cdot (4 \cdot 8)$
(c) $(5 \cdot 4) \cdot 8 = 8 \cdot (5 \cdot 4)$ (d) $a \cdot b = b \cdot a$
(e) $(x \cdot y) \cdot z = x \cdot (y \cdot z)$ (f) $3 \cdot 5 \cdot 7 \cdot 4 = (3 \cdot 5) \cdot (7 \cdot 4)$

2. Group the numbers 3, 5, 7, and 4 as a product in several different ways; carry out the multiplication and compare the results.

3. The following are examples of the application of one of the laws of multiplication or addition. Identify the law exemplified by each.

(a) $5 \cdot (3+7) = 5 \cdot (7+3)$
(b) $5 \cdot 3 + 2 = 3 \cdot 5 + 2$
(c) $5 \cdot (3+2) = (3+2) \cdot 5$
(d) $a \cdot b + a \cdot c = b \cdot a + c \cdot a$
(e) $a \cdot (b \cdot c) = (b \cdot c) \cdot a$
(f) $(a+b) \cdot (c+d) = (a+b) \cdot (d+c)$
(g) $3 \cdot (4+5+6) = 3 \cdot (4+(5+6))$
(h) $(a \cdot b) \cdot c = a \cdot b \cdot c$

4.11f The Multiplicative Identity

The number 1 plays a special role as a member of a number system. It is the element of the number system which when multiplied by any number gives a product identically equal to that number. It is called the *multiplicative identity*. That is, for all whole numbers a,

$$1 \cdot a = a \cdot 1 = a.$$

Not only do we have $1 \cdot a = a$ or $a \cdot 1 = a$ but by the symmetric property of equality we have $a = 1 \cdot a$ or $a = a \cdot 1$. This important fact is often overlooked and is the source of some common errors in arithmetic and algebra.

To show that $1 \cdot a = a \cdot 1 = a$, we use statement 3 about Cartesian products.

$$1 \cdot a = n(\{e\}) \cdot n(A) = n(\{e\} \times A) = n(A) = a.$$

4.11g The Distributive Law

In our discussion so far we have been considering two binary operations, addition and multiplication, on the set W. The distributive law concerns the behavior of these two binary operations when they are combined in an arithmetic operation. To illustrate:

$$\begin{matrix} XXX & XXXXX \\ XXX & XXXXX \\ XXX & XXXXX \end{matrix}$$

If we want to express the total number of X's, it can be done in two ways. We have 3 X's in each of $(3+5)$ columns, or $3(3+5)$ X's, which is equal to 24 X's. We also have 3 rows of 3 X's and 3 rows of 5 X's or $(3 \cdot 3 + 3 \cdot 5)$ X's. But this is also 24 X's. This tells us that

$$3(3+5) = (3 \cdot 3) + (3 \cdot 5).$$

Executing the operations indicated within the parentheses first in $3(3+5)$, we add and then multiply. In $(3 \cdot 3) + (3 \cdot 5)$ we would multiply first and then add. Since in our example the result is the same in either instance, it is immaterial whether we added first and then multiplied or multiplied first and then added.

In general, if a, b, and c are elements of W, it is always true that

$$a \cdot (b+c) = a \cdot b + a \cdot c.$$

This is the distributive law. We say that multiplication is distributive with respect to addition in the set W.

This law can be made reasonable by making use of fact 4 of Cartesian products. For sets A, B, and C,

$$A \times (B \cup C) = (A \times B) \cup (A \times C).$$

In particular, when B and C are disjoint, we have

$$n[A \times (B \cup C)] = n[(A \times B) \cup (A \times C)].$$

If $a = n(A)$, $b = n(B)$, and $c = n(C)$, on the left we have,

$$n[A \times (B \cup C)] = n(A) \cdot [n(B \cup C)] = n(A) \cdot [n(B) + n(C)] = a(b+c).$$

On the right we have

$$n[(A \times B) \cup (A \times C)] = n(A \times B) + n(A \times C) = n(A) \cdot n(B)$$
$$+ n(A) \cdot n(C).$$

That is,

$$a \cdot (b+c) = a \cdot b + a \cdot c, \quad \text{or}$$

$$a(b+c) = ab + ac.$$

Example 1

1. $2(3+4) = 2 \cdot 3 + 2 \cdot 4 = 14.$
2. $2(3+x) = 2 \cdot 3 + 2 \cdot x = 6 + 2x.$
3. $a(1+b) = a \cdot 1 + a \cdot b = a + ab.$

Recall that the equals relation has the property of being symmetric. As a consequence, the distributive law can be written as follows:

$$a \cdot b + a \cdot c = a(b+c).$$

In this form it reminds us that much of factoring in algebra is simply an application of the distributive law. Thus, for example,

$$2 \cdot x + a \cdot x = (2+a)x.$$

4.11h Properties of Zero in Multiplication

The special role of 0 as a number is particularly bothersome to many teachers. There are many people who still do not think of 0 as a number. Yet they have no trouble with 0 in situations involving addition or subtraction. Asking them, however, to multiply by 0 or divide into 0, gives rise to the wildest kind of responses.

For any number a,

$$0 \cdot a = a \cdot 0 = 0.$$

In terms of statement 5 about Cartesian products, this property of 0 follows simply. We have

$$0 \cdot a = n(\emptyset) \cdot n(A) = n(\emptyset \times A) = n(\emptyset) = 0.$$

and by the commutative law, $a \cdot 0 = 0 \cdot a = 0$.
In the special instance where $a = 0$ we have

$$0 \cdot 0 = 0,$$

Note that, from the point of view of multiplication as repeated addition, $3 \cdot 0$, for example, has more meaning than $0 \cdot 3$, since we can consider $3 \cdot 0$ as $0+0+0=0$.

Exercise 4.11h

Apply the distributive law to rename the following:

1. $a(b+2)$ **2.** $2a+ac$ **3.** $23(2+1)$

4. $2 \cdot 10^2 + 3 \cdot 10^2$ **5.** $30 \cdot 10 + 2 \cdot 10$

6. $(a+b)(c+d) = a(c+d) + b(c+d)$

7. $(a+b)(c+d)=(a+b)c+(a+b)d$

8. $(a+b)(a+b)=(a+b)a+(a+b)b$

9. $x(y+2)$ **10.** $2x+bx$ **11.** $2ax+3x$

12. $2ax+5a$ **13.** $3\cdot(10\cdot 10)+2\cdot 10$

14. $10\cdot 3\cdot 10+2\cdot 10$ **15.** $2xab+a$ **16.** $(20+3)2$

Assume that the operations symbolized by Δ and $\oplus$ are defined on a set S, and using a, b, and c as elements of S, symbolize that

17. Δ is commutative.

18. $\oplus$ is commutative.

19. Δ is associative.

20. $\oplus$ is associative.

21. Δ is distributive with respect to $\oplus$.

4.12 THE SYSTEM OF WHOLE NUMBERS

We summarize our previous statements in a formal definition.

Definition 4.12. By the system of whole numbers we mean the set
$W=\{0, 1, 2, 3, 4, \ldots\}$,
the binary operations, addition $(+)$ and multiplication $(\cdot)$, and the following laws. For m, n, and k in W,

Closure Laws

1. There is a uniquely determined sum which we write as $m+n$ in W.
2. There is a uniquely determined product which we write as $m\cdot n$ in W.

Associative Laws

3. $m+(n+k)=(m+n)+k$.
4. $m\cdot(n\cdot k)=(m\cdot n)\cdot k$.

Commutative Laws

5. $m+n=n+m$.
6. $m\cdot n=n\cdot m$.

Distributive Law

7. $m\cdot(n+k)=m\cdot n+m\cdot k$.

Identities

8. There is a unique element 0 such that for any m in W, $0+m=m+0=m$.
9. There is a unique element 1 such that for any m in W, $1\cdot m=m\cdot 1=m$.

Exercise 4.12

1. Show that the multiplicative identity is unique. (*Hint*: Assume $\hat{1}$ is any other multiplicative identity.)

2. Assuming that $m \cdot 0 = 0$, show that $(m+1) \cdot 0 = 0$. State the law used at each step.

3. $1 \cdot 5 = 5$ because 1 is the multiplicative identity. There are occasions when we also write $5 = 1 \cdot 5$. Give an example of a situation in which it is convenient to use $1 \cdot 5$ in place of 5. (*Hint*: Write $5 + 10m$ as a multiple of 5.)

4. (a) Can you write 4 as a multiple of 0?
(b) Can you write 0 as a multiple of 4? Explain.

5. (a) Does 5 "divide" 0? Explain.
(b) Does 0 "divide" 5? Explain.

6. Distinguish between 0 and 1 as cardinal numbers and as elements of a number system.

7. Show that $(a+b)(c+d) = ac + ad + bc + bd$, and justify each step.

8. Show that $(a+b)^2 = a^2 + 2ab + b^2$, and justify each step.

9. Which number is larger?
(a) $(2^3)^0$ or $2^{(3^0)}$
(b) $(2^0)^3$ or $2^{(0^3)}$

4.13 ORDER RELATIONS FOR THE WHOLE NUMBERS

In much of arithmetic, in fact, in much of our everyday life, we are interested in the biggest, the smallest, the least expensive, the most profitable, and so on. In the simplest instances these comparisons are easy to make, but we do meet situations where some work is needed before we can reach a conclusion.

Example 1

The symbol π denotes the number which expresses the ratio of the circumference of a circle to its diameter. In computations involving this symbol one is often told to use 22/7 and at other times to use 3.1416. How do these three numbers compare? Are any two equal? If not, which is the largest and which the smallest?

We do not intend to answer these questions at this time. It is our intention to make precise certain concepts which will in time enable the reader to answer these questions and supply sound reasons to support his answers. We begin by making precise the meaning of "less than" for the whole numbers.

We define the relation "less than" (denoted by $<$) in terms of the one-to-one correspondence for sets. We remarked earlier that the concepts of "more than" and "less than" may have had meaning in the earliest civilizations, even in the absence of counting systems. Preschool children comprehend these concepts in terms of pieces of candy. Their idea of sharing equally appears in the familiar "one for you and one for me," but this is the matching relation or one-to-one correspondence. If the "one for you and one for me" ends while "me" still has several unmatched pieces, the "less than" concept has dramtic meaning for one person, at least.

Definition 4.13. If A and B are finite sets and A can be matched to a *proper* subset of B under a one-to-one correspondence, we say that the cardinal number $n(A)$ of the set A is *less than* the cardinal number $n(B)$ of the set B.

We write this

$$n(A) < n(B).$$

We can also read this as "$n(B)$ is greater than $n(A)$." This is written

$$n(B) > n(A).$$

Example 2

The cardinal number of the set $\{a, b, c\}$ is 3. The cardinal number of the set $\{1, 2, 3, 4, 5\}$ is 5. The set $\{a, b, c\}$ can be matched with a proper subset of the set $\{1, 2, 3, 4, 5\}$ so $3 < 5$. The cardinal number of the empty set is 0. The cardinal number of the set $\{1\}$ is 1. Since the empty set is a subset of $\{1\}$, we have $0 < 1$.

The "less than" relation in the set of whole numbers is trivially a transitive relation. A transitive relation is called an *order*. It is not sufficient for an order relation to be transitive to make comparisons. It must be *linear* also. An order relation is linear if it satisfies the *trichotomy law.*

4.13a The Trichotomy Law

Definition 4.16a. The trichotomy law. If m and n are any two whole numbers, then one and only one of the following relations holds:

1. $m = n$.
2. $m < n$.
3. $m > n$.

The word "trichotomy" is suggestive of the three choices which are available. The word "linear" is suggestive of a "number line."

Although the concept of the number line is familiar, we review it briefly.

On a horizontal line, which we denote by L, we choose an arbitrary point and label this point 0. We then choose any other point on the line L to the right, with respect to the viewer, of the point labled 0 and label this point 1.

We use the line segment from the point labeled 0 to the point labeled 1 to mark off consecutive points to the right and we label these points with the natural numbers, consecutively. Each natural number corresponds to just one such point, and each point corresponds to just one such number. Disregard the point labeled 0 for the moment and consider the rest. There is a one-to-one correspondence between the natural numbers and the points obtained in this manner.

Hereafter we shall not distinguish between these points and their corresponding labels. This allows us to speak of the numbers between 5 and 17, for example, with the concept of "between-ness" suggested by the number line.

Given any two points on a horizontal line one is to the left or to the right of the other or they are identical, much as the numbers are related to each other by size. Intuitively, one number x is "less than" another number y if the number x is to the left of the number y on the number line. The number line is a useful device for depicting the very important and useful order properties of the numbers. We add, multiply, subtract, divide, and, in general, calculate with numbers, but equally important, we compare numbers.

It is important to note in the statement of the trichotomy law that one and only one of three statements is true of any two given numbers. The implication of this is that if we know that one of the statements is not true then one or the other of the two remaining must be true.

Example 1

Let $A = \{n|n \text{ is a whole number and } n > 5\}$. Think of a whole number which is not in the set A. The number under consideration must be not greater than 5, that is, it must be either 5 or any whole number less than 5. We indicate this as follows. Let k be the number, then k is one of the numbers in the set

$A' = \{m|m \text{ is a whole number and } m \leqslant 5\}$. (See figure on page 96.)

"Not greater than," written $\ngtr$, is equivalent to "less than or equal," written $\leqslant$.

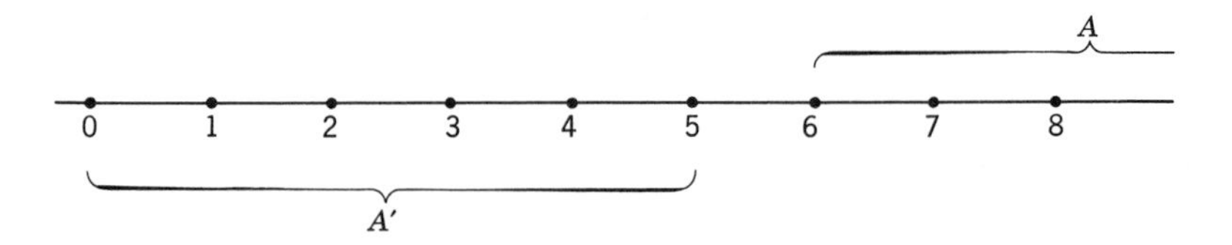

To say that two numbers m and n are distinct means that either $m < n$ or $m > n$. If $a < n$ and $n < b$, we abbreviate and write $a < n < b$.

Exercise 4.13

1. Let $A = \{n \in W | n > 7\}$. Use the symbol A' to denote the elements in W which are not in A. Write the set A' in two ways and indicate the sets A and A' on the number line.

2. Let $X = \{x \in W | 2x + 3 \geqslant 11\}$. Write the set X' in two ways and indicate the sets X and X' on the number line.

3. Let $Y = \{n \in W | 3 < n \leqslant 9\}$ and $Z = \{n \in W | 2n > 12\}$. List the elements in $Y \cap Z$. List the elements in $Y \cap Z'$.

4. If $3 < 5$, how is $3 + n$ related to $5 + n$ for any whole number n?

5. If $3 < 7$, how is $3 \cdot n$ related to $7 \cdot n$ for n a whole number?

6. Let $S = \{n \in W | 5 < n \leqslant 10\}$. Write S' using the set-builder notation.

4.13b Bounds on Sets

We can now introduce the idea of a bound on sets of numbers.

Definition 4.13b. Let S denote a set of whole numbers. We say that c is an *upper bound* of the set S if $a \leq c$ for each a in S. Similarly, we say that b is a *lower bound* for the set S if $b \leq a$ for each a in S.

Example 3

Let $A = \{2, 3, 19\}$. An upper bound of A is 20 and 1 is a lower bound of A. Also, 100 is an upper bound. Note that 19 is also an upper bound and 2 is a lower bound. Furthermore, 19 is an upper bound and is less than any other upper bound.

Definition 4.13c. (1) If c is an upper bound of a set A, (2) if d is any other upper bound, and (3) $c \leq d$, then we call c the *least upper bound* of the set A, that is, c is the *smallest* of the upper bounds.

Definition 4.13d. (1) If b is a lower bound of a set A, (2) if e is any other lower bound, and (3) $e \leq b$, then we call b the *greatest lower bound* of the set A, that is b is the *largest* of the lower bounds.

A set of whole numbers may have many upper bounds but it has only one least upper bound; it may have many lower bounds but it has only one greatest lower bound.

It may be of interest to remark that there is no biggest *number* less than 10. We shall say more about this later. Now it is sufficient to note that there is a biggest *whole number* less than 10. It is the number 9.

Exercise 4.13a

In the following, consider $n \in W$, and recall that the expression $a < n < b$ means that $n > a$ and $n < b$.

1. (a) Let $A = \{n|3 < n < 12\}$.

List the elements in the set A. List an upper and lower bound of the set A.

(b) Let $B = \{n|0 < n < 4\}$.

List the elements of the set B. List two distinct upper bounds of the set B.

2. If $C = \{n|3 < n\}$ and $D = \{n|n < 12\}$, then $C \cap D = ?$

3. List the elements in the set $A \cap B$ for the sets A and B of problem 1.

4. List the elements in the set $A \times B$ for the sets A and B of problem 1.

5. Let $C = \{n|6 \leq n \leq 12\}$.
List the elements of C. What is the least upper bound of C? List the elements of $B \cup C$ (set B of problem 1).

6. List the elements in the set $B \cap C$ (set B of problem 1 and set C of problem 5).

7. Describe the sets of whole numbers satisfying the following inequalities:
(a) $3+n < 10$.
(b) $n+2 < 6$.
(c) $3+n < 9$ and $2+n < 6$.

8. For whole numbers, a, b, and c, show that if $a = b$, then $a+c = b+c$. Is the converse true?

9. For whole numbers a, b, and c, show that if $a = b$, then $a \cdot c = b \cdot c$. Is the converse true?

10. State each of the properties of equality given in problems 8 and 9 in another way.

Show that the following equalities hold by beginning with the left side and, using the fundamental properties of the system of whole numbers, transforming it so that it is identical with the right side. Assume the letters are to represent whole numbers, for example,
to show $5+6+a = 6+a+5$

$$
\begin{aligned}
5+6+a &= 5+(6+a) && \text{by the associative law of addition} \\
&= (6+a)+5 && \text{by the commutative law of addition} \\
&= 6+a+5 && \text{by the associative law of addition}
\end{aligned}
$$

Hence, by the transitive property of equals, the given equality holds.

11. $x+2=2+x$

12. $(x+3)+(y+2)=(2+3)+(x+y)$

13. $3xy=x(3y)$

14. $(3b)^2=9b^2$

15. $3(x+2y)=6y+3x$

16. $3\cdot 10^0+4\cdot 10^0=7\cdot 10^0$

17. $2\cdot 10^2+8\cdot 10^2=1\cdot 10^3$

18. $13\cdot 10^2=1\cdot 10^3+3\cdot 10^2$

19. $(3+2)(4+1)=25$ (two ways)

20. $(3+x)(2+y)=6+3y+2x+xy$

21. $(a+b)(a+b)=a^2+2ab+b^2$

4.13b Open sentences in one variable

The expression $n<5$ is a sentence. It is read "n is less than 5." It is an *open sentence.* It becomes a statement when we substitute a whole number in place of n.

If we replace n by 2, we also use the expression "let n equal 2," we get the statement

$2<5$ "2 is less than 5"

If we let n equal 7 we get the statement

$7<5$ "7 is less than 5"

Both $2<5$ and $7<5$ are declarative statements. The former is true and the latter is false.

The numbers, which when substituted for n, give true statements constitute the *solution set* of the open sentence. The set $\{6, 7, 8, \ldots\}$ is the solution set of the open sentence $n>5$. (The solution set is sometimes called the Truth Set.)

Open sentences in elementary arithmetic books often look like the following:

1. $3+\square=10$
2. $2+5=\triangle$
3. $n+4=6$

The symbols $\square$, $\triangle$, and n are called *variables* (see section 2.2a). These open sentences become true or false statements when the variables are replaced by whole numbers. The sentences above are examples of *simple sentences*. They are simple sentences in the same sense as used in English grammar.

The following open sentences are called *compound* sentences:

1. $3 < n < 10$
2. $x \leq 5$

The first is read "n is greater than 3 *and* n is less than 10." It is the *conjunction* of two simple sentences. If we let $n = 4$, then both of the simple sentences $4 > 3$ and $4 < 10$ are true and the compound statement is also true.

If we let $n = 12$, we get $12 > 3$ which is a true statement and $12 < 10$ which is a false statement. In this case, the compound statement is false. The solution set of the first compound open sentence is the set $\{4, 5, 6, 7, 8, 9\}$.

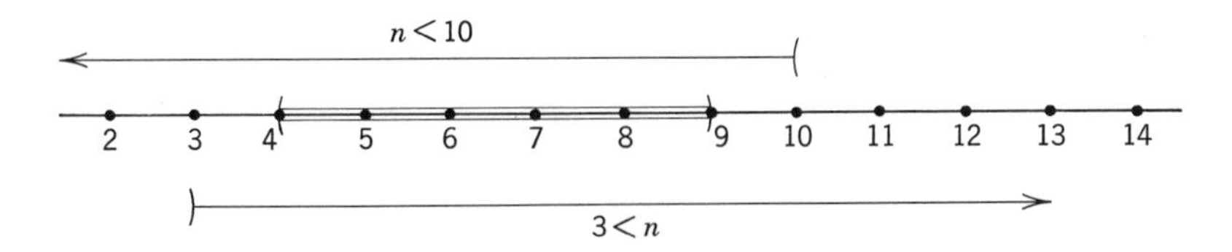

Notice that the solution set of the open compound sentence is the intersection of the solution sets of the two open simple sentences.

The second open sentence is read "x is less than 5 or x is equal to 5." It is the *disjunction* of the two simple sentences $x < 5$ and $x = 5$. In this case, the compound statement is true if either of the two simple statements is true. That is, the solution set in this case is the union of the solution sets of the two simple open sentences.

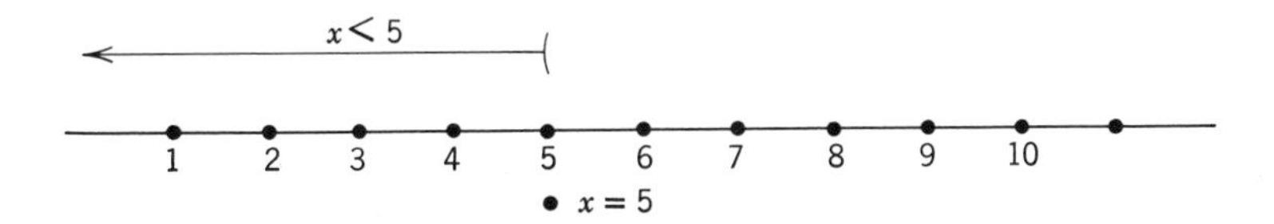

Exercise 4.13b

1. Identify each of the simple sentences in the following compound sentences.

(a) $2 < n \leq 12$
(b) $4 \leq n < 20$
(c) $75 < n < 77$

2. What is the least upper bound of the solution set in each of the problems in (1)?

3. A useful convention in plotting solution sets is to use square brackets through the point corresponding to the number when $\leq$ or $\geq$ is involved and parentheses when $<$ or $>$ is involved.

$5<n$ $n \leqq 17$

5 17

Plot the solution sets of the open sentences in problem 1.

4. Here are some mathematical sentences. Identify which are simple sentences and, if they are compound, identify the simple components.

(a) $a \in A$ (b) $a \notin A$
(c) $A \subseteq B$ (d) $A \not\subseteq B$
(e) $6 \mid 18$ (f) $6 \nmid 18$

Sentence (b) is the *negation* of sentence (a). Sentences (d) and (f) are the *negations* of sentences (c) and (e), respectively. Note that if sentence (a) is true, then its negation (b) is false. This is more obvious in statements (e) and (f).

4.14 FINGER COUNTING

The choice of the number ten as the base for numeration systems probably has an anatomical basis, as mentioned earlier, in that a man has ten fingers. Some historians refer to the choice of ten as the base as a "physiological accident." But equally plausible bases on these grounds are base *five* (the digits of one hand) or base *twenty* (the digits of both hands and both feet). Once one exponential-positional system is known, other bases present no problem. In fact, other bases have been and are being used.

Let us see how and why other bases might be used as well as base ten. We are interested in this for the simple reason that investigating place-value systems with bases other than ten leads to a better understanding of the basic concepts involved.

After years of experience you are so familiar with counting and computation in the decimal system of numeration (the base ten system) that you may be unaware of the difficulties a pupil experiences in his first attempts at counting and computation.

Let us suppose you are a beginner in order that you may realize better the problems of learning. Since you are not a "beginner" with the decimal system of numeration, we will turn to some other system. It may even be profitable to look at several other systems. We realize you do not quite have the status of a beginner, for you have the advantage of having some general knowledge about systems of numeration. In particular, the ideas of place value, the symbols themselves, and the meaning of the symbols are not strange to you. In spite of this advantage, however, you should be able to relive some of the difficulties inherent in learning to count and compute.

To introduce you to other methods of counting, let us describe a procedure we shall call "finger counting." Consider the open right hand as descriptive of zero—we have not begun to count. As we count we fold the fingers in, beginning with the little finger. We count—"one," "two," "three," "four"—and we have run out of fingers. We could fold in the thumb and make a fist to indicate five, but then we would have no more "digits" on that hand with which to keep a record of our counting. Instead, let us use the left hand to record the "fists." Now we have one finger folded in on our left hand to indicate one "fist" (five) and can open our right hand to continue the counting. Now, instead of saying "six, seven, eight," etc., which are decimal names for numbers, let us say, "one fist and one," "one fist and two," "one fist and three," "one fist and four"; here we must stop to consider again. If we fold in the right thumb, we really have "one fist and one fist" or "two fists." We can record this by folding in a second finger on our left hand and opening our right hand. Next in our counting comes "two fists and one," "two fists and two," "two fists and three," etc. Continuing in this manner, we soon come to "four fists and four," then "four fists and a fist," or a whole fistful of fists, and we need to look for another place to record "fistfuls of fists." We could turn to our toes as a new *place* for keeping this record, but it is rather difficult to control toes as one does fingers.

In any event, we have gone far enough with our discussion for you to recognize that what we are really doing is counting in a system of numeration with base *five*. We see that if we use the properties of a place-value system and if we borrow symbols from the decimal system, a numeral such as 12 would then mean "one fist and two," or, "one *five* and two," or, $1 \cdot 5 + 2$, or $1 \cdot 5^1 + 2 \cdot 5^0$. (Recall that $5^0 = 1$.) This is the expanded form of our numeral 12 using *base five.*

Let us compare counting in a base five system with the base ten, or decimal, system.

Base ten		*Base five*	
1	one	1	one
2	two	2	two
3	three	3	three
4	four	4	four
5	five	10	one five and zero
6	six	11	one five and one
7	seven	12	one five and two
8	eight	13	one five and three
9	nine	14	one five and four
10	ten	20	two fives and zero
11	eleven	21	two fives and one
12	twelve	22	two fives and two
.		.	
.		.	
.		.	

23 twenty-three	43 four fives and three
24 twenty-four	44 four fives and four
25 twenty-five	100 one five-fives, zero fives, and zero
.	.
.	.
.	.
33 thirty-three	113 one five-fives, one five, and three or $1 \cdot 5^2 + 1 \cdot 5^1 + 3 \cdot 5^0$
.	.
.	.
.	.

Since we are borrowing symbols from the decimal system of numeration, we are inclined to borrow names also, but there is a limitation. The numeral 12 is not "twelve" in the quinary (base five) system. We would read it as "one five and two," or "one-two, base five," and symbolize it 12_{five} to distinguish it from the decimal numeral 12. Similarly, 21_{five} is "two fives and one" or "two-one, base five," *not* "twenty-one."

We should point out here that it is not at all necessary to use the decimal symbols in our quinary system. Our familiarity with their meaning leads us to borrow decimal symbols. We could, for instance, take five arbitrary letters such as *O*, *E*, *T*, *F*, *M*, and give them, respectively, the meaning we associate with zero, one, two, three, and four. Then the numerals to represent the first fifteen (decimal language) counting numbers would look like this:

Decimal: 1, 2, 3, 4, 5, 6, 7, 8, 9, 10, 11, 12, 13, 14, 15

Quinary: *E*, *T*, *F*, *M*, *EO*, *EE*, *ET*, *EF*, *EM*, *TO*, *TE*, *TT*, *TF*, *TM*, *FO*

In spite of occasional confusions, such as mistaking 12_{five} for "twelve," it is easier to work with the familiar symbols 0, 1, 2, 3, 4, and we will continue to do so.

4.15 PLACE-VALUE SYSTEMS WITH BASES OTHER THAN TEN

In a quinary system the first place to the left of our reference point (is this a decimal point?) is the units position. This is true of all exponential-positional systems. We associate 5^0 ($5^0 = 1$) with this place as its "place value." The next place to the left represents the number of fives ($5^1 = 5$), the next place the number of five-fives, or twenty-fives ($5^2 = 25$), the next place the number of five (five-fives), or the one hundred twenty-fives ($5^3 = 125$), the next the number of six hundred twenty-fives ($5^4 = 625$), and so forth. We symbolize these place values by the powers of five, using the decimal language symbol 5^0, 5^1, 5^2, 5^3, 5^4, etc.

The first place to the right of the reference point has associated with it 5^{-1} ($5^{-1} = 1/5$), and a digit in this place represents the number of fifths. The next place to the right has associated with it 5^{-2} ($5^{-2} = 1/5^2 = 1/25$), and a digit in this place represents the number of twenty-fifths, etc.

Example 1

Consider 18_{ten}, the symbol for the number called eighteen in our decimal language. Expressing this in a base five system of numeration is essentially the regrouping of a representative set S where $n(S) = 18$.

set S	set S
[XXXXXXXXXX][XXXXXXXX]	[(XXXXX) (XXXXX) (XXXXX)] [XXX]
$1 \cdot 10^1 + 8 \cdot 10^0$	$3 \cdot 5^1 + 3 \cdot 5^0$

To accomplish this without the mechanics of regrouping, and using our knowledge of decimal computation, we ask "What is the highest power of five that is less than eighteen, and what is the greatest multiple of this power that is less than eighteen?" Of course, the answers are 5^1 and $3 \cdot 5^1$. This means that in our quinary numeral we can use a "3" in the 5^1 place. This, however, is not enough since we have accounted for only fifteen of the eighteen. This means that we need three more, so we can use a "3" in the 5^0, or units, place. Now our quinary numeral looks like this: 33_{five}. We conclude $18_{\text{ten}} = 33_{\text{five}}$.

Example 2

Similarly, for 333_{ten}, we see that

$$
\left.\begin{aligned}
333_{\text{ten}} &= 250 + 83 \\
&= 2 \cdot 5^3 + 75 + 8 \\
&= 2 \cdot 5^3 + 3 \cdot 5^2 + 5 + 3 \\
&= 2 \cdot 5^3 + 3 \cdot 5^2 + 1 \cdot 5^1 + 3 \\
&= 2 \cdot 5^3 + 3 \cdot 5^2 + 1 \cdot 5^1 + 3 \cdot 5^0
\end{aligned}\right\} \text{Regrouping}
$$

$$= 2313_{\text{five}}.$$

Example 3

$$
\begin{aligned}
2341_{\text{five}} &= 2 \cdot 5^3 + 3 \cdot 5^2 + 4 \cdot 5^1 + 1 \cdot 5^0 && \text{Expanded form with decimal numerals} \\
&= 250 + 75 + 20 + 1 \\
&= 346_{\text{ten}}.
\end{aligned}
$$

Example 4

$$\begin{aligned}42.321_{\text{five}} &= 4\cdot 5^1+2\cdot 5^0+3\cdot 5^{-1}+2\cdot 5^{-2}+1\cdot 5^{-3}\\ &= 20+2+\tfrac{3}{5}+\tfrac{2}{25}+\tfrac{1}{125}\\ &= 20+2+\tfrac{6}{10}+\tfrac{8}{100}+\tfrac{8}{1000}\\ &= 22.688_{\text{ten}}.\end{aligned}$$

This notion of counting in systems with bases other than ten can be extended to any choice of base. Table 1 gives the representation for numbers in base ten (decimal), base five (quinary), base two (binary, and base twelve (duodecimal).

Table 1

Symbols Base Ten: 0, 1, 2, 3, 4, 5, 6, 7, 8, 9
Base Five: 0, 1, 2, 3, 4
Base Two: 0, 1
Base Twelve: 0, 1, 2, 3, 4, 5, 6, 7, 8, 9, *T*, *E*

Base Ten (Decimal)	Base Five (Quinary)	Base Two (Binary)	Base Twelve (Duodecimal)
1	1	1	1
2	2	10	2
3	3	11	3
4	4	100	4
5	10	101	5
6	11	110	6
7	12	111	7
8	13	1000	8
9	14	1001	9
10	20	1010	*T*
11	21	1011	*E*
12	22	1100	10
13	23	1101	11
14	24	1110	12
15	30	1111	13
16	31	10000	14

Note that in the last row of Table 1 we have: "one ten and six units," "three fives and one unit," "one sixteen, zero eights, zero fours, zero twos, and zero units," and "one twelve and four units," all representing the same number. Is this possible? This is as though we call John Arthur Palmer, "John," "John Arthur," "Art," or "Palmer." Each is simply a different name for the same person.

$$16_{\text{ten}} = 31_{\text{five}} = 10000_{\text{two}} = 14_{\text{twelve}}.$$

Exercise 4.15

1. Write the following number in base five and then in base ten:
$2 \cdot 5^3 + 3 \cdot 5^2 + 4 \cdot 5^1 + 2 \cdot 5^0$

2. Write the first fifty (decimal language) numerals as quinary numerals, borrowing the symbols from the decimal system of numeration.

3. Write each of the following in expanded form and find the decimal (base ten) equivalent:

(a) 320_{five} (b) 121_{five} (c) 203_{five}
(d) 21.1_{five} (e) 1.11_{five}

4. Rename as quinary system numerals the following decimal system numerals (for example, $27_{\text{ten}} = 102_{\text{five}}$):

(a) 14_{ten} (b) 140_{ten} (c) 327_{ten}

5. The following line is the number line in a base *four* system. The first two points have been labeled.

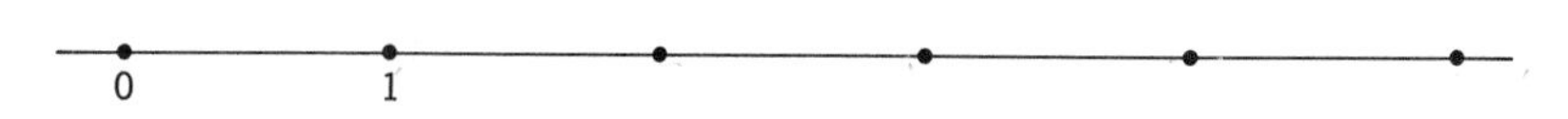

(a) Finish labeling the points using base four numerals.
(b) Plot the base four numerals 1.2; 2.3; 2.12; 10.1; and 20.2.

6. A group of four people want to divide evenly 2331.332_{four} ounces of petunia seeds. How much should each receive?

7. You move to a country that has a monetary system that consists of the following coins:

Coin name	Coin symbol	United States equivalent
Sens	ⓢ	1¢
Fens	ⓕ	5¢
Qens	ⓠ	25¢
Dens	ⓓ	$1.25
Mens	ⓜ	$6.25
Nens	ⓝ	$31.25

You want to convert your United States money to that of this new country so that you will have the least number of coins to carry. Convert the following amounts:

(a) $3.56 (b) $13.50
(c) $29.32 (d) $76.19

For example, $5.36 would be four dens, one quens, two fens, and one sens.

ⓓ ⓓ ⓓ ⓓ ⓠ ⓕ ⓕ ⓢ

8. What follows EE in base twelve?

9. What comes after TT in base twelve?

10. What number follows 99 in base twelve?

11. Is the base twelve number $4E$ even or odd? Explain why.

12. Is the base twelve number $7T$ even or odd? Explain why.

13. Which base twelve number is larger and by how much?

    (a) $9T$ or $T9$?
    (b) ET or EE?
    (c) $9E$ or $E1$?

14. What base five number follows 44?

15. What base five number follows 4?

4.16 THE ALGORITHMS

An algorithm is simply a procedure for performing an operation such as addition or multiplication. Whenever we perform operations with single-digit numbers, we recall the necessary facts from memory or use a table of elementary facts. When faced with computation involving multiple-digit numerals, however, we turn to an algorithm. The algorithms enable us to carry out complicated computations by using elementary facts and knowing the procedure. The place-value or exponential-positional system of numeration and the fact that the numbers are elements of a number system make the algorithm possible. The algorithms we use now are not the only possible ones. There have been changes in the past in the algorithms of arithmetic, and possibly there will be others in the future. Improvements in the algorithms lead to more efficient use of the system.

It will be noted as we proceed that the algorithms presented are procedures for naming the numbers determined by the binary operations of the system of whole numbers and do not depend on the *base* of the system but simply on the fact that we have a *place-value* system. This implies that the procedures will work as well in base five, base twelve, base two, or any other base, as they do in the base ten, or decimal, system.

4.17 COMPUTATION IN BASE FIVE

First let us consider some computations in a base five system of numeration. You will recall that after you learned to count in the decimal system the next step in the learning process was concerned with the elementary facts of addition and multiplication, the addition combinations and the multiplication tables. We will avoid having to memorize these facts by presenting tables of elementary facts as follows:

+	0	1	2	3	4
0	0	1	2	3	4
1	1	2	3	4	10
2	2	3	4	10	11
3	3	4	10	11	12
4	4	10	11	12	13

·	1	2	3	4
1	1	2	3	4
2	2	4	11	13
3	3	11	14	22
4	4	13	22	31

We read these tables as follows: to find $2+4$, we go down the left-hand column until we find the numeral 2, then across that row until we are in the column headed 4, where we find the numeral 11. This means that in base five arithmetic, $2+4=11$. We read this, "Two plus four is one-one, base five." It would be incorrect to say, "Two plus four is 'eleven.'" "Eleven" is a decimal word, and we are dealing in base five, or quinary, arithmetic.

The multiplication table is used similarly. Consider the product $3 \cdot 4$. We go down the left-hand column until we find the numeral 3, then across that row until we are in the column headed 4, where we find the numeral 22. This means that in quinary arithmetic $3 \cdot 4 = 22$. We read this, "Three times four is two-two, base five."

With the use of these tables, a little intuition, and an understanding of the system of whole numbers, we can carry out computations in base five. Let us try a few simple problems.

4.17a The Addition Algorithm

First, using properties of our system of whole numbers, and tables of elementary facts, we shall show that $13_{\text{five}}+21_{\text{five}}=34_{\text{five}}$. We shall then compare the complete algorithm with our usual procedure. (Note that in what follows, $10=10_{\text{five}}=5_{\text{ten}}$.)

(a) $13+21=(1\cdot 10^1+3\cdot 10^0)+(2\cdot 10^1+1\cdot 10^0)$ by our system of numeration.

(b) $(1\cdot 10^1+3\cdot 10^0)+(2\cdot 10^1+1\cdot 10^0)=1\cdot 10^1+(3\cdot 10^0+2\cdot 10^1)+1\cdot 10^0$ by the associative law of addition.

(c) $1\cdot 10^1+(3\cdot 10^0+2\cdot 10^1)+1\cdot 10^0=1\cdot 10^1+(2\cdot 10^1+3\cdot 10^0)+1\cdot 10^0$ by the commutative law of addition.

(d) $1\cdot 10^1+(2\cdot 10^1+3\cdot 10^0)+1\cdot 10^0=(1\cdot 10^1+2\cdot 10^1)+(3\cdot 10^0+1\cdot 10^0)$ by the associative law of addition.

(e) $(1\cdot 10^1+2\cdot 10^1)+(3\cdot 10^0+1\cdot 10^0)=(1+2)\cdot 10^1+(3+1)\cdot 10^0$ by the distributive law.

(f) $(1+2)\cdot 10^1+(3+1)\cdot 10^0=3\cdot 10^1+4\cdot 10^0$ by the table of elementary facts.

(g) $3\cdot 10^1+4\cdot 10^0=34$ by our system of numeration.

(h) $13+21=34$ by the transitive property of equals.

In our usual procedure we would first write the problem in the form

$$\begin{array}{r} 13 \\ \underline{21} \end{array}$$

Steps (a), (b), (c), (d), and (e) of the complete algorithm show that we are justified in such an arrangement of work in that we will get the same result if we perform column addition.

$$\begin{array}{r} 13 \\ \underline{21} \\ 34_{\text{five}} \end{array}$$

Steps (f), (g), and (h) indicate the performance of the addition.

Let us consider an example involving a "carry" and show how that appears in the complete algorithm. Consider the problem $13_{\text{five}} + 24_{\text{five}}$.

(a)	$13+24=(1\cdot 10^1+3\cdot 10^0)+(2\cdot 10^1+4\cdot 10^0)$	System of numeration
(b)	$=1\cdot 10^1+(3\cdot 10^0+2\cdot 10^1)+4\cdot 10^0$	Associative law of addition
(c)	$=1\cdot 10^1+(2\cdot 10^1+3\cdot 10^0)+4\cdot 10^0$	Commutative law of addition
(d)	$=(1\cdot 10^1+2\cdot 10^1)+(3\cdot 10^0+4\cdot 10^0)$	Associative law of addition
(e)	$=(1+2)\cdot 10^1+(3+4)\cdot 10^0$	Distributive law
(f)	$=3\cdot 10^1+12\cdot 10^0$	Table of elementary facts
(g)	$=3\cdot 10^1+(1\cdot 10^1+2\cdot 10^0)\cdot 10^0$	System of numeration
(h)	$=3\cdot 10^1+(1\cdot 10^1)\cdot 10^0+(2\cdot 10^0)\cdot 10^0$	Distributive law
(i)	$=3\cdot 10^1+1\cdot (10^1\cdot 10^0)+2\cdot (10^0\cdot 10^0)$	Associative law of multiplication
(j)	$=3\cdot 10^1+1\cdot 10^1+2\cdot 10^0$	Law of exponents
(k)	$=(3\cdot 10^1+1\cdot 10^1)+2\cdot 10^0$	Associative law of addition
(l)	$=(3+1)\cdot 10^1+2\cdot 10^0$	Distributive law
(m)	$=4\cdot 10^1+2\cdot 10^0$	Table of elementary facts
(n)	$=42$	System of numeration
(o)	$13+24=42$	Transitive property of equals

In this algorithm, the first five steps justify our arrangement of the problem for column addition, that is, column addition can be thought of as arranging the powers of the base five so that the distributive law can be applied. Step (f) is an indicated addition, and if this is written as column addition it would appear as follows:

$$\begin{array}{l} 13 \\ \underline{24} \\ 12 \\ 30 \quad \text{base five} \end{array}$$

Steps (g) through (k) justify the carry, and this is usually written as

$$\begin{array}{rl} \text{carry} \longrightarrow & 1 \\ & 13 \\ & \underline{24} \\ & 42 \quad \text{base five} \end{array}$$

Steps (e) through (o) complete the computation and justify our writing the sum as a column sum with the appropriate carry.

The addition algorithm applies to sums of more than two numbers. In writing out the complete algorithms for such problems we simply extend the ideas used in the simpler problems.

We wish to emphasize that our addition algorithm is based on elementary facts and properties of our system of whole numbers. Each step we take in carrying out a computation can be justified in terms of one or more of the fundamental concepts.

Example 1

Add 234_{five} and 432_{five}:

Procedure

$$\begin{array}{r} {\scriptstyle 11} \\ 234 \\ \underline{432} \\ 1221 \end{array}$$

1st column: From the table, $4+2=11$. Write 1 and carry 1 as in decimal arithmetic.

2nd column: From the table, $3+3=11$. $11+1$ (the carry) $=12$. Write 2, carry 1.

3rd column: From the table, $2+4=11$. $11+1$ (the carry) $=12$. Write 12.

We can easily check our result by converting the base five numerals to decimal numerals and computing in decimal arithmetic.

$$\begin{aligned} 234_{five} &= 2 \cdot 5^2 + 3 \cdot 5^1 + 4 \cdot 5^0 && \text{Expanded form with} \\ &= 50 + 15 + 4 && \text{decimal numerals} \\ &= 69_{ten} \end{aligned}$$

Similarly,

$$\begin{aligned} 432_{five} &= 117_{ten}. \\ 69 + 117 &= 186. && \text{Decimal arithmetic} \\ 1221_{five} &= 1 \cdot 5^3 + 2 \cdot 5^2 + 2 \cdot 5^1 + 1 \cdot 5^0. && \text{Expanded form with} \\ &= 125 + 50 + 10 + 1. && \text{decimal numerals} \\ &= 186_{ten}. \end{aligned}$$

This implies that our quinary arithmetic is correct. We obtained the same sum in both systems, and we feel quite confident that the decimal arithmetic is correct.

Example 2

Add 323_{five}, 413_{five}, and 343_{five}:

	Procedure
11	
323	1st column: From the table $3+3=11$; then $11+3=14$. Write 4,
413	carry 1.
343	2nd column: $4+1=10$, $10+2=12$, $12+1=13$. Write 3, carry 1.
2134	3rd column: $3+4=12$, $12+3=20$, $20+1=21$. Write 21.

Notice that in the procedure for this problem some of the additions performed in finding the sum of a column are problems in themselves. The answers are not found directly from the table. For example, $12+3=20$.

4.17b The Multiplication Algorithm

Using the tables of elementary facts and the properties of our system of whole numbers, we will show that $(14_{five})(2_{five}) = 33_{five}$.

(a)	$(14)(2) = (1 \cdot 10^1 + 4 \cdot 10^0)(2 \cdot 10^0)$	System of numeration
(b)	$= (1 \cdot 10^1)(2 \cdot 10^0) + (4 \cdot 10^0)(2 \cdot 10^0)$	Distributive law
(c)	$= 1 \cdot (10^1 \cdot 2) \cdot 10^0 + 4 \cdot (10^0 \cdot 2)\ 10^0$	Associative law of multiplication
(d)	$= 1 \cdot (2 \cdot 10^1) \cdot 10^0 + 4 \cdot (2 \cdot 10^0) \cdot 10^0$	Commutative law of multiplication
(e)	$= (1 \cdot 2)(10^1 \cdot 10^0) + (4 \cdot 2)(10^0 \cdot 10^0)$	Associative law of multiplication
(f)	$= (1 \cdot 2)(10^1) + (4 \cdot 2)(10^0)$	Law of exponents
(g)	$= 2 \cdot 10^1 + 13 \cdot 10^0$	Table of elementary facts
(h)	$= 2 \cdot 10^1 + (1 \cdot 10^1 + 3 \cdot 10^0) \cdot 10^0$	System of numeration
(i)	$= 2 \cdot 10^1 + (1 \cdot 10^1) \cdot 10^0 + (3 \cdot 10^0) \cdot 10^0$	Distributive law
(j)	$= 2 \cdot 10^1 + 1 \cdot (10^1 \cdot 10^0) + 3 \cdot (10^0 \cdot 10^0)$	Associative law of multiplication
(k)	$= 2 \cdot 10^1 + 1 \cdot 10^1 + 3 \cdot 10^0$	Law of exponents
(l)	$= (2 \cdot 10^1 + 1 \cdot 10^1) + 3 \cdot 10^0$	Associative law of addition
(m)	$= (2+1) \cdot 10^1 + 3 \cdot 10^0$	Distributive law
(n)	$= 3 \cdot 10^1 + 3 \cdot 10^0$	Table of elementary facts
(o)	$= 33$	System of numeration
(p)	$(14)(2) = 33$ base five	Transitive property of equals

Step (a) is usually shortened to

$$(10+4)2.$$

The distributive law (step b) allows us to multiply 10 by 2 and 4 by 2 first and then add

$$10 \cdot 2 + 4 \cdot 2.$$

Steps (f) through (m) show why we multiply 1 by 2 first and then add the 1 carried over from the previous multiplication. This might be written in the following way, which is suggestive of what is actually taking place:

```
14
 2
--
13
20
--
33   base five
```

The usual practice of finding products is a much abbreviated procedure as is indicated in the following example.

Example 1

In multiplying 123_{five} by 323_{five} we are actually finding the sum of the following products:

(a) $(100+20+3)3$
(b) $(100+20+3)20$
(c) $(100+20+3)300$

This represents several applications of the basic laws. If we carry out the indicated addition as column addition, it would appear as in Column I in the following. Columns II and III give some indication of the actual steps that are omitted in the usual procedure.

	I	II	III
	123	123	123
	323	323	323
(a)	14		
	110		
	3000	3124	3124
(b)	110		
	400		
	2000	3010	301
(c)	1400		
	11000		
	30000	42400	424
	104034	104034	104034

Notice in column III that we "indent" the 301 because we are actually multiplying by 20, not 2. We indent the 424 two places because we are actually multiplying by 300, not 3.

It is left to the student to check this example by converting to base ten arithmetic as in Example 1.

Example 2

Multiply 342_{five} by 23_{five}:

$\overset{1}{21}$	*Procedure*
342	Multiplying by 3: $3 \cdot 2 = 11$. Write 1, carry 1. $3 \cdot 4 = 22$, $22 + 1$
23	(the carry) $= 23$. Write 3, carry 2.
2131	$3 \cdot 3 = 14$, $14 + 2$ (the carry) $= 21$. Write 21.
1234	
20021	Multiplying by 2: same as for the 3 except that the 2 is in the second place, and we must set the partial products one place to the left as in decimal arithmetic.

The check by conversion to base ten:

$$342_{\text{five}} = 3 \cdot 5^2 + 4 \cdot 5^1 + 2 \cdot 5^0 = 75 + 20 + 2 = 97_{\text{ten}}.$$

$$23_{\text{five}} = 2 \cdot 5^1 + 3 \cdot 5^0 = 10 + 3 = 13_{\text{ten}}.$$

$$20021_{\text{five}} = 2 \cdot 5^4 + 0 \cdot 5^3 + 0 \cdot 5^2 + 2 \cdot 5^1 + 1 \cdot 5^0 = 1250 + 10 + 1.$$

$$= 1261_{\text{ten}}.$$

$13 \cdot 97 = 1261$ Decimal arithmetic.

Hence the quinary arithmetic is correct.

Subtraction and division can also be accomplished with the use of the addition and multiplication tables. The addition table answers the question $a + b = ?$ for a and b quinary digits, but it may also answer the question $a + ? = c$. For example, to find $11 - 3$ in quinary arithmetic, we consider the problem $3 + ? = 11$. To find the answer we go down the left-hand column until we find the numeral 3, then across the row until we find the numeral 11. The numeral that names the number that must be added to 3 to obtain 11 is at the top of this column. Hence $11 - 3 = 3$. Similarly, $12 - 4 = 3$, $10 - 2 = 3$, $11 - 2 = 4$, etc.

Example 3

Subtract 2341 from 4332 in quinary arithmetic:

$^{3}\,{}^{1}2_{1}$	*Procedure*	
~~4~~ ~~3~~ 3 2	1st column:	$1 + ? = 2, 1 + 1 = 2$. Write 1.
2 3 4 1	2nd column:	Borrow, then $4 + ? = 13, 4 + 4 = 13$. Write 4.
1 4 4 1	3rd column:	Borrow, then $3 + ? = 12, 3 + 4 = 12$. Write 4.
	4th column:	$2 + ? = 3, 2 + 1 = 3$. Write 1.

We can use our multiplication tables for division in the sense that the question $a \div b = ?$ can be interpreted as $b \cdot ? = a$.

Example 4

Divide 33011 by 4 in quinary arithmetic:

```
   4224
4)33011
  31
  --
   20
   13
   --
    21
    13
    --
     31
     31
     ==
```

Procedure

Same as for decimal arithmetic, using the base five multiplication table to find the partial quotients.

For example, $33 \div 4 = ?$; $4 \cdot ? = 33$; $4 \cdot 4 = 31$, which is as close as we can get to 33. So 4 is the first quotient figure. Multiply, subtract, bring down the next digit of the divident, then continue as before.

We have presented the traditional approach to this division problem. It is well to point out that the modern approach involves returning to an older format, or arrangement, of the work as follows:

```
4)33011|   4000
  31000|
  -----
   2011|    200
   1300|
   ----
    211|     20
    130|
    ---
     31|      4
     31|   ====
     ==    4224
```

This format indicates a more complete procedure. Here the partial quotients are obtained in essentially the same way as in the traditional approach, using the multiplication table to assist in making estimations.

Exercise 4.17b

1. Check Examples 2, 4, and 5 by converting the quinary numerals to decimal numerals and doing the arithmetic in the decimal system.

2. Carry out the following addition problems in base five arithmetic:

```
(a) 231    (b) 2342    (c) 2013
    413        1341        4002
    ---        3124        2144
               ----        3241
                           ----
```

3. Carry out the following multiplication problems in base five arithmetic:

```
(a) 2342    (b) 1341    (c) 2144
       3          32         234
    ----        ----        ----
```

4. Carry out the following subtraction problems in base five arithmetic:

(a) 413	(b) 2342	(c) 3000
231	1424	2143

5. Carry out the following division problems in base five arithmetic:

(a) 4)2033 (b) 3)1414 (c) 10)204320

6. Check your results for problem 2 by converting to the decimal system.

7. Construct the tables of elementary facts for base three (ternary) arithmetic and carry out the following computations.

(a) Add:		(b) Subtract:		(c) Multiply:	
212	222	212	2121	212	21022
220	111	121	212	22	2102

(d) Divide 21021 by 12. Divide 21020 by 12.

8. What is the minimum number of weights required to weigh objects weighing up to 41 ounces with a two-pan balance?

4.18 COMPUTER ARITHMETIC

In base two (binary) arithmetic we need but two symbols, and it is convenient to use the decimal symbols 0 and 1. The elementary facts needed for computation in base two are elementary indeed! Bearing in mind the special properties of zero and one, we see that the only other fact needed is $1+1=10$ in base two.

It may seem a waste of time even to consider a base two arithmetic, so you may be surprised to learn that the binary system has many useful and practical applications. Binary numbers are used in statistical investigations and in the study of probabilities, in analyzing games, and in the arithmetic units of some of the electronic computers. Whenever a situation may be described in terms of "on" or "off," "yes" or "no," "charge" or "no charge," or similar dual choices, the binary system of numeration is useful in its analysis.

Example 1

Add 1011_{two} and 1100_{two}.

	Procedure
1011	1st column: $0+1=1$. Write 1.
1100	2nd column: $0+1=1$. Write 1.
10111	3rd column: $1+0=1$. Write 1.
	4th column: $1+1=10$. Write 10.

Example 2

Add in base two, 1011, 1101, 1001.

Procedure

```
  1011   1st column:  1+1 = 10, 10+1 = 11. Write 1, carry 1.
  1101   2nd column:  0+0 = 0, 0+1 = 1, 1+1 (carry) = 10. Write 0,
  1001   carry 1.
------
100001   3rd column:  0+1 = 1, 1+0 = 1, 1+1 = 10. Write 0, carry 1.
         4th column:  1+1 = 10, 10+1 = 11, 11+1 = 100. Write 100.
```

Again, as in quinary arithmetic, some of the additions performed in finding the sum of a column are problems in themselves. For example, $11+1 = 100$ is not found directly from the table.

Example 3

Multiply the binary numbers: (1011)(101).

Procedure

```
  1011   We follow the same procedure as in decimal arithmetic. The
   101   problem of multiplying is much simpler, however. In multiply-
------
  1011   ing by a 1 all we need do is copy the multiplicand in proper
 10110   position.
------
110111
```

Subtraction can be carried out with the use of the addition facts as in Example 4 of the preceding section.

In dividing binary numbers it is not necessary to estimate a trial quotient. Because of the nature of the system we have but two choices at each step in the division process. Either the divisor "goes" or it does not. A simple example is sufficient to illustrate this.

Example 4

Divide the binary numbers: 1000010101 by 1101.

```
         101001
     ----------
1101)1000010101
     1101
     ----
       1110
       1101
       ----
          1101
          1101
          ====
```

4.18b Radix Fractions

In decimal arithmetic we have *decimal fractions* such as 0.25, 0.375, etc. We must also have a representation for such numbers in other bases. Since the word "decimal" is associated with a base ten system, we do not use it in arithmetic of other bases. In place of "decimal" point we use "reference" point and in place of "decimal" fraction we use "radix" fraction and we must identify the base in which we are working.

Let us consider some binary radix fractions.

Example 5

(a) 0.1_{two} in expanded form in decimal language is $1 \cdot 2^{-1}$ or $\frac{1}{2}$ which would be 0.5_{ten}.

(b) 0.11 in expanded form in decimal language is $1 \cdot 2^{-1} + 1 \cdot 2^{-2}$ or $\frac{1}{2} + \frac{1}{4} = \frac{3}{4} = 0.75_{ten}$.

(c) $0.875_{ten} = 0.500 + 0.250 + 0.125 = \frac{1}{2} + \frac{1}{4} + \frac{1}{8} = 1 \cdot 2^{-1} + 1 \cdot 2^{-2} + 1 \cdot 2^{-3} = 0.111_{two}$.

Note that $0.1_{two} = 0.5_{ten}$ and $0.01_{two} = 0.25_{ten}$. Moving the reference point one place to the left in a binary numeral divides the number being represented by two. It is true in general that moving the reference point one place to the left has the effect of dividing the number represented by the base. Moving the reference point one place to the right has the effect of multiplying the number represented by the base.

Example 6

(a) 11.1_{two} represents a number one-half as large as $111._{two}$

(b) 23.4_{five} represents a number one-fifth as large as $234._{five}$

(c) 21.4_{ten} represents a number one-tenth as large as $214._{ten}$

(d) $100._{ten}$ represents a number ten times as large as $10._{ten}$

(e) $101._{two}$ represents a number twice as large as 10.1_{two}

You might think that conversion of decimal fractions to binary radix fractions would be a difficult task to accomplish with a computer. The procedure can be illustrated as in Example 7.

Example 7

$$0.375_{ten} = (0.375)\left(\frac{2}{2}\right) = \frac{0.750}{2} = \frac{0}{2} + \left(\frac{0.750}{2}\right)\left(\frac{2}{2}\right) = \frac{0}{2} + \frac{1.5}{2^2}$$

$$= \frac{0}{2} + \frac{1}{2^2} + \left(\frac{0.5}{2^2}\right)\left(\frac{2}{2}\right) = \frac{0}{2} + \frac{1}{2^2} + \frac{1.0}{2^3} = \frac{0}{2} + \frac{1}{2^2} + \frac{1}{2^3}$$

But this is the expanded form (in decimal language) for the binary numeral 0.011. So, $0.375_{ten} = 0.011_{two}$.

This procedure can be shortened as indicated by the following example.

Example 8

$$\begin{array}{l|r}
 & 0.375 \\
 & \underline{2} \\
\text{binary radix} & 0.750 \\
\text{fraction} & \underline{2} \\
\text{digits} & 1.500 \\
 & \underline{2} \\
\downarrow & 1.000
\end{array} \qquad 0.375_{ten} = .011_{two}$$

	0.390625	
	2	
	0.781250	
	2	
	1.56250	
binary radix	2	$0.390625_{ten} = 0.011001_{two}$
fraction	1.1250	
digits	2	
	0.250	
	2	
	0.50	
	2	
	1.0	

	0.52	
	5	
quinary radix	2.60	$0.52_{ten} = 0.23_{five}$
fraction	5	
digits	3.0	

Exercise 4.18

1. Check Examples 1, 3, and 4 of Section 4.18 by converting the binary numerals to decimal numerals and carrying out the computation in decimal arithmetic.

2. Carry out the following addition problems in binary arithmetic:

(a) $101 + 111$ (b) $1101 + 1001 + 1111$
(c) $1001 + 1111 + 101$ (d) $10011 + 1001 + 100$
(e) $11011 + 11100 + 10111$ (f) $1111 + 1111 + 1111$

3. Compute in binary arithmetic

(a) (1011)(111) (b) (11010)(101) (c) (11011)(1101)
(d) $11011 - 1101$ (e) $10011 - 1010$ (f) $1100100 \div 1010$

4. An application of binary arithmetic is involved in the following number game. Consider the following cards:

Card A		Card B		Card C		Card D		Card E	
16	24	8	24	4	20	2	18	1	17
17	25	9	25	5	21	3	19	3	19
18	26	10	26	6	22	6	22	5	21
19	27	11	27	7	23	7	23	7	23
20	28	12	28	12	28	10	26	9	25
21	29	13	29	13	29	11	27	11	27
22	30	14	30	14	30	14	30	13	29
23	31	15	31	15	31	15	31	15	31

A student is given these cards and asked to indicate which ones have his age printed on them. He looks them over carefully and then answers that his age is printed on cards *A* and *E*. The owner of the cards quickly adds the upper left-hand numbers on these two cards and says, "You are 17 years old."

Try this with your friends. You will find the system infallible.

What is the system? Figure out the "why" behind this simple trick. As a beginning you might write the binary numerals from one to thirty-two and use these as a guide.

4.19 DUO-DECIMAL ARITHMETIC

Duo-decimal (base twelve) arithmetic can be carried out as in base five or base two. There is one difference, however—the symbols for the decimal digits will be insufficient in number. We need twelve symbols for a base twelve system of numeration, as indicated in Section 4.15. The tables of elementary facts for addition and multiplication would be as shown in Tables 2 and 3.

Table 2 Addition, Base Twelve

+	0	1	2	3	4	5	6	7	8	9	*T*	*E*
0	0	1	2	3	4	5	6	7	8	9	*T*	*E*
1	1	2	3	4	5	6	7	8	9	*T*	*E*	10
2	2	3	4	5	6	7	8	9	*T*	*E*	10	11
3	3	4	5	6	7	8	9	*T*	*E*	10	11	12
4	4	5	6	7	8	9	*T*	*E*	10	11	12	13
5	5	6	7	8	9	*T*	*E*	10	11	12	13	14
6	6	7	8	9	*T*	*E*	10	11	12	13	14	15
7	7	8	9	*T*	*E*	10	11	12	13	14	15	16
8	8	9	*T*	*E*	10	11	12	13	14	15	16	17
9	9	*T*	*E*	10	11	12	13	14	15	16	17	18
T	*T*	*E*	10	11	12	13	14	15	16	17	18	19
E	*E*	10	11	12	13	14	15	16	17	18	19	1*T*

In the duo-decimal system of numeration we do have some names for some of the powers of the base. For the name of the power of the base in the first place to the left of the reference point we can use "units," just as we do for the decimal system. The next place to the left could have the name "dozens," just as we say "tens" in the decimal system. The next place would be called "gross" (a dozen dozens) and the next "great loss" or "dozen gross," and so on.

The duo-decimal system has been advocated by many people as a standard system of numeration instead of the decimal system. It has certain advantages over the decimal system, but the problem of conversion on a national or international scale would be almost insurmountable. The student will be asked in the next set of exercises to name some of the advantages of the duo-decimal system over the decimal system.

Table 3 Multiplication, Base Twelve

·	1	2	3	4	5	6	7	8	9	T	E
1	1	2	3	4	5	6	7	8	9	T	E
2	2	4	6	8	T	10	12	14	16	18	$1T$
3	3	6	9	10	13	16	19	20	23	26	29
4	4	8	10	14	18	20	24	28	30	34	38
5	5	T	13	18	21	26	$2E$	34	39	42	47
6	6	10	16	20	26	30	36	40	46	50	56
7	7	12	19	24	$2E$	36	41	48	53	$5T$	65
8	8	14	20	28	34	40	48	54	60	68	74
9	9	16	23	30	39	46	53	60	69	76	83
T	T	18	26	34	42	50	$5T$	68	76	84	92
E	E	$1T$	29	38	47	56	65	74	83	92	$T1$

4.19a The Addition Algorithm

The complete algorithm for the sum $T4_{\text{twelve}} + 16_{\text{twelve}}$ is as follows:

$$\begin{aligned}
T4+16 &= (T \cdot 10^1 + 4 \cdot 10^0) + (1 \cdot 10^1 + 6 \cdot 10^0) && \text{System of numeration} \\
&= T \cdot 10^1 + (4 \cdot 10^0 + 1 \cdot 10^1) + 6 \cdot 10^0 && \text{Associative law of addition} \\
&= T \cdot 10^1 + (1 \cdot 10^1 + 4 \cdot 10^0) + 6 \cdot 10^0 && \text{Commutative law of addition} \\
&= (T \cdot 10^1 + 1 \cdot 10^1) + (4 \cdot 10^0 + 6 \cdot 10^0) && \text{Associative law of addition} \\
&= (T+1) \cdot 10^1 + (4+6) \cdot 10^0 && \text{Distributive law} \\
&= E \cdot 10^1 + T \cdot 10^0 && \text{Table of elementary facts} \\
&= ET && \text{System of numeration} \\
T4+16 &= ET && \text{Transitive property of equals}
\end{aligned}$$

Notice that the algorithm is independent of the base, that is, the procedure that is followed is the same in base twelve as in base five, or as in base ten. This also holds true for the algorithms of subtraction, multiplication, and division.

Example 1

Add $7ET45_{\text{twelve}}$ and 21372_{twelve}.

Procedure

$$\begin{array}{r} 7ET45 \\ \underline{21372} \\ T11E7 \end{array}$$

1st column: $5+2=7$. Write 7.
2nd column: $4+7=E$. Write E.
3rd column: $T+3=11$. Write 1, carry 1.
4th column: 1 (carry) $+E=10$, $10+1=11$. Write 1, carry 1.
5th column: 1 (carry) $+7=8$, $8+2=T$. Write T.

Example 2

Multiply $T45_{\text{twelve}}$ and 372_{twelve}.

	Procedure
$T45$	Multiplying by 2: $2 \cdot 5 = T$. Write T. $2 \cdot 4 = 8$. Write 8. $2 \cdot T = 18$.
$\underline{372}$	Write 18.
$188T$	Multiplying by 7: $7 \cdot 5 = 2E$. Write E, carry 2. $7 \cdot 4 = 24$, $24 + 2 =$
$606E$	26. Write 6, carry 2. $7 \cdot T = 5T$, $5T + 2 = 60$. Write 60. Multiply-
$\underline{2713}$	ing by 3: $3 \cdot 5 = 13$. Write 3, carry 1. $3 \cdot 4 = 10$, $10 + 1 = 11$.
$31367T$	Write 1, carry 1. $3 \cdot T = 26$, $26 + 1 = 27$. Write 27.

Subtraction and division can be carried out in a manner similar to that used for arithmetic in the other bases discussed. This type of computation will be left for the exercises.

Exercise 4.19

1. What are some of the advantages of the binary system over the decimal system?

2. What are some of the advantages of the duo-decimal system over the decimal system?

3. Carry out the following computations in duo-decimal arithmetic:

(a) $3E + T4$ (b) $204 + 60T$
(c) $1702 + 91TE6$ (d) $188TE + E7E03T$
(e) $60T - T4$ (f) $312 - 5T$
(g) $18E8T - 5TE4$ (h) $604T5 - 231E4$
(i) $(4T)(E3)$ (j) $(54)(214E)$
(k) $(2E4)(5467)$ (l) $(325)(TE41)$
(m) $293 \div 7$ (n) $11413 \div 5$
(o) $E468 \div 4E$ (p) $12E114 \div 214$

4. Use the addition algorithm and decimal arithmetic to find the following sums: Give reasons for each step.

(a) $27 + 9$ (b) $36 + 8$ (c) $379 + 96$ (d) $432 + 899$

5. Use the multiplication algorithm and decimal arithmetic to find the following products:

(a) $(36)(9)$ (b) $(47)(8)$ (c) $(36)(45)$ (d) $(57)(92)$

6. (a) What is the effect of multiplying a number by the base of the system of numeration?
(b) Multiply the base five number 342_{five} by the base.
(c) Multiply the base twelve number $TET7_{\text{twelve}}$ by the base.

7. Use the table of elementary facts in Section 4.19 and the addition algorithm to find the sum of each of the following:

(a) $9E7 + T$ (b) $TT + T$ (c) $EE + E$ (d) $TE + 9$

8. Use the table of elementary facts in Section 4.19 and the multiplication algorithm to find the following products:

(a) $(EE)(9)$ (b) $(TET)(8)$ (c) $(7E)(T5)$ (d) $(6T)(5E)$

9. The following exercises illustrate additional applications of base two arithmetic.

The question of multiplication in the Egyptian hieroglyphics has undoubtedly occurred to the reader. Addition is a simple process in this system of numeration and multiplication can be thought of as addition of one of the numbers to itself as many times as indicated by the second number. Thus $3 \cdot 6 = 6+6+6$. This is the way multiplication is carried out on a desk calculator. This method was probably the way multiplication was carried out for a long time by the Egyptians. However, their writing tools and working conditions dictated a more compact multiplication process. A very ingenious process called "doubling and summing" shortened their work considerably. This process was based on the fact that any number can be expressed as a sum of powers of 2. (This is the base two idea.) The base two notation was not used, however. Let us consider an example.

Find (39)(46).

Expressing 39 as a sum of powers of 2 we get $39 = 32+4+2+1$, or $39 = 2^5+2^2+2^1+2^0$. Hence multiplying by 39 is the same as multiplying by 1, 2, 4, and 32 and summing (use of the distributive law).

1	....	46
2	....	92
4	....	184
8	....	368
16	....	736
32	....	1472

Note that successive numbers in each column are obtained by "doubling" the preceding number.

(a) Which numbers in the right-hand column should be added to obtain the product of 39 and 46?
(b) Multiply (27)(49) by this method.
(c) Multiply (325)(202) by this method.

A variation of this method is called "halving and doubling." The work would be laid out as follows:

39	....	46*
19	....	92*
9	....	184*
4	....	368
2	....	736
1	....	1472*

Notice that in the process of "halving" we disregard remainders. To obtain the product, (39)(46), add the numbers in the right-hand column opposite the *odd* numbers in the left-hand column, that is,

$(39)(46) = 46 + 92 + 184 + 1472 = 1794.$

(d) Use this method for the problems in (b) and (c).

Consider the procedure for halving and doubling and this time write the remainder after each "halving" as follows:

39 1
19 1
9 1
4 0
2 0
1 1

Now write the remainders from left to right in the same order they appear from bottom to top, that is, 100111.

(e) What is the binary numeral 100111 as a decimal numeral?

4.20 THE "100" COUNTING TABLE

A useful and interesting device for teaching counting in any base is the "100" counting table. The counting table shown in Table 4 is for bases two through twelve.

Table 4 The "100" Counting Table

1	2	3	4	5	6	7	8	9	*T*	*E*	10
11	12	13	14	15	16	17	18	19	1*T*	1*E*	20
21	22	23	24	25	26	27	28	29	2*T*	2*E*	30
31	32	33	34	35	36	37	38	39	3*T*	3*E*	40
41	42	43	44	45	46	47	48	49	4*T*	4*E*	50
51	52	53	54	55	56	57	58	59	5*T*	5*E*	60
61	62	63	64	65	66	67	68	69	6*T*	6*E*	70
71	72	73	74	75	76	77	78	79	7*T*	7*E*	80
81	82	83	84	85	86	87	88	89	8*T*	8*E*	90
91	92	93	94	95	96	97	98	99	9*T*	9*E*	*T*0
*T*1	*T*2	*T*3	*T*4	*T*5	*T*6	*T*7	*T*8	*T*9	*TT*	*TE*	*E*0
*E*1	*E*2	*E*3	*E*4	*E*5	*E*6	*E*7	*E*8	*E*9	*ET*	*EE*	100

As shown, it is the "100" counting table for base twelve. By covering the columns headed T and E and the rows beginning with $T1$ and $E1$, (except the last entry 100), we obtain the base ten counting table. By covering the columns headed 7, 8, 9, T, and E, and the rows beginning with 71, 81, 91, $T1$, and $E1$, (except for the last entry 100), we have the counting table for base seven.

To make a "100" counting table for other bases, cover the appropriate rows and columns.

REVIEW EXERCISES I

1. What is the primary reason for studying place-value systems of numeration to bases other than ten?

2. Let us make up a base four place-value system of numeration using symbols other than the decimal numerals. The new symbols are as follows:

$n(\emptyset) = \Delta$
$n(\{\Delta\}) = -$
$n(\{\Delta, -\}) = \cap$
$n(\{\Delta, -, \cap\}) = \Sigma$

(a) Write the first fifteen numerals in this system.
(b) What is the successor of $\Sigma\Sigma\Sigma$?
(c) Which is larger, $\cap^{\Sigma}$ or $\Sigma^{\cap}$?
(d) What is the decimal numeral for $-\Delta.\cap\cap$?

3. Using the system of numeration of problem 2, carry out the indicated operations.

(a) $\cap\ \Sigma{-}\Sigma\ +\ \Sigma\Sigma\Delta\ \cap = ?$
(b) $\cap\ \Sigma{-}\Sigma\ \cdot\ -\Delta\Delta = ?$

4. Construct a number line for the whole numbers and label the points using the system of numeration in problem 2.

5. On the number line of problem 4, plot the following points: $-.\cap$, $\Delta.{-}\cap$, $\cap\,.\,\cap-$.

6. Let W denote the set of whole numbers and N denote the set of natural numbers. We define a relation in the set $W \times N$ as follows:

$(a, b) = (c, d)$ if and only if $ad = bc$.

(a) List three elements which are related to (3, 5).
(b) Which of the following elements do not belong to $W \times N$:
(2, 1) (1, 1) (6, 0) (1, 8) (0, 0) (0, 1)?
(c) Use numerical examples to convince yourself that this is an equivalence relation.

7. Define the binary operation $\oplus$ in $W \times N$ as follows:

$[(a, b), (c, d)] \xrightarrow{\oplus} (a, b) \oplus (c, d) = (ad + bc, bd).$

(a) What is $(7, 5) \oplus (6, 8)$?
(b) What is $(3, 4) \oplus (0, 4)$?

8. Does the binary operation defined in problem 7 satisfy the closure law? Illustrate.

9. Does the binary operation defined in problem 7 satisfy the commutative law? Illustrate.

10. Does the binary operation defined in problem 7 satisfy the associative law? Illustrate.

11. Is there an element which acts as an identity for the binary operation defined in problem 7? What is it?

12. In the set $W \times W$, define a relation as follows:

    $(a, b) = (c, d)$ if and only if $a + d = b + c$.

    (a) Use numerical examples to convince yourself that this is an equivalence relation.
    (b) List elements in the equivalence class to which (3, 3) belongs.

13. Define the binary operation $\ominus$ in $W \times W$ as follows:

    $[(a, b), (c, d)] \xrightarrow{\ominus} (a, b) \ominus (c, d) = (ac + bd, ad + bc)$

    (a) What is $(7, 5) \ominus (6, 8)$?
    (b) What is $(3, 8) \ominus (7, 7)$?

14. Does the binary operation defined in problem 13 satisfy the closure law? Illustrate.

15. Does the binary operation defined in problem 13 satisfy the commutative law? The associative law?

16. Is there an element which acts as an identity for the binary operation defiNed in problem 13? What is it?

REVIEW EXERCISES II

1. What do we mean by the statement, "The set B has cardinal 87, or cardinal number 87"?

2. Use the set-builder notation in any way you can to specify the empty set $\emptyset$.

3. The number 0 is a unique number in many ways. List some of these ways.

4. Show that 0 is unique as an element of the system of whole numbers (that is, show that there is one and only one additive identity in the system of whole numbers).

5. Show that $3 \cdot 0 = 0$, and justify each step.

6. If $A = \{2, 3, 5, 16\}$, what is the least upper bound of A?

7. In the indicated multiplication,

$$\begin{array}{r} 372 \\ \underline{39} \\ 3348 \\ \underline{1116} \\ 14508 \end{array}$$

explain why the fourth row of digits is indented one place to the left before the addition is performed.

Give precise definitions of each of the following:

8. The order relation "$<$" for the whole numbers.

9. The order relation "$\leq$" for the whole numbers.

10. The subset of a set.

11. The union of two sets.

12. The empty set.

13. The Cartesian product of two sets.

14. Equality of ordered pairs.

15. A relation: (a) intuitively, (b) in terms of sets.

16. Range of a relation.

17. The "equals" relation for numbers.

18. The cardinal of a set.

Give examples of relations possessing the following properties:

19. Reflexive and transitive, but not symmetric.

20. Transitive, but not reflexive and not symmetric.

21. Single-valued.

REFERENCE

Wolfe, Lee R., "Computer Concepts Possessed by 7th Grade Children," *The Arithmetic Teacher*, Vol. 15, No. 1, January 1968.

CHAPTER 5

The System of Integers

5.1 INTRODUCTION

Historically the natural numbers were invented for the purpose of counting. The rational numbers (fractions) were introduced in connection with problems of measuring. It took a very long time for zero to be accepted as a number. The same was true of negative numbers. Although some irrational numbers were known by the early Greek mathematicians, it was not until comparatively recently that a complete theory of irrational numbers was developed. Almost the same sequence of development is followed in the teaching of numbers in our present educational system. Numbers, when introduced in this sequence, are probably learned more efficiently than in any other order because it is easy to relate each new number concept to physical experiences. The concept of a half of something is learned long before $\frac{1}{2}$ is introduced as a number. By the time $\frac{1}{2}$ is introduced as a number it is easy to relate the fact that two halves of a cake are the same as a whole cake to the fact that $\frac{1}{2}+\frac{1}{2}=1$. These analogies must be selected carefully. Thus, to state that two halves of something are equal to the whole has meaning only in certain contexts. Not many people would be willing to accept two halves of an automobile tire for a whole tire!

We depart from tradition and treat the negative numbers before we treat the fractions (the rational numbers). This should present no difficulties, for this is not a first course in arithmetic. It does allow us to present the various number systems in what seems to us a reasonable order. This approach permits us to emphasize the power and limitations of each system and to gain insight into the *structure* of the various number systems. The idea of the structure of number systems is important in the understanding of the structure of mathematics. We hope, finally, to clarify the arithmetic processes in terms of the systems of numeration and the structure of the number systems.

For a discussion of the teaching of arithmetic of signed numbers to elementary students see, "A Rationale in Working with Signed Numbers," Louis S. Cohen, *The Arithmetic Teacher*, November, 1965.

5.2 THE SET OF INTEGERS

Recall that the *set* N of natural numbers is the set of "counting numbers," $\{1, 2, 3, \ldots\}$. The *set* W of whole numbers is the set $\{0, 1, 2, 3, 4, \ldots\}$. By the *system* of whole numbers we meant the set W, the two binary operations, $+$ and $\cdot$, and the rules governing the behavior of the numbers under the binary operations.

The system of whole numbers is a system in which we can answer the following simple questions.

First, what is the number $2+3$? We write this

$$2+3=?$$

or

$$2+3=n, \quad n=?$$

Second, "I am thinking of a number. If I add 3 to this number, I get 5. What is the number?" We write this

$$n+3=5, \quad n=?$$

(We are essentially doing subtraction.)

There are, however, severe limitations to the system of whole numbers. Questions similar in form to the second question just presented need not have answers. For example, what number must be added to 5 to get 4? What number added to 3 is 0?

$$5+?=4.$$

$$3+?=0.$$

In order to answer these simple questions and similar ones we define a new set, J, which we call the *integers*.

Let us say that

$^{-}3$ is the number which when added to 3 gives 0
and is the only such number.
$^{-}1$ is the number which when added to 1 gives 0
and is the only such number.
$^{-}15$ is the number which when added to 15 gives 0
and is the only such number.

In general, if n is any number, ^{-}n is the number which when added to n gives 0 and is the only such number. The new number, ^{-}n, is called the *additive inverse* of n.

The number n may be *any* element of J, and since 3 is the number which when added to $^{-}3$ gives 0 and the only such number, the *additive inverse* of n is a more appropriate name than "negative n."

Note that

$$^{-}3+3=0,$$

and

$$^{-}3+{}^{-}(^{-}3)=0,$$

and since $^{-}3$ has only one additive inverse

$$^{-}(^{-}3)=3$$

In general, from the definition of additive inverse it is clear that

$$^{-}(^{-}m)=m \text{ for any integer } m. \quad \textbf{(I-1)}$$

We shall use this freely in what follows and refer to this fact by the symbol **I-1**.

> ***Definition 5.2.*** The *set* J of integers consists of the set N of natural numbers, zero, and for each n in N, the number ^{-}n, such that $n+{}^{-}n=0$.

$$J=\{\ldots,{}^{-}n,\ldots,{}^{-}4,{}^{-}3,{}^{-}2,{}^{-}1,0,1,2,3,4,\ldots,n,\ldots\}.$$

The numbers $^{-}1$, $^{-}2$, $^{-}3$, etc., when associated with the number line are plotted to the left of the point labeled 0 in the same way as 1, 2, 3, etc., were plotted to the right (see Section 4.13a).

5.3 PROPERTIES OF THE SET OF INTEGERS

Notice that the set N of natural numbers is a subset of the set J of integers. As integers, we will refer to the natural numbers as *positive integers*. The additive inverses of the positive integers will be called *negative integers*.

The Set J

The set N

$$\underbrace{\ldots{}^{-}4,{}^{-}3,{}^{-}2,{}^{-}1,}_{\text{Negative integers}}\ \underbrace{0,}_{\text{Zero}}\ \underbrace{\overbrace{1,2,3,4,5,6,\ldots}^{\text{The set }N}}_{\text{Positive integers}}$$

When we say m is an integer, *only one* of the following statements must be true.

1. $m=0$, or
2. m is positive, or
3. ^{-}m is positive.

For example: $0 = 0$; 1 is positive; $^{-}1$ is such that $^{-}(^{-}1)$ is positive; etc. That is, the *set of integers* is partitioned into three mutually disjoint sets, and any integer can belong to one and only one of these three sets. This is a statement of the *trichotomy law.*

5.3a Properties of the Positive Integers

Recall that the natural numbers are the *positive integers* when considered as a subset of the integers. But the natural numbers are closed under the binary operations of addition and multiplication. We restate the closure laws (1 and 2 of the following), for they will be useful later in establishing properties of the order relation.

1. The sum of two positive integers is a positive integer.
2. The product of two positive integers is a positive integer.
3. If m is an integer and $m \neq 0$, then either m is positive or ^{-}m is positive.

We will extend the binary operations of addition and multiplication to the set of integers and require that the closure laws, the commutative laws, the associative laws, and the distributive law are satisfied. The new system, consisting of the set J of integers, the binary operations defined on J and the laws governing these operations, is called the *system of integers.*

Exercise 5.3

1. Is 0 a positive number? Why?

2. Is $^{-}0$ a negative number?

3. Is $0 = {}^{-}0$? Show why.

4. If a is an integer, is $^{-}(^{-}a)$ positive? What is the meaning of $^{-}(^{-}a)$? What is it equal to? What is $^{-}(^{-}3)$? What is ^{-}m if $m = {}^{-}2$?

5. State the trichotomy law for the integers in another way.

6. Is ^{-}m a negative integer if m is an integer?

7. What is $^{-}(m+n)$?

8. What is the additive inverse of $^{-}3$?

9. What is the additive inverse of ^{-}m?

10. What is the additive inverse of 0?

5.4 THE SYSTEM OF INTEGERS

The *set* of integers can be thought of as an enlarged set of numbers which contains the set of natural numbers and the singleton set $\{0\}$ as proper subsets. When we speak of the *system of integers,* we mean the *set* J defined in Section 5.2 and binary operations, addition and multiplication, satisfying certain laws. We want addition of the integers to be an extension of the binary operation of addition of the natural numbers in the sense that when we think of the natural numbers as positive integers, addition of the positive integers should be consistent with the addition of natural numbers. The same is true for multiplication.

In the system of integers, which is an enlargement of the system of whole numbers, we require the binary operations to satisfy the same laws as for the whole numbers and one new law concerning the existence and uniqueness of the additive inverse of any element.

Definition 5.4. By the *system of integers* we mean the set

$$J = \{\ldots, {}^{-}n, \ldots, {}^{-}4, {}^{-}3, {}^{-}2, {}^{-}1, 0, 1, 2, 3, 4, \ldots, n, \ldots\},$$

the binary operations, addition (+) and multiplication (·), and the following laws.

Closure Laws

1. For m and n in J there is a uniquely determined sum, which we write $m+n$, in J.
2. For m and n in J there is a uniquely determined product, which we write $m \cdot n$, in J.

Associative Laws. For m, n, and k in J,

3. $m+(n+k) = (m+n)+k$.
4. $m \cdot (n \cdot k) = (m \cdot n) \cdot k$.

Commutative Laws. For m and n in J,

5. $m+n = n+m$.
6. $m \cdot n = n \cdot m$.

Distributive Law. For m, n, and k in J,

7. $m \cdot (n+k) = m \cdot n + m \cdot k$.

Identities

8. There is a unique element 0 such that for any m in J, $m+0 = 0+m = m$.
9. There is a unique element 1 such that for any m in J, $1 \cdot m = m \cdot 1 = m$.

Additive Inverses

10. For each m in J there is a unique element ${}^{-}m$ in J such that $m + {}^{-}m = {}^{-}m + m = 0$.

5.4a Addition of Integers

The use of the foregoing laws enables us to extend the binary operation of addition to the set of integers, using what we already know about addition in the set of whole numbers.

The positive integers are the natural numbers renamed so that addition of the positive integers is addition of the natural numbers.

To extend the binary operation of addition to the negative integers we make use of our knowledge of the addition of positive integers. If m and n are any two integers, we assert that

$$^{-}m + ^{-}n = ^{-}(m+n). \qquad \textbf{(A-1)}$$

This assertion states that the binary operation, addition, assigns to the ordered pair $(^{-}m, ^{-}n)$ the number we write as $(^{-}m + ^{-}n)$, which has another name, $^{-}(m+n)$. This "other name" for the number is the one that enables us to find the sum in terms of addition of positive integers.

The following example indicates the procedure for establishing this assertion, using positive integers for m and n. The general proof is left as an exercise for the reader.

Example 1

We wish to show that $(^{-}2 + ^{-}3) = ^{-}(2+3)$.

$^{-}(2+3) + (2+3) = 0$ by the property of additive inverses.

$$\begin{aligned}
(^{-}2 + ^{-}3) + (2+3) &= (^{-}2 + ^{-}3) + (3+2) && \text{Why?} \\
&= ^{-}2 + (^{-}3 + 3) + 2 && \text{Why?} \\
&= ^{-}2 + 0 + 2 && \text{Why?} \\
&= ^{-}2 + 2 && \text{Why?} \\
&= 0 && \text{Why?}
\end{aligned}$$

We have

$$^{-}(2+3) + (2+3) = 0$$

and

$$(^{-}2 + ^{-}3) + (2+3) = 0.$$

Since the additive inverse of any number, in this case $(2+3)$, must be unique (one and only one), we conclude that

$$(^{-}2 + ^{-}3) = ^{-}(2+3).$$

To extend the binary operation of addition to the addition of a negative integer and a positive integer, using properties already developed, we assert that if m and n are integers, $(^{-}m + n)$ is the number which when added to m gives n.

This assertion states that the binary operation, addition, assigns to the ordered pair $(^{-}m, n)$ the number we write as $(^{-}m + n)$. It is the solution to the equation

$$m + ? = n.$$

We shall indicate the procedure for establishing this assertion and actually finding a new name for the number $(^{-}m + n)$ by several examples. For this purpose, recall the simple equation

$$3 + x = 7, \quad x = ?$$

What number must be added to 3 to get 7?

Let us proceed to "solve" this equation. Adding the additive inverse of 3 to each side of the equation we get

$$
\begin{aligned}
{}^{-}3+(3+x) &= 7+{}^{-}3 \\
({}^{-}3+3)+x &= (7+{}^{-}3) && \text{Why?} \\
0+x &= (7+{}^{-}3) && \text{Why?} \\
x &= (7+{}^{-}3) && \text{Why?}
\end{aligned}
$$

But from the table of elementary facts we know that $4+3=7$. Then by substitution we have

$$
\begin{aligned}
x &= (4+3)+{}^{-}3 \\
&= 4+(3+{}^{-}3) && \text{Why?} \\
&= 4+0 && \text{Why?} \\
&= 4 && \text{Why?}
\end{aligned}
$$

This tells us that $(7+{}^{-}3)$ is just another name for 4, or that $(7+{}^{-}3)=4$.

By convention we write this as $(7-3)$ instead of $(7+{}^{-}3)$.

This may seem like an unnecessarily complicated way of solving a very simple problem. We admit that it may seem so, but the object of this example is to illustrate the role of the basic laws in understanding arithmetic. The following problem is essentially the same as the last problem but one which is usually avoided at primary and intermediate levels because of the fact that the result is a negative integer.

$$9+x=4, \quad x=?$$

What number added to 9 gives 4?

Again, we proceed by "solving" this equation.

$$
\begin{aligned}
9+x &= 4 \\
{}^{-}9+(9+x) &= 4+{}^{-}9 && \text{Why?} \\
({}^{-}9+9)+x &= 4+{}^{-}9 && \text{Why?} \\
0+x &= 4+{}^{-}9 && \text{Why?} \\
x &= 4+{}^{-}9 && \text{Why?} \\
x &= 4+{}^{-}(4+5) && \text{Why?} \\
x &= 4+({}^{-}4+{}^{-}5) && \text{Why?} \\
x &= (4+{}^{-}4)+{}^{-}5 && \text{Why?} \\
x &= 0+{}^{-}5 && \text{Why?} \\
x &= {}^{-}5 && \text{Why?}
\end{aligned}
$$

It is essential to see $(7-3)$, that is, $(7+{}^{-}3)$, is the number which when added to 3 gives 7. Also, $(11-5)$, which is $(11+{}^{-}5)$, is the number which when added to 5 gives 11.

These remarks lead naturally to the generalization that $(n-m)$ is that number which when added to m gives n.

$$(n-m)+m=n.$$

Furthermore, since

$$m+({}^{-}m+n)=n,$$

then

$$({}^{-}m+n)=(n+{}^{-}m)=n-m. \qquad \textbf{(A-2)}$$

This expression may not have meaning in the system of whole numbers because $n-m$ may not be a whole number. In elementary arithmetic $n-m$ is usually interpreted as the *difference* of n and m or as m *subtracted* from n. "Subtraction" as a binary operation is neither associative nor commutative (see Sections 4.8b, 4.8c).

To summarize this section on the addition of integers, let us consider several numerical examples.

Example 2

(a) ${}^{-}5+{}^{-}9={}^{-}(5+9)={}^{-}14$
(b) ${}^{-}5+\ 9={}^{-}5+(5+4)=({}^{-}5+5)+4=0+4=4$
(c) $5+{}^{-}9=5+{}^{-}(5+4)=5+({}^{-}5+{}^{-}4)=(5+{}^{-}5)+{}^{-}4$
$\quad =0+{}^{-}4={}^{-}4$

Example 3

(a) ${}^{-}6+{}^{-}15={}^{-}(6+15)={}^{-}21$
(b) ${}^{-}6+\ 15={}^{-}6+(6+9)=({}^{-}6+6)+9=0+9=9$
(c) $6+{}^{-}15=6+{}^{-}(6+9)=6+({}^{-}6+{}^{-}9)=(6+{}^{-}6)+{}^{-}9$
$\quad =0+{}^{-}9={}^{-}9$

Exercise 5.4a

1. Add $9+{}^{-}3$, showing and justifying each step.

2. Add ${}^{-}7+4$, showing and justifying each step.

3. Rewrite the problems of Examples 2 and 3, justifying each step.

4. Show that $({}^{-}m+{}^{-}n)={}^{-}(m+n)$ for any integers m and n.

5. Write the additive inverses of each of the following:

(a) 12 (b) 2 (c) ${}^{-}3$ (d) ${}^{-}1$
(e) 0 (f) a (g) ${}^{-}a$ (h) $2+a$
(i) ${}^{-}2+a$ (j) ${}^{-}a+2$ (k) ${}^{-}3+3$ (l) ${}^{-}5+3$
(m) ${}^{-}(a+b)+2$ (n) ${}^{-}2+(a+b)$ (o) ${}^{-}a+b+3$ (p) ${}^{-}3+{}^{-}2$

6. Add

(a) ${}^{-}3+{}^{-}5$ (b) ${}^{-}2+{}^{-}1+{}^{-}6$
(c) $a+{}^{-}3+{}^{-}1$ (d) ${}^{-}13+{}^{-}8$
(e) ${}^{-}18+3+{}^{-}7$ (f) $(18+{}^{-}3)+{}^{-}7$
(g) $18+({}^{-}3+{}^{-}7)$ (h) ${}^{-}9+{}^{-}13$

7. Find the solution set for each of the following open sentences.

(a) $3+n=10$.
(b) $n+5=1$.
(c) $a+x=b$, consider x as the variable.
(d) $n+a=b$, consider n as the variable.

8. Show that $17-12$ is another name for 5.

9. Show that $372-176$ is another name for 196.

10. We break a yardstick into two pieces. If one piece is n in. long, how long is the other piece.

11. Where would you break the yardstick if the longer piece is to be 3 times the length of the shorter piece?

12. Is subtraction as a binary operation commutative? Use a numerical example to justify your answer.

13. Is subtraction as a binary operation associative? Use a numerical example to justify your answer.

14. Use the properties of the system of integers (see Definition 5.4) to prove that $m \cdot 0 = 0 \cdot m = 0$ for any integer m.

5.4b Multiplication of Integers

The set of positive integers is the same as the set of natural numbers. Thus multiplication of the positive integers will be the same as multiplication of the natural numbers.

In order to extend the operation of multiplication to the integers, we show that, for m and n integers,

$$({}^{-}m)(n) = {}^{-}(m \cdot n). \qquad \textbf{(M-1)}$$

Before attempting the general case, let us look at a numerical example. We shall use the fact that a number can have only one additive inverse to show that $({}^{-}4)(3) = {}^{-}12$.

Example 1

$${}^{-}12+12=0 \qquad \text{Why?}$$
$${}^{-}(4\cdot 3)+(4\cdot 3)=0 \qquad \text{Why?}$$

Thus ${}^{-}(4\cdot 3)$ is the additive inverse of $(4\cdot 3)$.
But

$$({}^{-}4)(3)+(4)(3)=({}^{-}4+4)3 \qquad \text{Why?}$$
$$({}^{-}4+4)3=0\cdot 3=0 \qquad \text{Why?}$$

Hence $({}^{-}4)(3)$ is also an additive inverse of $(4\cdot 3)$. But the additive inverse of any number is unique. Hence $({}^{-}4)(3)={}^{-}(4\cdot 3)={}^{-}12$.

Now let us show, in general, that $(^{-}m)(n) = {}^{-}(m \cdot n)$.

$^{-}(m \cdot n) + (m \cdot n) = 0$	by the properties of the additive inverse.
$(^{-}m)(n) + (m)(n) = (^{-}m+m)n$	by the distributive law.
$(^{-}m+m)n = 0 \cdot n = 0$	by the properties of the additive inverse and zero (see problem 14, Exercise 5.4a).

We have shown that $^{-}(m \cdot n) + (m \cdot n) = 0$. We have also shown that $(^{-}m)(n) + (m \cdot n) = 0$. But $(m \cdot n)$ has only one additive inverse, so $(^{-}m)(n) = {}^{-}(m \cdot n)$.

In particular, note that we have now shown how to find the product of a positive integer and a negative integer in terms of the product of two positive integers.

Similarly, for any integers m and n,

$$(m)(^{-}n) = {}^{-}(m \cdot n).$$

The proof of this will be left as an exercise for the reader.

The foregoing statements may be summarized by saying that the binary operation of multiplication assigns to any ordered pair which consists of an integer and the additive inverse of an integer, the additive inverse of the product of the integers. The statement, "A negative number times a positive number is a negative number," is seen to be a simple consequence of the basic laws.

We now consider the product of the additive inverses of any two integers. We will show that

$$(^{-}m)(^{-}n) = m \cdot n. \qquad \textbf{(M-2)}$$

To do this we proceed as follows:

$$\begin{aligned}
(-m)(-n) &= (-m)(-n) + 0 \\
&= (-m)(-n) + (0)(n) \\
&= (-m)(-n) + (-m+m)(n) \\
&= (-m)(-n) + [(-m)(n) + (m)(n)] \\
&= (-m)(-n) + (-m)(n) + (m)(n) \\
&= [(-m)(-n) + (-m)(n)] + (m)(n) \\
&= (-m)(-n+n) + (m)(n) \\
&= (-m)(0) + (m)(n) \\
&= 0 + (m)(n) \\
&= (m)(n)
\end{aligned}$$

Hence $(^{-}m)(^{-}n) = m \cdot n$. This implies the familiar rule, "A negative number times a negative number is a positive number." We have demonstrated this to be a fact which is a consequence of the basic laws.

Exercise 5.4b

Establish the following equalities, using the properties of the system of integers, **I-1**, **A-1**, **A-2**, **M-1**, **M-2**, and properties of zero:

1. $^-(1 \cdot 1) = {}^-1$.
   for example, $^-(1 \cdot 1) = (^-1)(1)$ by **M-1**.
   $= {}^-1$ by mult. identity.
2. $(^-1)(^-1) = 1$.
3. $(^-2)(^-3) = (2)(3)$.
4. $(^-5)(0) = 0$.
5. $(^-2)(^-3)(^-4) = {}^-24$.
6. $(^-2)(^-3 + {}^-4) = 14$, in two ways.
7. $(^-3)(^-4 + {}^-5) = 27$.
8. $8(7 - 3) = 32$, in two ways.
9. $8(3 - 7) = {}^-32$, in two ways.
10. $^-(7 - 3) = (3 - 7)$.
11. $^-(m - n) = (n - m)$.
12. $^-(^-3) = 3$.
13. $(^-2)(3) = {}^-(2 \cdot 3)$.
14. $(m)(^-n) = {}^-(m \cdot n)$.
15. $(^-n)(0) = 0$.

5.5 THE CANCELLATION LAWS

We have seen that $n \cdot 0 = 0$ in general. (See problem 14, Exercise 5,4a.) This is somewhat surprising to most people. In fact, most people confuse the number 0 with the meaning of the word "nothing." We repeat that 0 is not nothing, 0 *is a number*, a very important and useful number. Previously, the number 0 was used to define the additive inverses. We will use this number again when discussing *order* in the integers. For the present we consider another seemingly obvious arithmetical statement involving 0.

The system of integers has no zero divisors. By this we mean that the product of two integers is zero if and only if one of the factors is zero, that is, $a \cdot b = 0$ if and only if either $a = 0$ or $b = 0$. (The "or" here is the inclusive or.) This statement would undoubtedly mean more if one were familiar with a mathematical system where the product of two nonzero elements is zero. This is considered in Section 5.14b. For the present consider the following examples:

$$2 \cdot 0 = 3 \cdot 0 \quad \text{because both} \quad 2 \cdot 0 = 0 \quad \text{and} \quad 3 \cdot 0 = 0.$$

In general,

$$x \cdot 0 = y \cdot 0$$

Now consider the following:

$$x \cdot 2 = y \cdot 2.$$

Adding the additive inverse of $y \cdot 2$ to both sides of the equation we obtain

$$\begin{aligned} x \cdot 2 + {}^{-}(y \cdot 2) &= y \cdot 2 + {}^{-}(y \cdot 2) \\ &= 0. \end{aligned}$$

Then $(x + {}^{-}y) \cdot 2 = 0$ by the distributive law.
Notice that we are looking at the product of two terms, one being $(x + {}^{-}y)$ and the other, 2. Using the fact that the product of two terms is 0 if and only if one of the factors is 0, we see that, since 2 is not 0, $(x + {}^{-}y)$ must be 0, that is,

$$(x + {}^{-}y) = 0.$$

Then $(x + {}^{-}y) + y = 0 + y$

$x + ({}^{-}y + y) = y$	Why?
$x + 0 = y$	Why?
$x = y$	Why?

This is an example of the *cancellation law for multiplication.*

If a is an integer and $a \cdot x = a \cdot y$, can we conclude that $x = y$? We have seen that if $a = 0$, we cannot. If $a \neq 0$, then we can say that $x = y$.

The Cancellation Law for Multiplication. If $a \neq 0$ and $a \cdot x = a \cdot y$, then $x = y$.

We are not "dividing" by a. This is a subtle point which will be explained further when we discuss the system of rational numbers.

There is also a *cancellation law for addition.* However, this law does not need conditions on the terms.

Cancellation Law for Addition. If $a + x = a + y$, then $x = y$.

The cancellation law for addition can be proved as follows:

$a + x = a + y$	Hypothesis
${}^{-}a + (a + x) = {}^{-}a + (a + y)$	Why?
$({}^{-}a + a) + x = ({}^{-}a + a) + y$	Why?
$0 + x = 0 + y$	Why?
$x = y$	Why?

Exercise 5.5

1. What is the solution set of the open sentence $(x - 3)(x - 7) = 0$? (This type of argument is used in solving quadratic equations by factoring.)

2. If $x+3=y+3$, what can you say about x? Why?

3. If $y=6/(x-1)$, what must be true about x? Why?

4. Name the law that is exemplified by each of the following:

(a) $xy=yx$.
(b) $(x+y)+z=x+(y+z)$
(c) $xy+xz=x(y+z)$
(d) $x+a=a+x$
(e) $(2a)b=2(ab)$
(f) $a+0=a$
(g) $a=1\cdot a$
(h) $0=a+{}^{-}a$
(i) $({}^{-}a)(b+{}^{-}b)=({}^{-}a)(b)+({}^{-}a)({}^{-}b)$

5. State the reason that justifies each of the following steps in proving that $({}^{-}a)({}^{-}b)=(a)(b)$:

$$\begin{aligned}({}^{-}a)({}^{-}b) &= ({}^{-}a)({}^{-}b)+0\\ &= ({}^{-}a)({}^{-}b)+[{}^{-}(a)(b)+(a)(b)]\\ &= ({}^{-}a)({}^{-}b)+[({}^{-}a)(b)+(a)(b)]\\ &= [({}^{-}a)({}^{-}b)+({}^{-}a)(b)]+(a)(b)\\ &= ({}^{-}a)({}^{-}b+b)+(a)(b)\\ &= ({}^{-}a)(0)+(a)(b)\\ &= 0+(a)(b)\\ &= (a)(b)\\ ({}^{-}a)({}^{-}b) &= (a)(b)\end{aligned}$$

6. What is meant by ${}^{-}({}^{-}a)$?

7. What is ${}^{-}({}^{-}a)$? Can you prove your assertion?

8. What is the solution set of the open sentence $(x-3)(x-7)>0$?

9. If m is an integer, is ${}^{-}(m^2)$ necessarily a negative number?

10. If m and n are integers, is $({}^{-}m)({}^{-}n)$ necessarily a positive integer?

11. Justify each step in the proof of the cancellation law for addition.

12. What do we mean by zero divisors?

13. State the cancellation law for multiplication of integers.

5.6 PRIME NUMBERS AND COMPOSITE NUMBERS

We introduced the divisibility concept for the natural numbers earlier. Thus 12 is divisible by 3, 72 is divisible by 12, and 72 is also divisible by 9 and 8. We extend this concept to the integers.

Definition 5.6a. An integer n is divisible by an integer m, $m\neq 0$, if there is a single integer k such that $n=m\cdot k$.

Definition 5.6b. We say that m is a *divisor* of n if n is divisble by m. We also say that m is a *factor* of n.

It is obvious that 1 divides every integer n and also that n divides n if n is a nonzero integer, since $n = n \cdot 1$.

Definition 5.6c. We say that m is a *proper divisor* of n if m is a divisor of n and $m \neq 1$, $m \neq -1$, $m \neq n$, and $m \neq -n$. In this case m is called a *proper factor of n.*

(*Note:* Here and in what follows we use the traditional $-n$ for the additive inverse of n.)

Thus 2 is a proper divisor of 18; 9 is also a proper divisor of 18; and 2 and 9 are proper factors of 18.

Note that we cannot say that 0 divides any number. Suppose we ask, "Does 0 divide 2?" If so, we should be able to write 2 as some multiple of 0. But 0 times any number is 0, not 2. On the other hand, every nonzero integer divides 0, since $0 = 0 \cdot n$.

Now let us consider the following subset of the set of positive integers:

$$\{2, 3, 5, 7, \ldots\}$$

What do you think the next integer should be? It is 11. What is the next one? It is 13. What criterion are we using to determine whether an integer belongs to this set?

$$\{2, 3, 5, 7, 11, 13, \ldots\}$$

These numbers have no proper factors and are called *prime numbers.*

Definition 5.6d. An integer, p, $p > 1$, is a *prime* if it has no proper divisors.

The only divisors of a prime, p, are $1, -1, p$, and $-$p.

Definition 5.6e. A positive integer different from 1 which is not a prime is called a *composite.*

The negative integers may be classified similarly by considering the negative integer $(-n)$, $n \neq 1$, as $(-1)(n)$ and examining the positive integer n.

Exercise 5.6

1. List all the primes less than 50.

2. List *all* the positive divisors of 36. The *prime* divisors.

3. List *all* the positive divisors of 52. The *prime* divisors.

4. List *all* the positive divisors of 14. The *prime* divisors.

5. List *all* the positive divisors of 39. The *prime* divisors.

6. Find all primes less than 100 by first throwing away multiples of 2, then multiples of 3, then multiples of 5, and so on.

7. How can you tell whether a number is divisible by 2? By 3? By 4? By 5? By 9?

8. List the common divisors of 50 and 52; of 36 and 39; of 39 and 52.

9. Write the following numbers as products of prime factors: (a) 72, (b) 356, (c) 512, (d) 1000.

10. Is 1 divisible by 0? Why? Is 0 divisible by 1? Why?

5.7 PRIME FACTORIZATION

The set of positive integers greater than 1 is partitioned into two disjoint sets, the set consisting of the primes and the set consisting of the composites. The primes are, in a sense, the building blocks of the composites as indicated in the next statement.

The Fundamental Theorem of Arithmetic

Any integer, different from 0 and ± 1, can be written as a product of primes and ± 1 in one and only one way, except possibly for the order in which the factors occur.

Example 1

Consider the integer 72. We can think of this as $9 \cdot 8$, then factoring further as $3 \cdot 3 \cdot 8 = 3 \cdot 3 \cdot 2 \cdot 4 = 3 \cdot 3 \cdot 2 \cdot 2 \cdot 2 = 3^2 \cdot 2^3$. On the other hand, we could think as follows: $72 = 6 \cdot 12 = 2 \cdot 3 \cdot 12 = 2 \cdot 3 \cdot 3 \cdot 4 = 2 \cdot 3 \cdot 3 \cdot 2 \cdot 2 = 2 \cdot 2 \cdot 2 \cdot 3 \cdot 3 = 2^3 \cdot 3^2$. We have looked at the factorization of 72 into prime factors in two different ways but have arrived at a unique factorization except for order of the factors.

The fundamental theorem of arithmetic can be proved by using some of the elementary notions of number theory and mathematical induction. For our purposes we shall accept it as a fundamental principle.

The problem of finding the prime factors of a number is, in general, tedious. For large numbers the problem has been turned over to modern high-speed computers. There are a few divisibility facts which enable one to tell by inspection whether a given number is divisible by the first few small numbers. We present these, some with proof and some without, to assist in prime factorization of numbers. It should be especially noted that these tests for divisibility rely heavily on the fact that our numeration system is a place-value decimal system.

Divisibility by 2. A number is divisible by 2 if and only if the units digit of its numeral is even. The reason for this is that every power of 10 except 10^0 is divisible by 2. Hence the number is divisible by 2 if and only if the units digit of its numeral is divisible by 2.

Divisibility by 3. A number is divisible by 3 if and only if the sum of the digits of its numeral is divisible by 3. We make this seem reasonable by giving an example.

$$
\begin{aligned}
378 &= 3 \cdot 10^2 + 7 \cdot 10 + 8 \\
&= 3 \cdot 100 + 7 \cdot 10 + 8 \\
&= 3(99+1) + 7(9+1) + 8 \\
&= 3 \cdot 99 + 3 \cdot 1 + 7 \cdot 9 + 7 \cdot 1 + 8 \\
&= (3 \cdot 99 + 7 \cdot 9) + (3 + 7 + 8) \\
&= (3 \cdot 33 + 7 \cdot 3)3 + (3 + 7 + 8)
\end{aligned}
$$

We see that 3 divides the first term on the right in the last equality. If it also divides the second term $(3+7+8)$, then it must divide 378. Hence, if 3 divides $(3+7+8)$, 3 divides 378. Furthermore, if 3 divides 378, it must divide $(3+7+8)$. This argument is based on the distributive law and the meaning of "divides." Note that the same rule and a similar argument applies to divisibility by 9.

Divisibility by 5. A number is divisible by 5 if and only if the units digit of its numeral is 0 or 5.

Divisibility by 7. A technique for testing divisibility by 7 can best be illustrated by an example.

Example 1

5236 is divisible by 7 if $523 - 2 \cdot 6 = 511$ is divisible by 7.
511 is divisible by 7 if $51 - 2 \cdot 1 = 49$ is divisible by 7.
49 is divisible by 7, so 511 is divisible by 7, and 5236 is divisible by 7.

Example 2

25,252 is divisible by 7 if $2525 - 2 \cdot 2 = 2521$ is divisible by 7.
2521 is divisible by 7 if $252 - 2 \cdot 1 = 250$ is divisible by 7.
250 is divisible by 7 if $25 - 2 \cdot 0 = 25$ is divisible by 7.
25 is not divisible by 7, so 25,252 is not divisible by 7.

Divisibility by 11. A similar procedure can be used to check for divisibility by 11. The only change is that the units is subtracted from the whole number part of one-tenth of the number instead of twice the units digit.

Example 3

25,256 is divisible by 11 if $2525 - 6 = 2519$ is divisible by 11.
2519 is divisible by 11 if $251 - 9 = 242$ is divisible by 11.
242 is divisible by 11 if $24 - 2 = 22$ is divisible by 11.
But 22 is divisible by 11, so 25,256 is divisible by 11.

[See "A General Test for Divisibility by any Prime (except 2 and 5)" by Benjamin Bold.]

Exercise 5.7

1. List all the positive divisors of 72.

2. List the prime numbers less than 100.

3. Express each of the following as a product of prime factors: (a) 84, (b) 198, (c) 975, (d) 144, (e) 4455, and (f) 10^{12}.

An integer d is a *common divisor* of a set of integers if it is a divisor of each of them.

4. List all common divisors of 198 and 144.

5. List all common divisors of 84, 198, and 405.

6. Write 21,489 as a product of prime factors.

7. Write 4408 as a product of prime factors.

8. List the common divisors of 4408 and 72.

9. Test the following for divisibility by 2, 3, 4, 5, 6, 7, 8, 9, 10: (a) 627,433, (b) 2,288,817, (c) 324,244, (d) 625,530.

10. Remembering that the multiplication table for 9 seems to be difficult for some people, consider the following:

$9\times 4=?$ $\quad 10-4=\mathbf{6},\quad 9-6=\mathbf{3},\quad$ hence $9\times 4=36$.
$9\times 8=?$ $\quad 10-8=\mathbf{2},\quad 9-2=\mathbf{7},\quad$ hence $9\times 8=72$.
$9\times 5=?$ $\quad 10-5=\mathbf{5},\quad 9-5=\mathbf{4},\quad$ hence $9\times 5=45$.

(a) Why does this give the correct answer?
(b) Can you extend this idea to 9 times any nonnegative number less than ten?

5.8 THE DIVISION ALGORITHM

Although the "divides" relation holds only between certain ordered pairs of integers, we can still say something about *any* given pair. For example, given the integers 16 and 7, we can express 16 as a multiple of 7 plus a remainder of 2.

$$16=7\cdot 2+2.$$

This very obvious arithmetical statement illustrates the division algorithm. *The Division Algorithm.* If m and n are any two integers, such that n is greater than 0, then there is a unique pair of integers, q and r, such that

$$m=n\cdot q+r$$

where r is less than n and greater than or equal to 0. If $r=0$, then n divides m.

The division algorithm is a statement deducible from other basic assumptions; however, we accept it here as a fundamental principle that is intuitively plausible.

The division algorithm is a comparative statement about the pair of integers m and n. If we think of n as some fixed positive integer (say 3), then the division algorithm says that any integer m can be written as a multiple of 3 with only 0, 1, or 2 as possible remainders. If we let $n=7$, the remainders of integers when divided by 7 are 0, 1, 2, 3, 4, 5, 6. We shall refer to this again. For the present we shall see how it can be applied in finding the greatest common divisor and the least common multiple of a pair of integers.

5.9 THE GREATEST COMMON DIVISOR

In later sections of this book we will need to "reduce" fractions. We may find it convenient to write $\frac{3}{4}$ instead of $\frac{39}{52}$. The idea of "reducing" involves the greatest common divisor.

Definition 5.9a. A positive integer, d, is the *greatest common divisor* of the integers a and b if d is a common divisor of a and b and is a multiple of every other common divisor.

The abbreviation "g.c.d." designates greatest common divisor.

Example 1

g.c.d. $(36, 60) = 12$.
g.c.d. $(-10,35) = 5$.
g.c.d. $(6, 12) = 6$.
g.c.d. $(5, 7) = 1$.

Definition 5.9b. If g.c.d. $(a, b) = 1$, we say a and b are *relatively prime*.

5.9a The G.C.D. Using Prime Factorization

The problem of finding the g.c.d. of two integers is simple when the integers are small, and can usually be done by inspection. There are systematic procedures for determining the g.c.d. of two integers. We shall examine two methods. The first method is illustrated with numerical examples.

Example 2

We wish to find the g.c.d. (6, 15). Factoring, we have

$6 = 2 \cdot 3$,

and

$15 = 3 \cdot 5$.

We note that 3 is a divisor both of 6 and of 15 and is the only positive common divisor other than 1. Thus g.c.d. $(6, 15) = 3$.

Example 3

We wish to find the g.c.d. (72, 90). Factoring, we have

$72 = 2 \cdot 2 \cdot 2 \cdot 3 \cdot 3 = 2^3 \cdot 3^2$,

and

$90 = 2 \cdot 3 \cdot 3 \cdot 5 = 2 \cdot 3^2 \cdot 5$.

The only power of the prime 2 which divides both 72 and 90 is 2^1. The highest power of the prime 3 which divides both 72 and 90 is 3^2. These numbers can be expressed as

$$72 = (2 \cdot 3^2) \cdot 2^2,$$

and

$$90 = (2 \cdot 3^2) \cdot 5,$$

which shows that $(2 \cdot 3^2)$ is a common divisor of 72 and 90. It is the greatest common divisor of 72 and 90. It is the product of the highest powers of the primes *common to both* numbers.

In general, the g.c.d. of two numbers, m and n, is the product of the highest powers of the primes *common to* the factorizations of *both* m and n.

Exercise 5.9a

1. Find the g.c.d. of 84 and 198; of 36 and 54.

2. Find the g.c.d. of 975 and 144; of 17 and 51.

3. Find the g.c.d. of -84 and 144; of -96 and 336.

4. Find the g.c.d. of 198 and 975.

5. Find the g.c.d. of -198 and -144.

6. Find the g.c.d. of 84, 198, and 144.

7. Find the g.c.d. of 144, 198, and 975.

8. 11 divides $(10+1)$, (10^2-1), (10^3+1), (10^4-1), etc. Using a procedure similar to the check for divisibility by 9, determine a check for divisibility by 11.

9. If p is a prime and n is any integer, what is the g.c.d. (p, n)?

5.9b The G.C.D. Using the Division Algorithm

The second method of finding the g.c.d. of two integers involves the division algorithm. Suppose we are interested in finding the g.c.d. of 58 and 16. The division algorithm allows us to write

$$58 = 16 \cdot 3 + 10, \quad \text{where} \quad 0 \leqslant 10 < 16.$$

Notice that any number that divides 58 and 16 must also divide 10 because we can write

$$10 = 58 - 16 \cdot 3.$$

In particular, the g.c.d. of 58 and 16 must divide 10. This is a consequence of the distributive law and the meaning of "divides." It implies that this g.c.d.—let us call it d—is a common divisor of 16 and 10. Further, it must be the greatest common divisor of 16 and 10 because if there were another common divisor greater than d, this divisor would also have to divide 58 and d would not be the g.c.d. of 58 and 16. Furthermore, any number which divides 16 and 10 must divide 58 so

$$\text{g.c.d.}\ (58, 16) = \text{g.c.d.}\ (16, 10).$$

Now we have reduced our problem to that of finding the g.c.d. of 16 and 10. Applying the division algorithm again, we have

$$16 = 10 \cdot 1 + 6.$$

Again, any number that divides 16 and 10 must also divide 6 because we have

$$6 = 16 - 10.$$

In particular, the g.c.d. of 16 and 10 must divide 6. Also, any number which divides 10 and 6 must divide 16, so

$$\text{g.c.d.}\ (16, 10) = \text{g.c.d.}\ (10, 6).$$

By using the transitivity of equals we have reduced the problem to that of finding the g.c.d. of 10 and 6. Applying the division algorithm and reasoning as before, we have

$$10 = 6 \cdot 1 + 4,$$

and $6 = 4 \cdot 1 + 2,$
and $4 = 2 \cdot 2.$

The last statement, which identifies d, states that g.c.d. $(4, 2) = 2$. Following back up the chain of equations, we have $2|2$ and $2|4$, so $2|6$; $2|4$ and $2|6$, so $2|10$; $2|6$ and $2|10$, so $2|16$; $2|10$ and $2|16$, so $2|58$; $2|16$ and $2|58$, and it has been identified as the g.c.d. at each step; hence g.c.d. $(58, 16) = 2$.

Example 1

Find the g.c.d. (84, 198):

$198 = 84 \cdot 2 + 30$	g.c.d. (198, 84) = g.c.d. (84, 30)
$84 = 30 \cdot 2 + 24$	g.c.d. (84, 30) = g.c.d. (30, 24)
$30 = 24 \cdot 1 + 6$	g.c.d. (30, 24) = g.c.d. (24, 6)
$24 = \mathit{6} \cdot 4$	g.c.d. (24, 6) = 6.

Hence 6 is the g.c.d. (84, 198).

Example 2

Find the g.c.d. (198, 144):

$198 = 144 \cdot 1 + 54$	g.c.d. (198, 144) = g.c.d. (144, 54)
$144 = 54 \cdot 2 + 36$	g.c.d. (144, 54) = g.c.d. (54, 36)
$54 = 36 \cdot 1 + \mathit{18}$	g.c.d. (54, 36) = g.c.d. (36, 18)
$36 = \mathit{18} \cdot 2$	g.c.d. (36, 18) = 18.

Hence g.c.d. (198, 144) = 18.

The computation for finding the g.c.d. by the division algorithm may be performed as follows. We want the g.c.d. (84, 198).

$$
\begin{array}{ll}
84\overline{)198}\ \ 2 & 198 = 84 \cdot 2 + 30 \\
\quad 168 & \\
30\overline{)84}\ \ 2 & 84 = 30 \cdot 2 + 24 \\
\quad 60 & \\
24\overline{)30}\ \ 1 & 30 = 24 \cdot 1 + 6 \\
\quad 24 & \\
6\overline{)24}\ \ 4 & 24 = 6 \cdot 4
\end{array}
$$

g.c.d. (84, 198) → 6

This procedure for finding the g.c.d. is called the *Euclidean algorithm.*

The greatest common divisor of three or more numbers, say a, b, and c, can be found by pairing. First find g.c.d. $(a, b) = d$; then g.c.d. $(d, c) = g$; then g.c.d. $(a, b, c) = g$ (see problem 3c, Exercise 5.9b).

Example 3

Find the g.c.d. (12, 18, 33):

g.c.d. (12, 18) = 6.
g.c.d. (6, 33) = 3.

Hence g.c.d. (12, 18, 33) = 3.

Exercise 5.9b

1. Find the g.c.d. of the following sets of numbers by the prime factorization method:

(a) 84, 198 (b) 252, 144
(c) 36, 54 (d) 12, 30, 42

2. Find the g.c.d. of the following sets of numbers by the Euclidean algorithm:

(a) 14, 198 (b) 210, 126
(c) 735, 858 (d) 84, 210, 126

3. We define the binary operation $\odot$ on the set J of integers as follows. For integers m and n, not both zero,
$m \odot n =$ g.c.d. (m, n).

(a) Is the set J closed with respect to this operation?
(b) Is this operation commutative? Illustrate numerically.
(c) Is this operation associative? Illustrate numerically.
(d) Is there an identity element? If so, what is it?

4. In which type of arithmetic computation is the g.c.d. used? Illustrate numerically.

5. Find the g.c.d. of the following pairs: (a) 0, 9, (b) 0, -12, (c) -9, 0, (d) 0, 1.

6. What can you say about g.c.d. $(0, n)$ for any integer n?

7. What is g.c.d. (p, q) if p and q are both prime numbers?

5.9c Special Properties of the G.C.D.

An interesting and useful property of the g.c.d. of a pair of numbers, a and b, which is also a consequence of the division algorithm, is that the g.c.d. (a, b) can be written as the sum of some multiples of a and b.

Example 1

g.c.d. $(7, 17) = 1$
Multiples of 7: 7, 14, 21, 28, **35**, 42, . . .
Multiples of 17: 17, **34**, 51, 68, . . .

$1 = (5)(7) + (-2)(17).$

Example 2

g.c.d. $(6, 15) = 3$
Multiples of 6: 6, **12**, 18, 24, 30, . . .
Multiples of 15: **15**, 30, 45, 60, 75, . . .

$3 = (1)(15) + (-2)(6).$

Theorem I. In general, if g.c.d. $(a, b) = d$, then there are multiples $s \cdot a$ and $t \cdot b$ of a and b such that

$$d = s \cdot a + t \cdot b.$$

The proof of this theorem is not difficult but we consider it unnecessary for our purposes.

We now prove some simple theorems concerning prime numbers that serve as a review of some of the definitions and concepts we have introduced.

Theorem II. If p is a prime and p divides the product $a \cdot b$, then either p divides a or p divides b.

Proof. If a is a multiple of p, then p divides a. If a is not a multiple of p, since p is a prime whose only divisors are $\pm p$ and ± 1, g.c.d. $(a, p) = 1$ (see problem 9, Exercise 5.9a).

Hence

$1 = s \cdot a + t \cdot p.$ Theorem I

Multiplying by b we have

$b = s(ab) + t(pb).$

We assumed that p divides ab, so p divides $s(ab)$. Also, p divides

$t(pb)$, since this is a multiple of p. The distributive law implies that p divides $s(ab)+t(pb)$. Hence p divides b.

Theorem III. If m and n are relatively prime and m divides $n \cdot a$, then m must divide a.

Proof.	$1 = s \cdot m + t \cdot n$	Why?
	$a = s \cdot m \cdot a + t \cdot n \cdot a$	Why?
	Then m divides a.	Why?

Theorem IV. If a and b are relatively prime and a divides m and b divides m, then ab divides m,

Proof.	$m = a \cdot k$.	Why?
	b divides ak.	Why?
	Hence b divides k.	Theorem III
	Hence $k = b \cdot e$, for some e.	Why?

Substituting $b \cdot e$ for k in the first equation we obtain

$m = a \cdot b \cdot e.$
Hence $a \cdot b$ divides m.

5.10 THE LEAST COMMON MULTIPLE

The *least common multiple*, l.c.m., of a pair of positive integers m and n, is the smallest positive integer that is divisible by both m and n. The l.c.m. of 6 and 5 is 30; the l.c.m. of 12 and 18 is 36.

> ***Definition 5.10.*** The positive integer d is the l.c.m. of positive integers m and n, (1) if m divides d and n divides d and (2) if k is any other multiple of m and n, then d divides k.

The l.c.m. of three or more positive integers can be found by finding the l.c.m. of pairs in the same way that the g.c.d. of three or more numbers was found; that is, to find the l.c.m. (a, b, c), first find l.c.m. $(a, b) = k$; then l.c.m. $(k, c) = m$. Then l.c.m. $(a, b, c) = m$ (see problem 5c, Exercise 5.10).

5.10a The L.C.M. by Prime Factorization

The problem of finding the l.c.m. of a set of positive integers is an essential step in handling rational numbers (fractions). Here again there is more than one way of determining the l.c.m. We first examine the method involving prime factorization.

Example 1

Find the l.c.m. of 12 and 18:

$12 = 2^2 \cdot 3,$
$18 = 2 \cdot 3^2.$

The l.c.m. evidently must contain factors of 2 and 3. Furthermore, the

2 must be of the second power. The same is true for the 3. Then l.c.m. $(12, 18) = 2^2 \cdot 3^2 = 36$.

In general, the l.c.m. of two positive integers m and n is the product of the highest powers of all the *different* factors that occur in the prime factorization of either integer.

Example 2

We wish to find the l.c.m. of 60 and 36. Factoring, we have

$60 = 2^2 \cdot 3 \cdot 5,$

and

$36 = 2^2 \cdot 3^2.$

We see that the *different* prime factors that occur in either factorization are 2, 3, and 5. The highest power of 2 that occurs is 2^2, the highest of 3 that occurs is 3^2, and the highest of 5 that occurs is 5^1.

Thus the l.c.m. $(60, 36) = 2^2 \cdot 3^2 \cdot 5 = 180$.

5.10b Finding the L.C.M. from the G.C.D.

Another method of finding the l.c.m. of a *pair* of positive integers m and n is to divide their product by their g.c.d. This is intuitively evident. If d is the g.c.d. of m and n, then $m = d \cdot a$; $n = d \cdot b$; and g.c.d. $(a, b) = 1$. Hence $m \cdot n = d \cdot a \cdot d \cdot b$. But $a \cdot d \cdot b$ is a multiple of both m and n, since $a \cdot n = adb = m \cdot b$, and it is the l.c.m. (m, n). That is,

$$\text{l.c.m.}\ (m, n) = adb = \frac{dadb}{d} = \frac{m \cdot n}{\text{g.c.d.}\ (m, n)}.$$

When the numbers are large, this is the most practical way of finding the l.c.m.

Example 1

Find the l.c.m. (285, 76):

$285 = 76 \cdot 3 + 57$
$76 = 57 \cdot 1 + 19$
$57 = 19 \cdot 3$

Hence, g.c.d. $(285, 76) = 19$.

Then l.c.m. $(285, 76) = \dfrac{(285)(76)}{19} = (285)(4) = 1140$.

Exercise 5.10

1. Find the l.c.m. of each of the following, using the prime factorization method:

(a) 32, 40 (b) 8, 18, 27
(c) 36, 56 (d) 17, 7

2. Find the l.c.m. of each of the following using the g.c.d.:

(a) 18, 84 (b) 36, 27
(c) 96, 84 (d) 252, 33

3. Find the g.c.d. of each of the following pairs of numbers:

(a) (9, 16) (b) (22, 46)
(c) $(2^3 \cdot 3^5 \cdot 7^2, 3^2 \cdot 5^3 \cdot 7^4)$ (d) $(3^{10} \cdot 7^2 \cdot 11^4, 3^3 \cdot 7^3 \cdot 11^3)$
(e) (42, 90) (f) (74, 111)
(g) $(10^{19}, 10^{10})$ (h) $(10^{10}, 10^{15})$
(i) (806, 1116) (j) (1936, 3630)
(k) (72, 1000000) (l) (54, 1000000)

4. Find the l.c.m. of each of the pairs of problem 3.

5. We define the binary operation $*$ on the set J of integers as follows: for integers m and n, neither of which is zero,

$$m * n = \text{l.c.m.}\ (m, n).$$

(a) Is the set J closed with respect to this operation?
(b) Is this operation commutative? Illustrate numerically.
(c) Is this operation associative? Illustrate numerically.
(d) Is there an identity element? If so, what is it?

6. In which type of arithmetic computation is the l.c.m. used? Illustrate numerically.

5.11 ORDER RELATIONS FOR THE INTEGERS

We defined a "less than" relation and a "less than or equal to" relation for the whole numbers in terms of one-to-one correspondence between sets. Since this is not very meaningful for negative integers, we define these relations for *integers* in terms of the positive integers.

> ***Definition 5.11a.*** If m and n are any two *integers*, we say m "is less than" n if $(n-m)$ is a positive integer.

We denote this $m < n$. Expressions involving the relation $<$ are called *inequalities*, or *strict inequalities*.

> ***Definition 5.11b.*** If m and n are any two integers, we say m is "less than or equal to" n if either $(n-m)$ is a positive integer or $(n-m)$ is 0.

We denote this $m \leq n$. The relation, $\leq$, is also called an *inequality*, or a *weak inequality*.

Example 1

$5 < 7$, since $7-5=2$, a positive integer.
$0 < m$ for any positive integer m, since $m-0=m$, a positive integer.
$-7 < -5$, since $-5-(-7)=2$, a positive integer.

We indicate hereafter that a number m is positive by the symbol $m > 0$.

The order relation $<$ for the integers does not have the reflexive property because $n - n = 0$ and 0 is not a positive integer. On the other hand, $\leq$ is reflexive.

Neither the $<$ nor the $\leq$ relation is symmetric.

Both relations are transitive, that is, if $m < n$ and $n < k$, then $m < k$, also, if $m \leq n$ and $m \leq k$, then $m \leq k$. In establishing this for the "less than" relation, we observe that

$$m < n \text{ means } n - m > 0.$$

and

$$n < k \text{ means } k - n > 0.$$

It is necessary to show that these two statements imply $m < k$, that is, that $k - m > 0$. Note that

$$\begin{aligned}(k-n) + (n-m) &= [k + (-n)] + [n + (-m)] \\ &= k + [(-n) + n] + (-m) \\ &= k + (-m) \\ &= k - m.\end{aligned}$$

Since the sum of two positive integers is a positive integer, and $(k - m)$ and $(m - n)$ are positive, $k - m$ is a positive integer. Then, by definition, $m < k$.

The proof for $\leq$ can be made by replacing $<$ by $\leq$ throughout.

Defining order for the integers in terms of the positive elements of the system has certain algebraic advantages; for example, the trichotomy law can be stated as follows.

The Trichotomy Law for Integers

If $m \neq 0$, then either $m > 0$ or $-m > 0$.

This can be stated alternately in the following way:

If m and n are any two integers, then one and only one of the following statements is true:

(1) $m = n$.
(2) $m < n$.
(3) $m > n$.

The trichotomy law states that any two integers can be compared. This, after all, is one of the most important things we do with numbers.

We emphasize the trichotomy law because a precise statement of it is extremely useful. Students in higher mathematics sometimes have difficulty because they are not fully aware of the importance of this idea and its relation to the order properties of the numbers we use.

The order relations have some additional properties that are both interesting and useful.

Properties of the Inequalities $<$ and $\leq$

For a, b, and c integers

1. If $a < b$, then $a + c < b + c$.
2. If $a < b$ and $c > 0$, then $ac < bc$.
3. If $a < b$ and $c < 0$, then $ac > bc$.

These statements can also be written with $\leq$ in place of $<$. The last statement says that multiplying both members of an inequality by the same negative number reverses the sense of the inequality.

Exercise 5.11

1. Describe the following sets:

(a) $A = \{n | n$ is an integer and $-3 < n \leq 5\}$.
(b) $B = \{n | n$ is an integer and $0 \leq n \leq 7\}$.
(c) $C = \{n | n$ is an integer and $n < 0\}$.
(d) $N = \{n | m$ is an integer and $0 < n\}$.
(e) $O = \{n | n$ is an integer and $n = -n\}$.
(f) $J = \{n | n$ is an integer$\}$.

2. Write the simple open sentences for each of the compound open sentences of problem 1.

3. Describe the complements of the sets of problem 1.

4. Specify each of the sets in problem 3 in the manner of problem 1.

5. Describe $N \times N$ (set N of problem 1).

6. Describe $J \times J$ (set J of problem 1).

7. Is there a smallest integer? Is there a smallest positive integer? If so, what is it?

8. Is there a smallest non-negative integer? If so, what is it?

9. Is there a largest negative integer? If so, what is it?

10. How many integers satisfy the inequalities $-5 < n < 5$? List them and indicate them on the number line.

11. Show that if $a < b$, then $a + c < b + c$.

12. Show that if $a < b$ and $c > 0$, then $ac < bc$.

13. Show that if $a < b$ and $c < 0$, then $ac > bc$.

14. How many solutions does the inequality $3 + n < 10$ have in the set of natural numbers? In the set of integers?

15. What is the sum of the first 5 positive integers?

16. What is the sum of the first 10 positive integers? Here is an easy way to add the first 5 positive integers:

$$\begin{array}{l} 1+2+3+4+5 \\ \underline{5+4+3+2+1} \\ 6+6+6+6+6 \end{array}$$

Five 6's are 30, but we added each of the numbers twice, so we must divide our answer by 2. Hence the sum of the first 5 positive integers is 5(6)/2.

17. Write out the sum of the first 10 positive integers in the same way and find the sum.

18. Find the sum of the first 20 positive integers by the method suggested in problem 16.

19. Find the sum of the first 50 positive integers.

20. Write an expression for the sum of the first n positive integers in the form suggested by problem 16.

21. What is the sum of the first 3 positive odd integers?

22. What is the sum of the first 4 positive odd integers?

23. What is the sum of the first 5 positive odd integers?

24. Write an expression for the kth positive odd integer.

25. How are the answers to problems 21, 22, and 23 related?

26. What is the sum of the first 10 positive odd integers?

27. What is the sum of the first 11 positive odd integers?

28. Write an expression for the sum of the first n positive odd integers in the form suggested by problems 21 through 27.

5.12 ABSOLUTE VALUE

We use the familiar concept of number line to introduce absolute value (see Sections 4.13a).

5.12a Distances on the Number Line

If we wish to know the number of intervals between pairs of points, it is simple enough to count them. This, however, is not necessary if we use the names we have attached to the points.

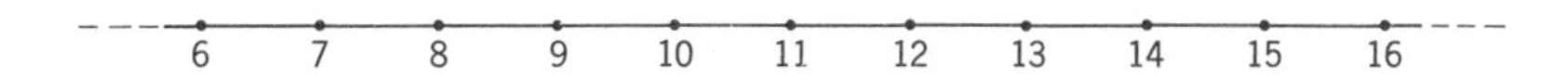

For example, we might ask how many units it is from the point labeled 7 to the point labeled 13. The result can be obtained if we notice that the 7 label actually can be interpreted to mean that there are 7 intervals between that point and the point labeled 0. Also, the 13 label indicates that there are 13 intervals between 0 and 13. We see immediately that there are $13-7=6$ intervals from the point labeled 7 to the point labeled 13.

We might be tempted to say, in general, that the number of intervals from the point labeled m to the point labeled n is $n-m$. However, if we ask how many intervals there are between the point labeled 13 and the point labeled 7, our generalization would yield -6, which is meaningless as an answer to the question, "How many?"

The *number* of intervals between points on a line is called the *distance* between points on the line. To define properly the distance between points in terms of the labels attached to the points we introduce the concept of *absolute value* of a number.

Definition 5.12a. If m is a number, the *absolute value* of m, denoted by $|m|$, is defined as follows:

If $m > 0$, then $|m| = m$.
If $m = 0$, then $|m| = 0$.
If $m < 0$, then $|m| = -m$.

Example 1

$|3| = 3$, since $3 > 0$.
$|-7| = 7$, since $-7 < 0$ and $-(-7) = 7$.
$|0| = 0$.

5.12b Properties of Absolute Value

1. The absolute value of the product of two numbers is the product of the absolute values of the numbers.

$$|m \cdot n| = |m| \cdot |n|.$$

2. The absolute value of the sum of two numbers is not always the same as the sum of the absolute values. It may be less but never greater. We indicate this as follows:

$$|m+n| \leq |m| + |n|.$$

Definition 5.12b. The distance between points labeled m and n on the number line is $|m-n|$.

5.12c Properties of Distance

1. The distance between two points is a nonnegative number. It may be 0. This happens when the two points are the same.
2. The distance from point A to point B is the same as the distance from point B to point A.
3. The distance from A to B plus the distance from B to C is greater than or equal to the distance from A to C. This property is referred to as the *triangular inequality*. Why?

Heretofore we have been speaking of distances along a line. We can also speak of distances in a plane. Here there are two kinds of distances familiar to most people, the "street" distance and the distance "as the crow flies." Consider a portion of a city map as shown in Figure 1.

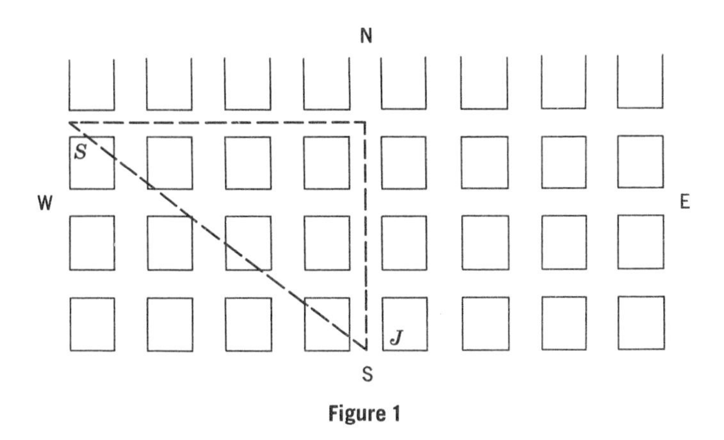

Figure 1

How far is it from the Smith's house (S) to the Jones' house (J)? If you drive, the answer is 4 blocks east and 3 blocks south, or a total of 7 blocks. We might also say 3 blocks south and 4 blocks east. This is one meaning of "distance" in the plane. We could also be concerned with how far it was between houses "as the crow flies." Here we mean the straight-line distance from house to house. You may recall that we compute this as shown in Figure 2.

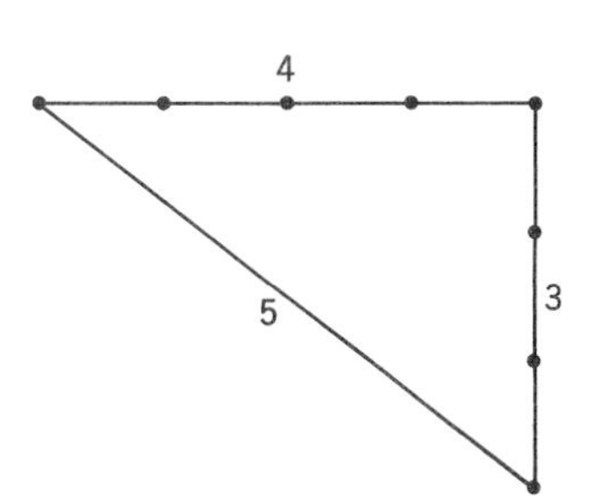

Figure 2. Distance (Smith to Jones): $\sqrt{4^2+3^2} = \sqrt{25} = 5$.

We have two distances from Smith's to Jones', by following the streets (7 blocks) and as the crow flies (5 blocks). We assert that both "distances" satisfy the properties of distances listed at the beginning of this section.

5.12d Interpretation of Absolute Value Statements

The expression $|m-4| < 3$ can be interpreted in two useful ways. One way to read it is "the number m whose distance from 4 is less than 3 on the number line." If we draw the number line, plot the number 4, then mark off a distance of 3 units in both directions from the number 4 we have a picture of the expression.

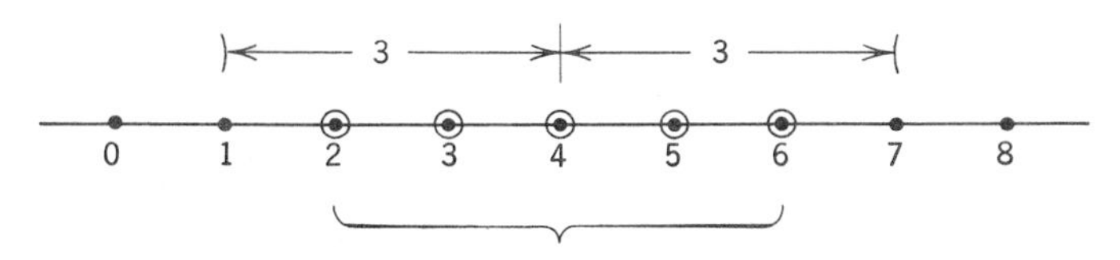

The numbers 2, 3, 4, 5, 6 are those whose distance from 4 is less than 3.

The expression $|m| < 3$ can be written as $|m-0| < 3$ and interpreted in the same way.

The other way of interpreting the expressions $|m| < 3$ and $|m-4| < 3$ is as open sentences. The open sentence $|m| < 3$ is a compound open sentence. It is the same as

$$-3 < m < 3.$$

It is the conjunction of the two simple sentences

$$-3 < m \qquad \text{and} \qquad m < 3.$$

The solution set or truth set of this compound sentence is the intersection of the solution sets of the simple open sentences.

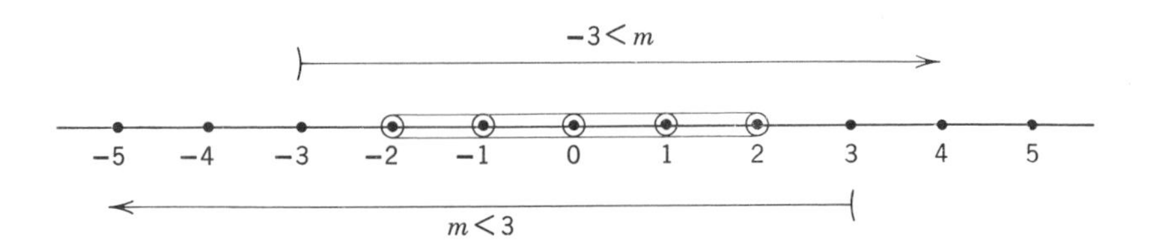

The open sentence $|m-4| < 3$ is the same as

It is the conjunction of the two simple sentences

$$-3 < m-4$$

and $\quad m-4 < 3.$

Using property (1), Section 5.11, we see that we can add 4 to both terms of the inequalities.

$$\begin{aligned} -3 &< m-4 \\ -3+4 &< m-4+4 \\ 1 &< m \end{aligned}$$

and

$$\begin{aligned} m-4 &< 3 \\ m-4+4 &< 3+4 \\ m &< 7 \end{aligned}$$

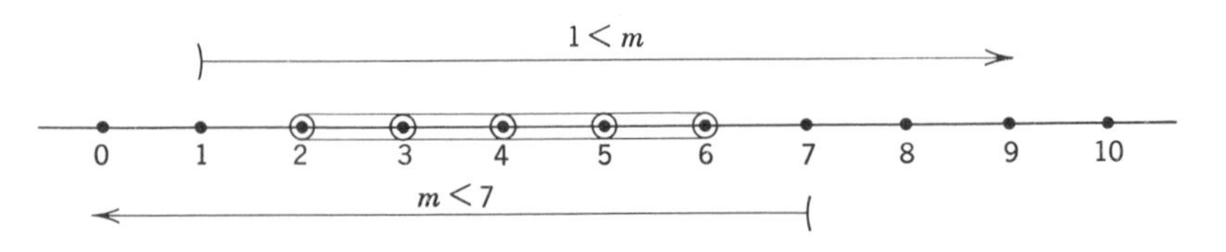

The expression $|m| > 3$ can be written $|m-0| > 3$ and interpreted as before. The solution set consists of all numbers m whose distance from 0 is greater than 3 on the number line.

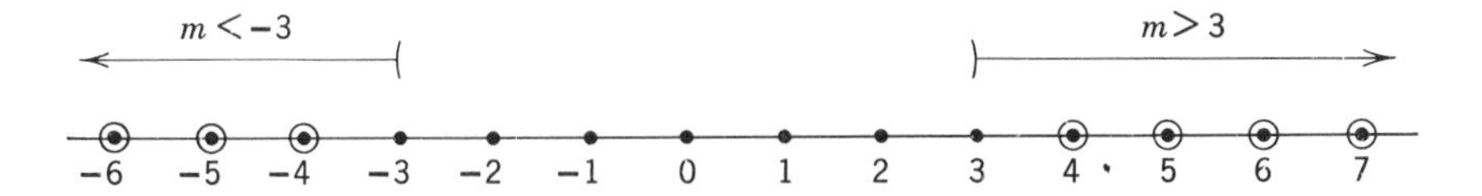

This suggests that $|m| > 3$ can be interpreted as the disjunction of the two simple open sentences $m > 3$ or $m < -3$. The solution set may be described by

$$\{m \in J | m > 3 \text{ or } m < -3\}.$$

The expression $|m+2| > 5$ can be written as $|m-(-2)| > 5$ and interpreted as the set of all numbers m whose distance from -2 is greater than 5 on the number line.

m + 2 < −5
m + 2 > 5
−9 −8 −7 −6 −5 −4 −3 −2 −1 0 1 2 3 4 5

This suggests that $|m+2| > 5$ can be written as the disjunction of the two simple open sentences, $m+2 > 5$ and $m+2 < -5$. That is,

$$\begin{array}{ccc} m+2 > 5 & \text{or} & m+2 < -5 \\ m+2+{}^{-}2 > 5+{}^{-}2 & & m+2+{}^{-}2 < -5+{}^{-}2 \\ m > 3 & & m < -7 \end{array}$$

The solution set may be described by

$$\{m \in J | m > 3 \text{ or } m < -7\}.$$

Exercise 5.12

1. Find the following sets and plot the sets and their complements on the number line:

(a) $\{m \in J \mid |m| < 5\}$
(b) $\{m \in J \mid |m| < 4\}$
(c) $\{m \in J \mid m < 5 \text{ and } m > -2\}$
(d) $\{m \in J \mid |m-5| < 2\}$

2. Find the following sets and plot the sets and their complements on the number line:

(a) $\{n \in J \mid |n| \leq 3\}$
(b) $\{n \in J \mid |n-8| \leq 2\}$
(c) $\{n \in J \mid |n| \leq 1\}$
(d) $\{n \in J \mid |n+5| \leq 3\}$
(e) $\{n \in J \mid |n-6| = 2\}$

3. If $a < b$, how is ac related to bc?

4. Use numerical examples to illustrate $|m+n| \leq |m|+|n|$.

(a) Under what conditions does the equality hold?
(b) Under what conditions does the strict inequality hold?

5. (a) Write an expression for the distance from the point 3 to the point 9 on the number line.
(b) Write an expression for the distance from the point 11 to the point 8.

6. (a) Write an expression for the distance from the point 4 to the point n.
(b) Write an expression for the distance from the point n to the point 18.

7. On the number line for the integers, which integers correspond to points whose distance from 7 is 3?

8. Which integers correspond to points whose distance from 7 is less than or equal to 3?

9. If m and n are integers, is $(-m)(n)$ a negative integer?

5.13 CLOCK ARITHMETIC

Suppose it is 9 o'clock and we had breakfast two hours ago. We plan to have a meeting in four hours (see Figure 3).

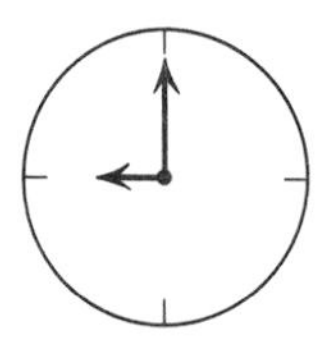
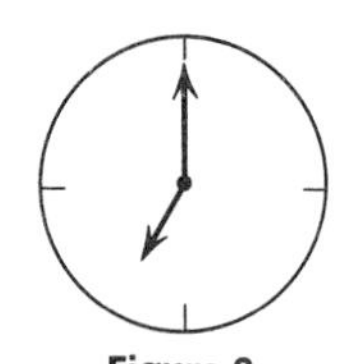
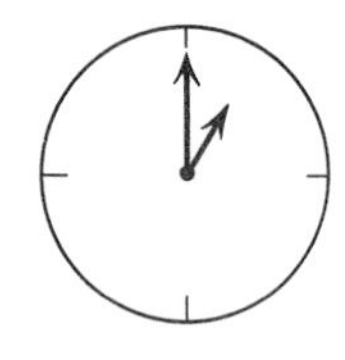

Figure 3

Two hours before 9 o'clock is 7 o'clock. Four hours after 9 o'clock is 1 o'clock. If we think of adding and subtracting hours, we have

$$9-2=7.$$
$$9+4=1.$$

If we did not know we were speaking about "time," this would be a strange arithmetic indeed. What time is 17 hours after 6 o'clock?

$$6+17=11.$$

Because one needs to know whether the time is morning or evening, the time is sometimes expressed in terms of a 24-hour clock. Thus 2200 hours is 10 o'clock P.M. 0300 hours is 3 o'clock A.M. 1200 hours is noon. Note that in these four-digit numerals related to time, the first pair range from 00 to and including 24 and the second pair range from 00 to and including 59.

In working time problems on the 12-hour clock it does not take long to realize that the correct time is obtained by adding the clock hours to the given hours, dividing by 12, and taking the remainder as the result.

Example 1

$$9+4=13=12\cdot 1+1\equiv 1.$$
$$9+17=26=12\cdot 2+2\equiv 2.$$
$$8+8=16=12\cdot 1+4\equiv 4.$$

Whenever we add or subtract h hours to a given time t, we obtain the "sum," using the division algorithm as follows:

$t+h=12\cdot g+r$, and r is the clock time.

Example 2

Add 26 hours to 9 o'clock.

$$9+26=35=12\cdot 2+11.$$

Hence the clock time is 11 o'clock.

Exercise 5.13

1. What is the clock time in each of the following:

(a) 75 hours after 3:00 A.M.? Will it be A.M. or P.M.?
(b) You plan to serve a roast which takes 15 hours to cook. Dinner is to be at 7:00 P.M. What time do you put the roast in the oven?
(c) It takes 7 hours to climb Harding Peak. You want to reach the summit at 3:00 P.M. What time should you start the climb?

2. Add the following in "clock arithmetic": (a) $9+5$, (b) $11+6$, (c) $8+15$, (d) $4+22$.

3. What is the clock time of the following:

(a) 112 hours after 3:00 P.M.?
(b) 64 hours after 7:00 A.M.?
(c) 42 hours after 5:00 P.M.?

5.14 THE CONGRUENCE RELATION

The congruence relation is a familiar one although we have not called it by this name before; in fact, we didn't give it a special name (see problem 5, Exercise 3.5b and problem 11, Exercise 4.4). In these problems we defined the relation, denoted by $\equiv$, as follows: For integers m and n, $m\equiv n$ if $m-n$ is a multiple of 4 (or 5), or, alternately, $m\equiv n$ if they each give the same remainder when divided by 4 (or 5). Let us define this relation more specifically.

Definition 5.14a. If a and b are integers and m is a positive integer, *a is congruent to b*, modulo m, if $(a-b)$ is a multiple of m.

Symbolically,

$$a\equiv b \pmod{m} \quad \text{if} \quad a-b=k\cdot m, \text{ for some integer } k.$$

The definition given here is equivalent to the definition that a is congruent to b, modulo m, if they each give the same remainder when divided by m. It is also equivalent to the definition that a is congruent to b, modulo m, if $a-b$ is divisible by m. The student will be asked to verify this in Exercise 5.14a.

Example 1

$39 \equiv 3 \pmod{12}$ because $39-3=36$, a multiple of 12.
$65 \equiv 113 \pmod{12}$ because $65-113=-48$, a multiple of 12.

Or we could say

$39 \equiv 3 \pmod{12}$ because $39=3 \cdot 12+3$, $3=0 \cdot 12+3$; they both give the same remainder when divided by 12.

Similarly, 65 and 113 both give a remainder of 5 when divided by 12.

Recall the definition of an equivalence relation. (See Section 3.4). Let us show that the congruence relation possesses the necessary properties of an equivalence relation.

The Reflexive Property.

$$a \equiv a \pmod{m} \quad \text{because} \quad a-a=0=0 \cdot m.$$

The Symmetric Property. If $a \equiv b \pmod{m}$, then $b \equiv a \pmod{m}$. To prove this we note that $a \equiv b \pmod{m}$ means $a-b=k \cdot m$, for some k. Then $b-a=(-k)m$, but this is also a multiple of m. Hence, if $a \equiv b \pmod{m}$, then $b \equiv a \pmod{m}$.

The Transitive Property. If $a \equiv b \pmod{m}$ and $b \equiv c \pmod{m}$, then $a \equiv c \pmod{m}$. To prove this we note that $a \equiv b \pmod{m}$ means $a-b=k \cdot m$, and $b \equiv c \pmod{m}$ means $b-c=j \cdot m$, for integers k and j.

Adding

$$\begin{aligned}(a-b)+(b-c) &= k \cdot m+j \cdot m \\ a-b+b-c &= (k+j)m \\ a-c &= (k+j)m.\end{aligned}$$

But $(k+j)$ is an integer, so $a-c$ is a multiple of m. This means that if $a \equiv b \pmod{m}$ and $b \equiv c \pmod{m}$, then $a \equiv c \pmod{m}$.

Since the congruence relation as just defined has the reflexive, symmetric, and transitive properties, it is an equivalence relation. Recall again the effect of an equivalence relation defined on a set (see Section 3.4). The relation partitions the set into disjoint subsets called equivalence classes.

Let us see what the congruence relation, modulo 12, does to the set J, the integers.

We first ask, "What numbers are congruent to 0, mod 12?" Obviously, they are multiples of 12. We call this the 0-class.

We next ask, "What numbers are congruent to 1, mod. 12?" These are 1, 13, 25, etc., and, in general, any number in the 0-class + 1. How do we obtain the numbers congruent to 2, mod 12? How many classes will we have? We answer the last question by asking, "How many remainders are possible when integers are divided by 12?"

We have 12 class s as follows:

...	−24	−12	0	12	24	36	...
...	−23	−11	1	13	25	37	...
...	−22	−10	2	14	26	38	...
...	−21	−9	3	15	27	39	...
...	−20	−8	4	16	28	40	...
...	−19	−7	5	17	29	41	...
...	−18	−6	6	18	30	42	...
...	−17	−5	7	19	31	43	...
...	−16	−4	8	20	32	44	...
...	−15	−3	9	21	33	45	...
...	−14	−2	10	22	34	46	...
...	−13	−1	11	23	35	47	...

Notice that any number in any row has the same remainder when divided by 12, as any other number in the same row. These rows, where the dots indicate that these are unending sequences of integers, are the *equivalence classes*, modulo 12, of the set of integers. Notice that each integer occurs in one and only one class. The classes are disjoint subsets of the set of integers, i.e., the set of integers is partitioned into disjoint equivalence classes.

Let us name the classes the 0-class, 1-class, etc., and use the following symbols:

$$[0], [1], [2], [3], [4], [5], [6], [7], [8], [9], [10], [11].$$

We call the system consisting of the set of equivalence classes, modulo 12, with addition and multiplication as defined in the following sections, J_{12}.

5.14a Addition in J_{12}

Now we observe an interesting phenomenon. Consider the numbers 27 and 38, whose sum is 65. Notice that 27 is in the class labeled [3] and 38 is in the class labeled [2], and their sum is in the class labeled [5]. Now try any number in [2] and any number in [3]. Their sum will always be in [5]. We indicate this by writing

$$[2]+[3]=[5].$$

We interpret this to mean that any number in the 2-class added to any number in the 3-class is a number in the 5-class.

Let us try some more "class" addition.

$$[5]+[9]=?$$

5 is in [5], and 9 is in [9]; $5+9=14$; 14 is in [2].

Hence

$$[5]+[9]=[2].$$

If this seems like a strange kind of arithmetic, recall that in our "clock" arithmetic 9 hours added to 5 o'clock was 2 o'clock.

We now consider the new set, which we have labeled J_{12}, and whose elements are [0], [1], [2], [3], [4], [5], [6], [7], [8], [9], [10], [11]. Previous examples lead us to define a binary operation in J_{12}, which we call "addition." If we let $[a]$ and $[b]$ be any elements in J_{12}, then

$$[a]+[b]=[a+b],$$

where $[a+b]$ is the class of $a+b$ reduced modulo 12. To show that this "operation" is "well defined," we must show that the "operation" is independent of the representatives of the classes; that is, regardless of which representative of the classes $[a]$ and $[b]$ are chosen, the "sum" will be in the class labeled $[a+b]$. Some numerical examples illustrate this. The formal proof is not difficult and is left as a challenge to the student.

Table 1 shows "addition" for this system.

Table 1 Addition in J_{12}

+	[0]	[1]	[2]	[3]	[4]	[5]	[6]	[7]	[8]	[9]	[10]	[11]
[0]	0	1	2	3	4	5	6	7	8	9	10	11
[1]	1	2	3	4	5	6	7	8	9	10	11	0
[2]	2	3	4	5	6	7	8	9	10	11	0	1
[3]	3	4	5	6	7	8	9	10	11	0	1	2
[4]	4	5	6	7	8	9	10	11	0	1	2	3
[5]	5	6	7	8	9	10	11	0	1	2	3	4
[6]	6	7	8	9	10	11	0	1	2	3	4	5
[7]	7	8	9	10	11	0	1	2	3	4	5	6
[8]	8	9	10	11	0	1	2	3	4	5	6	7
[9]	9	10	11	0	1	2	3	4	5	6	7	8
[10]	10	11	0	1	2	3	4	5	6	7	8	9
[11]	11	0	1	2	3	4	5	6	7	8	9	10

We observe from the addition table the following interesting properties of "addition" as it is defined in the set J_{12}. (Note that in the body of the table we have omitted the square brackets.)

The commutative property of "addition" results in the symmetry about the upper-left to lower-right diagonal; that is, $[a]+[b]=[b]+[a]$.

The first row and the first column in the body of the table indicate that [0] is the *additive identity*.

The element [0] occurs once and only once in each row or column. This means that the pair of elements whose sum is [0] are *additive inverses* of each other; for example, $[5]+[7]=[0]$. Hence [7] is the additive inverse of [5] and [5] is the additive inverse of [7]. The element [6] is its own additive inverse.

Not immediately evident but still something that can be shown is the fact that "addition" in this system is associative.

Exercise 5.14a

1. Use numerical examples to illustrate the fact that "addition" in J_{12} is associative.

2. Solve the following equations in J_{12}:

(a) $[3]+[x]=[7]$ (b) $[4]+[5]=[x]$
(c) $[8]+[x]=[0]$ (d) $[8]+[8]=[x]$
(e) $[x]+[x]=[0]$ (f) $[x]+[x]=[6]$

3. Show that the alternate methods of defining the congruence relation are equivalent (see Section 5.14).

4. How would you interpret $-[3]$?

5. Explain the following:

(a) $-[3]=[-3]$ (b) $-[3]=[9]$
(c) $-[6]=[6]$ (d) $[-2]=[10]$

5.14b Multiplication in J_{12}

We now define another binary operation, which we call "multiplication," in much the same way that we defined "addition."

If we let $[a]$ and $[b]$ be any elements in J_{12}, then

$$[a]\cdot[b]=[a\cdot b],$$

where $[a\cdot b]$ is the class of $a\cdot b$ reduced, modulo 12. This "operation" is also well defined. The proof is again left as a challenge to the reader. Table 2 shows "multiplication" for this system.

Table 2 Multiplication in J_{12}

	[1]	[2]	[3]	[4]	[5]	[6]	[7]	[8]	[9]	[10]	[11]
[1]	1	2	3	4	5	6	7	8	9	10	11
[2]	2	4	6	8	10	0	2	4	6	8	10
[3]	3	6	9	0	3	6	9	0	3	6	9
[4]	4	8	0	4	8	0	4	8	0	4	8
[5]	5	10	3	8	1	6	11	4	9	2	7
[6]	6	0	6	0	6	0	6	0	6	0	6
[7]	7	2	9	4	11	6	1	8	3	10	5
[8]	8	4	0	8	4	0	8	4	0	8	4
[9]	9	6	3	0	9	6	3	0	9	6	3
[10]	10	8	6	4	2	0	10	8	6	4	2
[11]	11	10	9	8	7	6	5	4	3	2	1

Examining the table closely, we observe that "multiplication" is commutative. Notice the symmetry about the main diagonal.

The first row and the first column in the body of the table indicate that [1] is the *multiplicative identity.*

The system has many "zero divisors"; that is, $[3] \neq [0]$ and $[4] \neq [0]$ but $[3] \cdot [4] = [0]$. By our definition of "divides," [3] and [4] are both nonzero divisors of zero.

We note that J_{12} is a mathematical system. It is closed under two binary operations, which are commutative and associative, and the distributive law of mutliplication with respect to addition holds. The system possesses an additive identity and a multiplicative identity, and there are inverses under addition. This system contains zero divisors and the cancellation law for multiplication does not hold.

5.14c Congruence Modulo 2

For integers a and b, $a \equiv b \pmod 2$ if $b - a$ is a multiple of 2. This is equivalent to saying a and b have the same remainders when divided by 2. The division algorithm tells us that the only possible remainders when dividing integers by 2 are 0 and 1. The congruence relation divides the set of integers into two disjoint subsets—those that have remainder 0 (the *even* integers) and those that have remainder 1 (the *odd* integers).

Equivalence classes, mod 2

$\{\cdots -14, -12, -10, -8, -6, -4, -2, 0, 2, 4, 6, 8, 10, 12, 14, \ldots\}$
$\{\cdots -13, -11, -9, -7, -5, -3, -1, 1, 3, 5, 7, 9, 11, 13, 15, \ldots\}$

Addition table mod 2

+	[0]	[1]
[0]	0	1
[1]	1	0

Multiplication table mod 2

·	[1]
[1]	1

Exercise 5.14c

1. Interpret the addition table mod 2 in terms of addition of even and odd integers.

2. Use numerical examples to illustrate the fact that multiplication is associative in J_{12}.

3. Can you always find a solution for the equation

$[a] \cdot [x] = [b]$? (modulo 12)

4. Solve the equation (modulo 12):

(a) $[x] \cdot [x] = [0]$ (b) $[x] \cdot [x] = [x]$
(c) $[3] \cdot [x] = [3]$ (d) $[3] \cdot [x] = [5]$

5. Does the cancellation law for multiplication hold in J_{12}? Illustrate with a numerical example.

6. If $[x+3] \cdot [x+4] = [0]$ in J_{12}, what can you say about x? List some solutions of this equation.

7. Let J_3 denote the equivalence classes, modulo 3. Use [0], [1], and [2] to denote the elements. Construct the addition and multiplication tables for this system.

8. Does J_3 have zero divisors?

9. What is the additive inverse of [2] in J_3?

10. What is the multiplicative inverse of [2] in J_3?

11. What is the g.c.d. (7980, 2310)? The l.c.m. (7980, 2310)?

12. What is the g.c.d. of 300, 210, and 230? The l.c.m.?

13. What is the fallacy in the following argument? Let $a = b$. Then

$$\begin{aligned} a^2 &= ab \\ a^2 - b^2 &= ab - b^2 \\ (a-b)(a+b) &= b(a-b) \\ a+b &= b \\ 2a &= a \\ 2 &= 1 \end{aligned}$$

REVIEW EXERCISES

1. An integer $p, p > 1$, is a ——— if it has no proper divisors.

2. ——— states that an integer p, $p \neq 0$ and $p \neq \pm 1$, can be written as a product of primes and ± 1 in one and only one way except possibly for the order in which the factors occur.

3. If a, b, and m are integers, and $m \neq 0$, ——— if $a-b$ is a multiple of m.

4. If a binary operation defined on a set S assigns to each ordered pair of elements of S a uniquely determined element which is an element of S, we say the operation satisfies the ———.

5. If W is the set of whole numbers, $+$ is a binary operation defined on W, and for all elements x in W there is an element a in W such that $a+x = x+a = x$, then a is called the ———.

6. If J is the set of integers, $+$ is a binary operation defined on J, and for each element x in J there is an element x' in J such that $x+x' = x'+x = 0$, the additive identity, then x' is called ———.

7. The ——— is the set J closed with respect to the operations of addition and multiplication, with each operation commutative and associative, with multiplication distributive with respect to addition, with identity elements for each operation, and with an additive inverse for each element in J.

8. The ——— of a set of integers is the largest positive integer that divides each integer of the set.

9. The ——— of a set of integers is the smallest positive integer such that each of the given integers divides it.

10. Identify each of the simple open sentences in the following compound open sentences:

(a) $|n-7| > 4$
(b) $|n-5| \leq 3$
(c) $|n| \leq 2$

11. Find the solution set for each of the problems in Exercise 10.

12. Find the solution set of each of the following compound sentences:

(a) $|x+9| < 3$
(b) $-3 < n < 4$

13. What is the additive inverse of 5?

14. What is the additive inverse of $^{-}5$?

15. What is $^{-}2$?

16. Designate the fundamental property of the system of integers that is illustrated by each of the following:

(a) $a+b$ is an integer
(b) $a+b=b+a$
(c) $a(b+c) = ab+ac$
(d) ab is an integer
(e) $a+{}^{-}a=0$
(f) $a+0=a$
(g) $(a+b)+c=a+(b+c)$
(h) $a(bc) = (ab)c$
(i) If $a+c=b+c$, then $a=b$
(j) $ab = ba$
(k) $a \cdot 1 = 1 \cdot a = a$

17. Give reasons for each step in the following proof that

$a+ac+b+bc = (a+b)(1+c)$:

(a) $a+ac+b+bc = a+(ac+b)+bc$
(b) $\quad = a+(b+ac)+bc$
(c) $\quad = (a+b)+(ac+bc)$
(d) $\quad = (a+b)+(a+b)c$
(e) $\quad = (a+b) \cdot 1+(a+b)c$
(f) $\quad = (a+b)(1+c)$
(g) $a+ac+b+bc = (a+b)(1+c)$

18. Show that $^{-}3+8=5$, justifying each step.

19. (a) List *all* positive divisors of 30.
(b) List *all* positive divisors of 75.
(c) Underline the prime divisors of 30 and the prime divisors of 75.
(d) Circle the g.c.d. of 30 and 75.

20. State which of the following numbers are prime. For those that are composite give the prime factorization:

(a) 147 (b) 251 (c) 493 (d) 1,000,000

21. Find the g.c.d. and l.c.m. of 180 and 10^{12}.

22. Find the g.c.d. and l.c.m. of 1116 and 806 using the Euclidean algorithm.

23. (a) Make a partial listing of each of the equivalence classes modulo 7.
(b) Construct addition and multiplication tables for J_7.
(c) What is the additive inverse of [3] in J_7?
(d) What is the multiplicative inverse of [5] in J_7?
(e) Are there zero divisors in J_7? If so, list them.

REFERENCES

Bold, Benjamin, "A General Test For Divisibility by Any Prime (except 2 and 5)," *The Mathematics Teacher*, April 1965, Vol. LVIII, No. 4.

Cohen, Louis S., "A Rationale in Working with Signed Numbers," *The Arithmetic Teacher*, Vol. 12, Number 7, November 1965.

CHAPTER 6

The System of Rational Numbers

6.1 INTRODUCTION

Just as an individual may be a teacher, a baseball player, a father, and a homeowner, the symbols which we call numerals have more than one interpretation. The natural numbers can be thought of as *elements* of the *system of whole numbers*. When considered as a subsystem of the *system of integers*, they are called the *positive integers*. We are making ordinal use of the natural numbers in answering the question, "Which one?" We are making cardinal use of the natural numbers in answering the question, "How many?"

6.2 INTERPRETATIONS OF NUMBER PAIRS

We now consider those numbers commonly referred to as *fractions* and written in the form $\frac{2}{5}$, $\frac{7}{11}$, $\frac{16}{3}$, etc. Again we have several seemingly unrelated *interpretations* of the symbols for these number pairs. We distinguish four principal meanings or interpretations of these number pair symbols. They are familiar to the reader but may not have been distinguished and fully appreciated.

The interpretations are:

1. The "element of a mathematical system" interpretation.
2. The "division" interpretation.
3. The "fraction" or "partition" interpretation.
4. The "ratio" or "rate pair" interpretation.

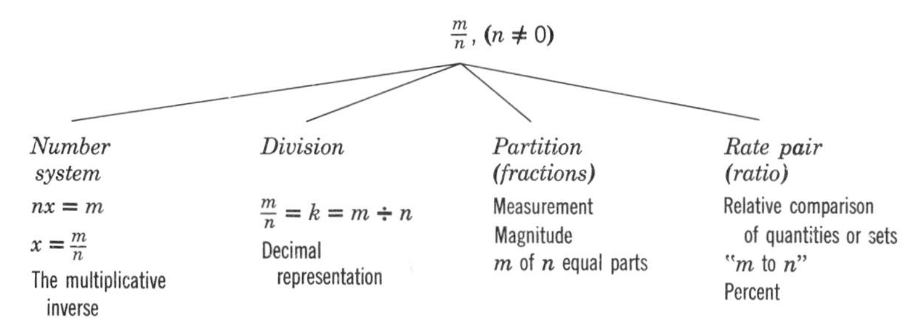

Figure 1

Each of these interpretations is much used, important, and in no danger of being made obsolete in any modern approach to arithmetic. Any restriction to a single interpretation can be as misleading and narrow as the interpretation of one of the six blind men who examined the elephant. Each interpretation is associated with a reasonably well-defined problem situation. The schematic diagram of Figure 1 presents some of the ideas associated with the various interpretations.

We shall consider the number pairs first as elements of that mathematical system called the *system of rational numbers*. The other interpretations will be treated where appropriate.

Exercise 6.2

1. Indicate the ordinal or cardinal use of the numbers in the following:
   (a) The numbers used to indicate the ranking of the baseball teams in the American League.
   (b) The score in the championship basketball game.
   (c) The number at the top of this page.

2. Discuss the cardinal use of 1 and 0.

3. Discuss the numbers 1 and 0 as elements of the system of integers.

4. Give a precise definition of the additive identity; the multiplicative identity.

5. Numbers are often used for the sole purpose of naming objects, events, or even people. In this sense the only property of the numbers used is that they are an endless source of new, distinct names. Social Security No. 517-16-1722 identifies one and only one person. Give other examples of the use of numbers as names.

6.3 THE SET OF RATIONAL NUMBERS

The number systems discussed so far can be considered in terms of the type of mathematical questions that can be answered in a particular system. The system of whole numbers is adequate for the following questions:

$$m + n = ?$$
$$m \cdot n = ?$$

In these questions m and n represent any elements in the system of whole numbers. The arithmetic operations are used to find names for these numbers.

The following questions may or may not have answers in the system of whole numbers:

$$m + ? = n.$$
$$n \cdot ? = m.$$

For example,

$$9 + ? = 4.$$
$$3 \cdot ? = 2.$$

Since there is no *whole number* that when added to 9 gives 4, we define new numbers $-1, -2, -3, \ldots, -n, \ldots$ so that $-1+1=0, -2+2=0, \ldots, -n+n=0, \ldots$. These numbers, together with the whole numbers and with addition and multiplication defined appropriately, constitute the system of integers. The system of integers can be thought of as an enlargement of the system of whole numbers in which the question

$$m + ? = n$$

can be answered for any integers m and n.

This enlarged system is still inadequate for answering the question

$$n \cdot ? = m.$$

(Note that if n is zero, so is m, since $0 \cdot a = 0$ for all a. We therefore rule out zero as a candidate for n.)

We are now interested in a mathematical system in which not only $m + ? = n$ is solvable, but also in which

$$a \cdot ? = b$$

has solutions in the system for any a and b, $a \neq 0$. The question has *integer* solutions for some pairs. For example,

$$3 \cdot ? = 15$$

has the solution 5. (Recall that we say that 3 "divides" 15.) For other pairs there is no integer solution. For example,

$$3 \cdot ? = 2.$$

We are looking for a number that when multiplied by 3 is the number 2. The fact that this number is intimately related to the numbers 2 and 3 is recognized and incorporated into the symbol representing that number. We must turn to "ordered pairs" of integers to answer our question.

> ***Definition 6.3.*** A *rational number* is a class of ordered pairs of integers. The ordered pairs are written in the form m/n, with the restriction that n is never 0.

That rational numbers are defined as *classes* will be made clear in what follows. Initially we will treat them as ordered pairs of integers, and the reader should be cautioned that in so doing we are assuming the identification of a particular ordered pair with a *class*.

That we require $n \neq 0$ is an acknowledgment of the "division" interpretation. In the ordered pair m/n, m is called the *numerator* and n is called the *denominator*. This terminology has its roots in the "fraction" interpretation.

Up to this point we have discussed natural numbers, positive integers, negative integers, and now we are introducing rational numbers. We emphasize that the terms are used as *names of sets*. Later we will speak of irrational numbers, real numbers, imaginary numbers, and complex numbers. Again, the terms are used strictly in the sense that they are *names* of various *sets of numbers*. The choice of names is very unfortunate because the literal meanings which the terms carry prejudice the student and tend to obstruct the learning of mathematics. The terms reflect the attitudes and suspicions which the number concepts have had to survive. There is no reason why this distrust and nonacceptance should be promulgated and the teaching of arithmetic made to suffer as a consequence. Hence, we repeat, the term negative integers does not mean nonintegers, but is the name of a definite set of numbers. The term rational numbers is the name of a definite set of numbers. The term irrational numbers is the name of a definite set of numbers.

There is a mathematical distinction between integers and rational numbers. Eventually we want to interpret the integers as rational numbers.

When we consider the set of all possible ordered pairs of integers, m/n, $n \neq 0$, there are many ordered pairs that appear different but that we are accustomed to regard as the same.

Example 1

$$\frac{1}{2}, \frac{2}{4}, \frac{3}{6}, \frac{4}{8}, \frac{5}{10}, \frac{6}{12}, \ldots$$

$$\frac{2}{3}, \frac{4}{6}, \frac{6}{9}, \frac{8}{12}, \frac{10}{15}, \frac{12}{18}, \ldots$$

$$\frac{2}{1}, \frac{4}{2}, \frac{6}{3}, \frac{8}{4}, \frac{10}{5}, \frac{12}{6}, \ldots$$

For simple ordered pairs it is easy to recognize this relationship. It is not so obvious for the numerals

$$\frac{27{,}946}{845{,}692} \quad \text{and} \quad \frac{307{,}406}{9{,}302{,}612}.$$

These numerals may be names for the same number, but how do you determine whether they are or not? To answer this question we define a relation in the set of ordered pairs of integers, m/n, $n \neq 0$.

6.4 EQUIVALENCE RELATION FOR ORDERED PAIRS OF INTEGERS

Definition 6.4. We say two ordered pairs of integers, m/n and p/q, are equivalent and write

$$\frac{m}{n} \doteq \frac{p}{q} \quad \text{if and only if} \quad mq = np.$$

We use the symbol $\doteq$ instead of $=$ to emphasize that this is a *new* relation defined on the set of ordered pairs of integers in terms of the relation, "equals," on the set of integers. This relation will eventually be written as "=" with the "names of the same number" meaning except when the ordered pairs are interpreted as ratios or rate pairs. The connection between $\doteq$ and $=$ is clarified in Section 6.7.

Exercise 6.4

1. Write five numerals equivalent to each of the following:

(a) $\frac{2}{3}$ (b) $\frac{-17}{19}$ (c) $\frac{10}{10}$

(d) $\frac{0}{5}$ (e) $\frac{17}{-19}$ (f) $\frac{4}{1}$

(g) $\frac{9}{1}$ (h) $\frac{0}{10}$ (i) $\frac{3}{4}$

(j) $\frac{-2}{3}$ (k) $\frac{4}{-6}$ (l) $\frac{-1}{-1}$

2. Which of the following are equivalent?

$$\frac{27{,}946}{854{,}692}, \quad \frac{11}{45}, \quad \frac{3{,}157{,}898}{96{,}580{,}196}, \quad \frac{3333}{13{,}329}, \quad \frac{63{,}327}{253{,}251}.$$

3. Which of the following are equivalent?

$$\frac{33}{29}, \quad \frac{33-2}{29-2}, \quad \frac{2 \cdot 33}{2 \cdot 29}, \quad \frac{2+33}{2+29}.$$

4. Show that each of the following pairs are equivalent:

(a) $\frac{-2}{3}$ and $\frac{2}{-3}$ (b) $\frac{-4}{-1}$ and $\frac{8}{2}$

(c) $\frac{-5}{3}$ and $\frac{10}{-6}$ (d) $\frac{-m}{-m}$ and $\frac{m}{m}$

(e) $\frac{0}{7}$ and $\frac{0}{6}$ (f) $\frac{1}{1}$ and $\frac{3}{3}$

(g) $\frac{0}{-1}$ and $\frac{0}{1}$ (h) $\frac{-m}{n}$ and $\frac{m}{-n}$

5. Find the rational number equivalent to each of the following for which the g.c.d. of the numerator and denominator is 1:

(a) $\frac{3964}{87{,}258}$ (b) $\frac{57}{76}$ (c) $\frac{144}{504}$

6. There are some open sentences of the type $ax + b = c$ that are understandable to elementary-level students if they are stated in a reasonable manner. "I am thinking of a number. By doubling it and adding 1, I get 9. What is the number?" Try this with fourth or fifth graders and obtain their response.

State the following problems in words:

(a) $3x + 1 = 7$ (b) $2x + 3 = 13$ (c) $3x + 1 = 10$

7. Make a partial list, including both positive and negative integers, of each of the equivalence classes of the integers modulo 13.

6.4a Properties of the Equivalence Relation $\doteq$

The relation $\doteq$ is transitive. The proof of this assertion proceeds as follows. We must show that if $a/b \doteq c/d$ and if $c/d \doteq e/f$, then $a/b \doteq e/f$, for a/b, c/d, and e/f ordered pairs of integers, b, d, and f not zero.

1. $\frac{a}{b} \doteq \frac{c}{d}$ means $ad = bc$.

2. $\frac{c}{d} \doteq \frac{e}{f}$ means $cf = de$.

If we show that $af = be$, then $a/b \doteq e/f$.
Multiplying (1) by f and (2) by b we get

$$adf = bcf \qquad \text{and} \qquad bcf = bde.$$

By the transitivity of "equals" we have $adf = bde$.
Using the cancellation law of multiplication for the integers, we have

$$af = be \qquad \text{or} \qquad a/b \doteq e/f.$$

The relation $\doteq$ is symmetric. We need to show that if $a/b \doteq c/d$, then $c/d \doteq a/b$. We leave this as an exercise.

The relation $\doteq$ is reflexive. This is also left as an exercise.

If a/b is an ordered pair of integers and x is an integer, $x \neq 0$, then $a/b \doteq ax/bx$.

Proof. $(ab)x = (ab)x$ by the reflexive property of equals in the system of integers.

$(ab)x = (ba)x$ by the commutative property of multiplication in the system of integers.

$a(bx) = b(ax)$ by the associative property of multiplication in the system of integers.

$\frac{a}{b} \doteq \frac{ax}{bx}$ by the definition of $\doteq$ in the set of ordered pairs.

6.5 EQUIVALENCE CLASSES OF ORDERED PAIRS OF INTEGERS

We have shown that the relation $\doteq$ is reflexive, symmetric, and transitive. This relation is an equivalence relation. An equivalence relation partitions the set in which it is defined into disjoint subsets.

Recall how the congruence relation modulo m, defined in the set of integers, partitioned the set into equivalence classes. The relation $\doteq$ has a similar effect on the set of ordered pairs of integers. We shall tabulate a few of the classes.

The class to which $\frac{1}{2}$ belongs is as follows:

$$\ldots, \frac{-5}{-10}, \frac{-4}{-8}, \frac{-3}{-6}, \frac{-2}{-4}, \frac{-1}{-2}, \frac{1}{2}, \frac{2}{4}, \frac{3}{6}, \frac{4}{8}, \frac{5}{10}, \frac{6}{12}, \frac{7}{14}, \ldots$$

We denote this class by $[\frac{1}{2}]$, although any other pair of the class could be used. The bracket notation denotes the class to which the object enclosed by the brackets belongs; $[\frac{10}{20}]$ denotes the same equivalence class as $[\frac{1}{2}]$, $\frac{1}{2}$ is simply a representative of the class $[\frac{1}{2}]$.

The class to which $\frac{-2}{3}$ belongs is as follows:

$$\ldots, \frac{8}{-12}, \frac{6}{-9}, \frac{4}{-6}, \frac{2}{-3}, \frac{-2}{3}, \frac{-4}{6}, \frac{-6}{9}, \frac{-8}{12}, \frac{-10}{15}, \frac{-12}{18}, \frac{-14}{21}, \ldots$$

We denote this class by $\left[\frac{-2}{3}\right]$. It is the same as $\left[\frac{-6}{9}\right]$ or $\left[\frac{4}{-6}\right]$.

The class to which $\frac{1}{1}$ belongs is as follows:

$$\ldots, \frac{-6}{-6}, \frac{-5}{-5}, \frac{-4}{-4}, \frac{-3}{-3}, \frac{-2}{-2}, \frac{-1}{-1}, \frac{1}{1}, \frac{2}{2}, \frac{3}{3}, \frac{4}{4}, \frac{5}{5}, \frac{6}{6}, \ldots$$

It is denoted by $[\frac{1}{1}]$ and will play a special rôle in operations with rational numbers.

Similarly, the class to which $\frac{0}{1}$ belongs plays a special role. The class is denoted by $[\frac{0}{1}]$ and is as follows:

$$\ldots, \frac{0}{-6}, \frac{0}{-5}, \frac{0}{-4}, \frac{0}{-3}, \frac{0}{-2}, \frac{0}{-1}, \frac{0}{1}, \frac{0}{2}, \frac{0}{3}, \frac{0}{4}, \frac{0}{5}, \frac{0}{6}, \ldots$$

Exercise 6.5

1. Show that $\doteq$ has the symmetric property.

2. Show that $\doteq$ has the reflexive property.

3. Indicate as in Section 6.5 the class to which each of the following numerals belong:

(a) $\frac{54}{81}$ (b) $\frac{16}{16}$ (c) $\frac{19}{2}$

(d) $\frac{0}{100}$ (e) $\frac{0}{-3}$ (f) $\frac{7}{36}$

4. Indicate as in Section 6.5 the class to which the following numerals belong:

(a) $\frac{-1}{2}$ (b) $\frac{2}{-1}$ (c) $\frac{-2}{3}$

(d) $\frac{-7}{6}$ (e) $\frac{0}{-3}$ (f) $\frac{-22}{2}$

(g) $\frac{2}{-3}$ (h) $\frac{4}{5}$ (i) $\frac{-4}{-5}$

(j) $\frac{1}{-2}$ (k) $\frac{12}{-3}$ (l) $\frac{12}{3}$

5. (a) Show that $\frac{2}{3}$ and $\frac{2 \cdot 2}{2 \cdot 3}$ belong to the same class.

(b) Show that $\frac{2}{3}$ and $\frac{2 \cdot n}{3 \cdot n}$ are related by $\doteq$ for any integer $n \neq 0$.

(c) Show that $\frac{m}{n}$ and $\frac{2m}{2n}$ are equivalent.

(d) Show that $\frac{a}{b}$ and $\frac{na}{nb}$ are equivalent. $n \neq 0$.

6. Use the ordered pairs $\frac{3}{5}, \frac{9}{15}$, and $\frac{-3}{-5}$ and illustrate that the $\doteq$ relation is transitive.

7. What can you say about $\left[\frac{-2}{3}\right]$ and $\left[\frac{2}{-3}\right]$?

8. What do we mean when we write

$\left[\frac{0}{11}\right] = \left[\frac{0}{4}\right]$?

9. What do we mean when we write

$\left[\frac{6}{6}\right] = \left[\frac{-3}{-3}\right]$?

6.6 RATIONAL NUMBERS AS EQUIVALENCE CLASSES

In the previous section we constructed several equivalence classes. We saw how any member of a class might act as a representative of the class to which it belongs. This is the reason for identifying particular ordered pairs with a *class*. The ability to think of a rational number as an equivalence class of ordered pairs of integers will be helpful in understanding fractions in arithmetic. This idea is not new. It is, in fact, a familiar concept. What comes to mind when you see $\frac{2}{2}$? What comes to mind when you see $\frac{8}{8}$? What comes to mind when you see $\frac{12}{6}$ or $\frac{6}{3}$? When one adds $\frac{2}{3}$ and $\frac{5}{6}$, the usual procedure is to add $\frac{4}{6}$ and $\frac{5}{6}$. That is, $\frac{4}{6}$ is the same, in some sense, as $\frac{2}{3}$.

The objects in the system of rational numbers will be these equivalence classes. We will define the binary operations on these *classes.*

The binary operations are defined in terms of *representatives of the classes.* Addition and multiplication as we define them may not yield directly the most convenient methods of finding the sum and product of two rational numbers, but the usual (convenient) way of performing these operations can be made reasonable in terms of our definitions.

6.7 ADDITION OF RATIONAL NUMBERS

Let $\frac{3}{4}$ and $\frac{1}{6}$ be representatives of the equivalence classes $[\frac{3}{4}]$ and $[\frac{1}{6}]$, respectively. The usual procedure for handling rational numbers does not make the distinction between the representative of a class and the class itself. For clarity of presentation and ease in understanding computations involving "fractions," we wish to keep this distinction for the present.

To the ordered pair $(\frac{3}{4}, \frac{1}{6})$ the binary operation "addition" assigns the numeral which we write as $(\frac{3}{4}+\frac{1}{6})$. We want this to be an ordered pair of integers to satisfy the *closure law.* Furthermore, we want this to be consistent with the addition of integers when we interpret the integers as a subset of the rational numbers.

There are several ways of determining which ordered pair of integers this should be. We discuss the usual procedure involving the "least common denominator" later. For the present, we find the ordered pair represented by $\frac{3}{4}+\frac{1}{6}$ in the following way:

Example 1

$$\frac{3}{4}+\frac{1}{6}=\frac{3\cdot 6+4\cdot 1}{4\cdot 6}=\frac{18+4}{24}=\frac{22}{24}$$

Notice that we are using the "names for the same number" interpretation of $=$. Since the relation $\doteq$ as defined for the ordered pairs has the same properties as $=$, when we wish to use the "names for the same number" interpretation, we shall drop the new symbol and revert to the symbol $=$.

In general, we define addition of a/b and c/d as follows:

Definition 6.7. For a/b and c/d representatives of rational numbers,

$$\frac{a}{b}+\frac{c}{d}=\frac{a\cdot d+b\cdot c}{b\cdot d}.$$

The expression $(ad+bc)/bd$ represents an ordered pair of integers because the set of integers is closed with respect to addition and multiplication. This procedure is not the usual way of adding numbers such as $\frac{3}{4}$ and $\frac{1}{6}$. However, it is consistent with the procedure for addition of integers when the ordered pairs are interpreted as integers (see problem 9, Exercise 6.7).

There are at present only a few schools that are teaching secondary school students computer languages, but we shall see much more of this in each succeeding year. When these students commence programming, they will discover that the method of addition of fractions by use of the least common denominators is not the easiest method to program, but rather that the method of Definition 6.7 is the easiest. One of the reasons for this is the fact that writing a program for the computer to find the least common denominator is quite involved.

The usual procedure uses the least common denominator with considerable economy of manipulation as follows:

Example 2

$$\frac{3}{4}+\frac{1}{6}=\frac{9}{12}+\frac{2}{12}=\frac{9+2}{12}=\frac{11}{12}.$$

The two methods involve essentially the same amount of work when the denominators are relatively prime.

Example 3

$$\frac{3}{4}+\frac{1}{5}=\frac{3\cdot 5+4\cdot 1}{4\cdot 5}=\frac{15+4}{20}=\frac{19}{20}.$$

$$\frac{3}{4}+\frac{1}{5}=\frac{15}{20}+\frac{4}{20}=\frac{15+4}{20}=\frac{19}{20}.$$

In either case, the important thing to note is that the sums may not look the same but they will be related by "equals" in the sense of "names for the same number." From the foregoing Examples 1 and 2, $\frac{22}{24}=\frac{11}{12}$ because these numerals are simply different names for the same number. Thinking of the rational numbers as equivalence classes, we see that $\frac{22}{24}$ and $\frac{11}{12}$ are different representatives of the class $[\frac{11}{12}]$.

Let us recapitulate. We wanted to define addition of "rational numbers." We wanted addition to satisfy the closure law. We also wanted the sum to be uniquely determined, that is, we wanted the binary operation to assign to any particular pair of numbers one and only one number. This is the reason we define the operation on the *classes*. Thus a representative of the class $[\frac{3}{4}]$ added to a representative of the class $[\frac{1}{6}]$ is a representative of the class $[\frac{3}{4}+\frac{1}{6}]$.

We have seen that this class can be determined in two ways. One way identified this class as $[\frac{22}{24}]$, another way as $[\frac{11}{12}]$. But $\frac{22}{24} \doteq \frac{11}{12}$, so $[\frac{22}{24}=\frac{11}{12}]$.

$$\frac{3}{4} \text{ is in } \left\{\ldots, \frac{-6}{-8}, \frac{-3}{-4}, \frac{3}{4}, \frac{6}{8}, \frac{9}{12}, \frac{12}{16}, \frac{15}{20}, \ldots\right\}$$

$$\frac{1}{6} \text{ is in } \left\{\ldots, \frac{-2}{-12}, \frac{-1}{-6}, \frac{1}{6}, \frac{2}{12}, \frac{3}{18}, \frac{4}{24}, \frac{5}{30}, \ldots\right\}$$

$$\frac{3}{4}+\frac{1}{6} \text{ is in } \left\{\ldots, \frac{-22}{-24}, \frac{-11}{-12}, \frac{11}{12}, \frac{22}{24}, \frac{33}{36}, \ldots\right\}$$

The circled numbers indicate addition as defined. The numbers in squares indicate addition in the usual way.

We state, but do not prove, the fact that *addition as defined on the classes is independent of the representatives of the classes.* Any representative of $[\frac{3}{4}]$ added to any representative of $[\frac{1}{6}]$ is in the class $[\frac{3}{4}+\frac{1}{6}]$. In other words, the *class* of the sum is uniquely determined. This is essentially the reason why the procedure as defined produces the "same" result as the method of least common denominators. The latter method merely uses a more convenient set of representatives of the classes. This usually results in less work in determining the class of the sum.

Exercise 6.7

1. Verify that $(ad+bc)/bd$ is an ordered pair of integers with nonzero denominator if b and d are nonzero integers.

2. Add $\frac{12}{16}$ and $\frac{4}{24}$ and show that your result is equivalent to $\frac{33}{36}$.

3. Add $\frac{-3}{-4}$ and $\frac{1}{6}$ and show that your result is equivalent to $\frac{11}{12}$.

4. Explain the following: $\left[\frac{3}{4}\right]+\left[\frac{1}{6}\right]=\left[\frac{3}{4}+\frac{1}{6}\right]$.

5. Explain the following: $\left[\frac{3}{4}+\frac{1}{6}\right]=\left[\frac{3\cdot 6+4\cdot 1}{4\cdot 6}\right]$.

6. Add the following:

(a) $\frac{2}{3}+\frac{0}{1}$ (b) $\frac{0}{3}+\frac{4}{3}$

(c) $\frac{2}{3}+\frac{-4}{6}$ (d) $\frac{1}{2}+\frac{1}{2}$

7. Add the following:

(a) $\frac{39}{27}+\frac{54}{144}$ (b) $\frac{369}{288}+\frac{39}{306}$

8. Show that addition as defined on the classes is independent of the representatives chosen.

9. Let the integer m correspond to the rational number $\left[\frac{m}{1}\right]$, and the integer n correspond to the rational number $\left[\frac{n}{1}\right]$. Then show that addition as defined for rational numbers produces a sum that corresponds to $(m+n)$.

6.8 MULTIPLICATION OF RATIONAL NUMBERS

Let $\frac{1}{2}$ and $\frac{3}{5}$ be representatives of the equivalence classes to which they belong. The binary operation, multiplication, assigns to the pair $(\frac{1}{2}, \frac{3}{5})$ the numeral $(\frac{1}{2} \cdot \frac{3}{5})$. Again, we want this product to be an ordered pair of integers so that multiplication will satisfy the closure law. This ordered pair is determined as follows:

$$\frac{1}{2} \cdot \frac{3}{5} = \frac{1 \cdot 3}{2 \cdot 5} = \frac{3}{10}.$$

Bear in mind that the numerals $\frac{1}{2}$, $\frac{3}{5}$, and $\frac{3}{10}$ are representatives of the classes $[\frac{1}{2}]$, $[\frac{3}{5}]$, and $[\frac{3}{10}]$.

In general, we define multiplication of rational numbers as follows:

Definition 6.8. For m/n and r/s, representatives of rational numbers

$$\frac{m}{n} \cdot \frac{r}{s} = \frac{m \cdot r}{n \cdot s}.$$

The product represents an ordered pair of integers.

The binary operations "addition" and "multiplication" for rational numbers could more properly be written as follows:

$$\left[\frac{a}{b}\right] + \left[\frac{c}{d}\right] = \left[\frac{ad+bc}{bd}\right]$$

$$\left[\frac{a}{b}\right] \cdot \left[\frac{c}{d}\right] = \left[\frac{a \cdot c}{b \cdot d}\right].$$

The brackets denote the equivalence class to which the numeral between the brackets belongs. We repeat, it is only in terms of the equivalence classes that the binary operations are *well defined.*

Exercise 6.8

1. Multiply: $\left(\frac{7}{3}\right)\left(\frac{2}{9}\right)$.

2. Multiply: $\left(\frac{4}{6}\right)\left(\frac{6}{4}\right)$.

3. Multiply: $\left(\frac{8}{12}\right)\left(\frac{6}{4}\right)$.

4. Multiply: $\left(\frac{3}{3}\right)\left(\frac{1}{2}\right)$.

5. Write each of the following as a product, for example, $\frac{14}{10} = \frac{2 \cdot 7}{2 \cdot 5} = \left(\frac{2}{2}\right)\left(\frac{7}{5}\right)$.

(a) $\frac{35}{49}$ (b) $\frac{33}{39}$

6. Show that multiplication is well defined; that is, show that the product is independent of the representative chosen.

7. Let the integer m correspond to $\left[\frac{m}{1}\right]$ and the integer n correspond to $\left[\frac{n}{1}\right]$. Show that multiplication as defined for rational numbers gives a product that corresponds to $m \cdot n$.

8. Carry out the indicated binary operation as defined.

(a) $\frac{5}{7}+\frac{5}{9}$ (b) $\frac{2}{3}+\frac{4}{3}$ (c) $\frac{4}{3}+\frac{0}{2}$

(d) $\frac{0}{5}+\frac{2}{5}$ (e) $\frac{16}{4}+\frac{12}{4}$ (f) $\frac{0}{1}+\frac{2}{1}$

(g) $\frac{-3}{2}+\frac{3}{2}$ (h) $\frac{3}{2}+\frac{3}{-2}$ (i) $\frac{4}{9}+\frac{0}{6}$

(j) $\frac{4}{9}+\frac{4}{-9}$ (k) $\frac{2994}{876{,}658}+\frac{779{,}922}{9{,}197{,}245}$

9. Carry out the indicated binary operation as defined.

(a) $\frac{2}{3}\cdot\frac{6}{5}$ (b) $\frac{9}{3}\cdot\frac{10}{7}$ (c) $\frac{13}{1}\cdot\frac{1}{13}$

(d) $\frac{19}{4}\cdot\frac{3}{3}$ (e) $\frac{7}{6}\cdot\frac{1}{1}$ (f) $\frac{4}{7}\cdot\frac{0}{2}$

(g) $\frac{6}{3}\cdot\frac{8}{2}$ (h) $\frac{5}{4}\cdot\frac{4}{4}$ (i) $\frac{7}{1}\cdot\frac{1}{1}$

(j) $\frac{2}{1}\cdot\frac{1}{2}$ (k) $\frac{2994}{876{,}658}\cdot\frac{779{,}922}{9{,}197{,}245}$

10. Explain the meaning of each of the following:

(a) $\left[\frac{6}{8}\right]+\left[\frac{4}{8}\right]=\left[\frac{5}{4}\right]$ (b) $\left[\frac{4}{7}\right]+\left[\frac{0}{6}\right]=\left[\frac{4}{7}\right]$

(c) $\left[\frac{7}{12}\right]+\left[\frac{-7}{12}\right]=\left[\frac{0}{1}\right]$ (d) $\left[\frac{6}{9}\right]+\left[\frac{2}{-3}\right]=\left[\frac{0}{-1}\right]$

(e) $\left[\frac{0}{100}\right]+\left[\frac{1}{1}\right]=\left[\frac{7}{7}\right]$

11. Explain the meaning of each of the following:

(a) $\left[\frac{9}{18}\right]\cdot\left[\frac{6}{3}\right]=\left[\frac{1}{1}\right]$ (b) $\left[\frac{4}{4}\right]\cdot\left[\frac{5}{7}\right]=\left[\frac{5}{7}\right]$

(c) $\left[\frac{0}{7}\right]\cdot\left[\frac{1}{1}\right]=\left[\frac{0}{1}\right]$ (d) $\left[\frac{18}{18}\right]\cdot\left[\frac{27}{36}\right]=\left[\frac{3}{4}\right]$

(e) $\left[\frac{-2}{3}\right]\cdot\left[\frac{4}{5}\right]=\left[\frac{-8}{15}\right]$

12. (a) Give a numerical example illustrating the fact that addition of rational numbers is commutative.
(b) Give a numerical example illustrating the fact that addition of rational numbers is associative.

13. (a) Give a numerical example illustrating the fact that multiplication of rational numbers is commutative.
(b) Give a numerical example illustrating the fact that multiplication of rational numbers is associative.

14. (a) $\left[\frac{5}{3}\cdot\frac{4}{7}\right]+\left[\frac{5}{3}\cdot\frac{9}{4}\right]=?$ (b) $\left[\frac{4}{7}+\frac{9}{4}\right]\cdot\left[\frac{5}{3}\right]=?$

15. (a) $\left[\frac{5}{5}\cdot\frac{16}{9}\right]+\left[\frac{7}{5}\cdot\frac{16}{9}\right]=?$ (b) $\left[\frac{5}{5}+\frac{7}{5}\right]\cdot\left[\frac{16}{9}\right]=?$

16. (a) $\left[\frac{0}{1}+\frac{5}{5}\right]\cdot\left[\frac{7}{7}\right]=?$ (b) $\left[\frac{6}{3}+\frac{2}{-1}\right]\cdot\left[\frac{17}{19}\right]=?$

17. (a) $\frac{784{,}327}{2{,}994{,}463}+\frac{9{,}548{,}012}{1{,}988{,}544}=?$ (b) $\frac{784{,}327}{2{,}994{,}463}\cdot\frac{9{,}548{,}012}{1{,}988{,}544}=?$

6.9 NAMING THE CLASSES (REDUCING FRACTIONS)

Any of the number pairs in an equivalence class can be used to name the class. However, among all the number pairs in an equivalence class, there is always one that is in some sense the simplest or most convenient, for example, $\frac{8}{12}$ is in the following class:

$$\ldots, \frac{-14}{-21}, \frac{-12}{-18}, \frac{-10}{-15}, \frac{-8}{-12}, \frac{-6}{-9}, \frac{-4}{-6}, \frac{-2}{-3}, \frac{2}{3}, \frac{4}{6}, \frac{6}{9}, \frac{8}{12}, \frac{10}{15}, \frac{12}{18}, \ldots$$

When one encounters $\frac{8}{12}$ one naturally thinks of $\frac{2}{3}$, so we tend to identify $\frac{8}{12}$ with $\frac{2}{3}$. We do not mean to imply that the two numerals are identical. We mean simply that $\frac{8}{12}$ and $\frac{2}{3}$ are in the same class, namely the class $[\frac{2}{3}]$ which we call $\frac{2}{3}$. We call this class $\frac{2}{3}$ because $\frac{2}{3}$ is the representative that is in *simplified* or *reduced* form.

Definition 6.9. A representative $\frac{m}{m}$ of a rational number $\left[\frac{m}{n}\right]$, is in *reduced* form if the greatest common divisor of the integers m and n is 1, n positive.

Notice that $\frac{2}{3}$ is in reduced form, $\frac{7}{12}$ is in reduced form, $\frac{14}{24}$ is not in reduced form. If the equivalence class is written out as done previously, the reduced form is easy to pick out. It is that element whose denominator is the least positive integer in the set of denominators.

Without writing out the equivalence class, there are essentially two approaches to "reducing" fractions. One approach uses the fundamental theorem of arithmetic. The other is based on finding the g.c.d. of the two integers. We discuss both approaches because both are used.

Let us simplify $\frac{6}{8}$ and $\frac{1{,}439{,}900}{3{,}496{,}900}$. When we simplify, or reduce, $\frac{6}{8}$, it is easily recognized as $\frac{3}{4}$. To a beginner, however, the process appears as $\frac{6}{8} = \frac{3 \cdot 2}{4 \cdot 2}$. We saw in Section 7.5 that $\frac{3 \cdot 2}{4 \cdot 2}$ is equivalent to $\frac{3}{4}$; that is, we factor the two integers and use the definition of our relation. This same technique will work for reducing $\frac{1{,}439{,}900}{3{,}496{,}900}$ but involves a great deal of computation. An alternate method for reducing $\frac{1{,}439{,}900}{3{,}496{,}900}$ is by finding the greatest common divisor of the number pair using the Euclidean algorithm. The reduced form is easily recognized as soon as it is put in the form $\frac{md}{nd}$, where d is the g.c.d. of the two integers. The problem of finding the reduced form of this particular number is left as an exercise. One gains a healthy respect for the Euclidean algorithm in such applications.

Exercise 6.9

Find the name of the class to which each of the following belong. That is, *simplify* each of the following:

1. $\frac{72}{144}$

2. $\frac{727}{1441}$

3. $\frac{163{,}264}{2{,}754{,}108}$

4. $\frac{1{,}439{,}900}{3{,}496{,}900}$

5. $\frac{39 \cdot 27}{52 \cdot 26}$

6. $\frac{228 \cdot 19}{204 \cdot 19}$

7. $\frac{39}{52} \cdot \frac{26}{27}$

8. $\frac{72}{38} + \frac{16}{19}$

9. $\frac{-54}{81}$

10. $\frac{108}{-162}$

11. $\frac{0}{5}$

12. $\frac{27}{27}$

6.10 THE SYSTEM OF RATIONAL NUMBERS

We review briefly the development of the system of rational numbers as far as we have progressed. We considered all possible ordered pairs of integers of the form m/n, where $n \neq 0$. We defined a criterion for determining when these ordered pairs are "related." This relation, $\doteq$, was shown to be an equivalence relation. This brought a certain amount of orderliness to the collection of the ordered pairs of integers; that is, the set of ordered pairs of integers was partitioned into disjoint classes of related elements. We denoted these classes by $\left[\frac{m}{n}\right]$. The binary operations were defined on these classes in terms of representatives of the classes.

Example 1

$$\left[\frac{2}{3}\right]+\left[\frac{5}{6}\right]=\left[\frac{2}{3}+\frac{5}{6}\right]=\left[\frac{2\cdot 6+3\cdot 5}{3\cdot 6}\right].$$

$$\left[\frac{2}{3}\right]\cdot\left[\frac{5}{6}\right]=\left[\frac{2}{3}\cdot\frac{5}{6}\right]=\left[\frac{2\cdot 5}{3\cdot 6}\right]\text{--.}$$

Recall the meaning of these symbols. They are to be interpreted to mean that any representative in the class $\left[\frac{2}{3}\right]$ added to any representative in the class $\left[\frac{5}{6}\right]$ is an element of the class of the ordered pairs obtained by carrying out the indicated operations in the last brackets. Simplifying fractions, then, amounted to finding the usual name of the class of the resultant; that is, the name of the class $\left[\frac{2\cdot 6+3\cdot 5}{3\cdot 6}\right]$ is found by reducing $\frac{12+15}{18}=\frac{27}{18}$ to $\frac{3}{2}$. We write this $\left[\frac{2}{3}\right]+\left[\frac{5}{6}\right]=\left[\frac{3}{2}\right]$. For convenience and economy of symbols, this can be shortened to $\frac{2}{3}+\frac{5}{6}=\frac{3}{2}$ without loss of meaning.

The same applies to multiplication.

6.10a Identities and Inverses

The *additive identity* for the system of rational numbers is that element which when added to any other element leaves it unchanged.

Example 1

$$\frac{0}{1}+\frac{3}{7}=\frac{0\cdot 7+1\cdot 3}{1\cdot 7}=\frac{0+3}{7}=\frac{3}{7}.$$

$$\frac{a}{b}+\frac{0}{1}=\frac{a\cdot 1+b\cdot 0}{b\cdot 1}=\frac{a+0}{b}=\frac{a}{b}.$$

It appears that $\frac{0}{1}$ is the *additive identity*. But $\frac{0}{1}$ is just one representative of the class $[\frac{0}{1}]$. Another element of this class is $\frac{0}{5}$.

Example 2

$$\frac{3}{7}+\frac{0}{5}=\frac{3\cdot 5+7\cdot 0}{7\cdot 5}=\frac{3\cdot 5+0}{7\cdot 5}=\frac{3\cdot 5}{7\cdot 5}$$

But

$$\frac{3\cdot 5}{7\cdot 5}\doteq\frac{3}{7}.$$

It is more appropriate to write

$$\left[\frac{3}{7}\right]+\left[\frac{0}{1}\right]=\left[\frac{3}{7}\right].$$

If we interpret this statement correctly, any member of $[\frac{0}{1}]$ can act as an additive identity, that is,

$$\frac{a}{b}+\frac{0}{n}\doteq\frac{a}{b}, \qquad n\neq 0.$$

The *multiplicative identity* is defined in much the same way with respect to multiplication. The *multiplicative identity* for the system of rational numbers is that element which when multiplied by any rational number gives a product which is identically equal to that number.

Example 3

$$\frac{4}{5}\cdot\frac{1}{1}=\frac{4\cdot 1}{5\cdot 1}=\frac{4}{5}.$$

Remembering that $\frac{4}{5}$ and $\frac{1}{1}$ are representatives of the classes to which they belong, we should write

$$\left[\frac{4}{5}\right]\cdot\left[\frac{1}{1}\right]=\left[\frac{4}{5}\right].$$

We are saying that any member of the class $[\frac{1}{1}]$ can act as the multiplicative identity, that is

$$\frac{a}{b}\cdot\frac{n}{n}\doteq\frac{a}{b} \qquad \text{for any integer } n\neq 0.$$

By the *additive inverse* of the element m/n in the system of rational numbers we mean that element which when added to m/n gives us the additive identity of the system of rational numbers. Just as the additive inverse of n in the system of integers is denoted by $-n$, the additive inverse of m/n is denoted by $-(m/n)$. Thus

$$\frac{m}{n}+\left(-\frac{m}{n}\right)=\frac{0}{1}.$$

Note, however, that

$$\frac{m}{n}+\frac{-m}{n}=\frac{mn+(-mn)}{nn}=\frac{0}{n^2}\doteq\frac{0}{1},$$

and

$$\frac{m}{n}+\frac{m}{-n}=\frac{-mn+mn}{-nn}=\frac{0}{-n^2}\doteq\frac{0}{1}.$$

That is, $\frac{-m}{n}$ and $\frac{m}{-n}$ behave the same as $\left(-\frac{m}{n}\right)$. We found earlier that

$\left[\frac{-m}{n}\right] = \left[\frac{m}{-n}\right]$. (See problem 4h, Exercise 6.4.) We are now saying that

$$\left[-\frac{m}{n}\right] = \left[\frac{-m}{n}\right] \quad \text{and} \quad \left[-\frac{m}{n}\right] = \left[\frac{m}{-n}\right].$$

In general $-\frac{m}{n} = \frac{-m}{n} = \frac{m}{-n}$, where we are using the "names of the same number" meaning of the "equals" relation. We adopt the convention here, as we did with the integers, and write, for example, $\frac{7}{5} - \frac{2}{3}$ instead of $\frac{7}{5} + \left(-\frac{2}{3}\right)$.

Recall that we introduced the negative integers so that we would have a mathematical system rich enough to provide a solution to the equation

$$m + x = n,$$

for any pair of whole numbers m and n. This system was called the system of integers.

We constructed the rational numbers in order to have a number system rich enough to provide a solution to the equation

$$n \cdot x = m,$$

for any pair of integers m and n, except when $n = 0$. More generally, we constructed a number system that could provide a solution to the equation

$$a \cdot x = b,$$

for any pair of numbers a and b in our system, where a is not in the zero class. Since the elements of our system of rational numbers can be written in the form $\left[\frac{m}{n}\right]$, we are saying that our number system is complete enough to provide solutions to the equation

$$\left[\frac{m}{n}\right] \cdot x = \left[\frac{r}{s}\right],$$

where $\left[\frac{m}{n}\right]$ and $\left[\frac{r}{s}\right]$ are any rational numbers and $\left[\frac{m}{n}\right] \neq \left[\frac{0}{1}\right]$.

Example 4

Find x so that

$$\frac{3}{1} \cdot x = \frac{1}{1}.$$

We have

$$x = \frac{1}{3},$$

since

$$\frac{3}{1} \cdot \frac{1}{3} = \frac{3 \cdot 1}{1 \cdot 3} = \frac{3}{3} \doteq \frac{1}{1}.$$

The *multiplicative inverse* of any rational number $\left[\frac{m}{n}\right] \neq \left[\frac{0}{1}\right]$, is that element which when multiplied by $\left[\frac{m}{n}\right]$ gives the multiplicative identity.

The multiplicative inverse of $\left[\frac{m}{n}\right]$ will be denoted by $\left[\frac{m}{n}\right]^{-1}$.

If we use the letter a to represent a rational number, we will write the multiplicative inverse of a as a^{-1}. The multiplicative inverse is also called the *reciprocal.*

We defined $\left[\frac{m}{n}\right]^{-1}$ as that element which when multiplied by $\left[\frac{m}{n}\right]$ is $\left[\frac{1}{1}\right]$.

$$\left[\frac{m}{n}\right] \cdot \left[\frac{m}{n}\right]^{-1} = \left[\frac{1}{1}\right].$$

Example 5

$$\frac{3}{11} \cdot \left[\frac{3}{11}\right]^{-1} = \frac{1}{1}.$$

Notice, however, that

$$\frac{3}{11} \cdot \frac{11}{3} = \frac{33}{33} = \frac{1}{1}.$$

The numbers $\left(\frac{3}{11}\right)^{-1}$ and $\frac{11}{3}$ both give the multiplicative identity when multiplied by $\frac{3}{11}$. Thus we write $\left[\frac{3}{11}\right]^{-1} = \left[\frac{11}{3}\right]$ in the sense that these are both names of the same number.

In general,

$$\left[\frac{m}{n}\right]^{-1} = \left[\frac{n}{m}\right].$$

That is,

$$\left[\frac{m}{n}\right] \cdot \left[\frac{n}{m}\right] = \left[\frac{1}{1}\right].$$

In the special case when $n = 1, \frac{m}{n} = \frac{m}{1}$, and we have

$$\left[\frac{m}{1}\right]^{-1} = \left[\frac{1}{m}\right].$$

Exercise 6.10

1. Write the additive inverse of each of the following and carry out the addition to verify the correctness of your choice.

(a) $\frac{5}{1}$ (b) $\frac{137}{329}$ (c) $\frac{-19}{3}$ (d) $\frac{2}{-10}$

(e) $\frac{6}{-7}$ (f) $\frac{5}{8}$ (g) $\frac{0}{-1}$ (h) $\frac{-3}{-4}$

(i) $\frac{0}{10}$ (j) $\frac{239}{316}$ (k) $\frac{m}{n}$ (l) $-\frac{m}{n}$

2. Write the multiplicative inverse of each of the following rational numbers in two ways and carry out the computation to verify the correctness of your choice.

(a) $\frac{2}{5}$ (b) $\frac{9}{14}$ (c) $\frac{17}{6}$ (d) $\frac{100}{19}$

(e) $\frac{-2}{11}$ (f) $\frac{8}{-3}$ (g) $\frac{-5}{-4}$ (h) $\frac{6}{1}$

(i) $\frac{9}{1}$ (j) $\frac{1}{23}$ (k) $\frac{33}{3}$ (l) $\frac{-13}{26}$

(m) $\frac{0}{1}$ (n) $\frac{2}{2}$ (o) $\frac{4}{1}$ (p) $\frac{1976}{1967}$

3. Carry out the indicated operations and simplify.

(a) $\frac{2}{5}+\frac{5}{2}$ (b) $\left(\frac{7}{3}+\frac{7}{3}\right)+\frac{7}{3}$

(c) $\frac{5}{3}\cdot\left(\frac{3}{17}+\frac{9}{4}\right)$ (d) $\frac{1}{3}+\left(\frac{1}{5}\right)^{-1}$

(e) $\left(\frac{1}{3}\right)^{-1}+\left(\frac{1}{5}\right)^{-1}$

4. What is the additive inverse of each of the numbers in problem 2?

5. What is the multiplicative inverse of each of the numbers in problem 1?.

6. Carry out the indicated operations.

(a) $\frac{2}{9}-\left(\frac{9}{2}\right)^{-1}$ (b) $\frac{5}{7}-\left(\frac{7}{5}\right)^{-1}$

(c) $\frac{11}{5}-\frac{4}{2}$ (d) $\frac{3}{4}-\frac{0}{2}$

The reader was asked in Exercise 6.8 to illustrate the commutative, associative, and distributive properties of addition and multiplication of rational numbers through the use of numerical examples. We still state these as laws which govern the behavior of operations with rational numbers, even though they can be proved to be satisfied by using the properties of the integers and the definitions of equality, addition, and multiplication of rational numbers.

Definition 6.10a. By the *system of rational numbers* we mean the set

$$R = \left\{x | x = \left[\frac{a}{b}\right], a \text{ and } b \text{ integers}, b \neq 0\right\},$$

the binary operations, addition (+) and multiplication (·), the relation $\frac{a}{b} = \frac{c}{d}$ if and only if $ad = bc$, and the following laws:

(Hereafter, for simplicity, we will omit the use of the square brackets.)

Closure Laws

1. For $\frac{a}{b}$ and $\frac{c}{d}$ in R there is a uniquely determined sum, which we write $\frac{a}{b} + \frac{c}{d} = \frac{ad + bc}{bd}$ in R.

2. For $\frac{a}{b}$ and $\frac{c}{d}$ in R there is a uniquely determined product which we write $\frac{a}{b} \cdot \frac{c}{d} = \frac{a \cdot c}{b \cdot d}$ in R.

Associative Laws. For $\frac{a}{b}, \frac{c}{d}$, and $\frac{e}{f}$ in R

3. $\frac{a}{b} + \left(\frac{c}{d} + \frac{e}{f}\right) = \left(\frac{a}{b} + \frac{c}{d}\right) + \frac{e}{f}$.

4. $\frac{a}{b} \cdot \left(\frac{c}{d} \cdot \frac{e}{f}\right) = \left(\frac{a}{b} \cdot \frac{c}{d}\right) \cdot \frac{e}{f}$.

Commutative Laws. For $\frac{a}{b}$ and $\frac{c}{d}$ in R

5. $\frac{a}{b} + \frac{c}{d} = \frac{c}{d} + \frac{a}{b}$.

6. $\frac{a}{b} \cdot \frac{c}{d} = \frac{c}{d} \cdot \frac{a}{b}$.

Distributive Law. For $\frac{a}{b}, \frac{c}{d}$, and $\frac{e}{f}$ in R

7. $\frac{a}{b} \cdot \left(\frac{c}{d} + \frac{e}{f}\right) = \frac{a}{b} \cdot \frac{c}{d} + \frac{a}{b} \cdot \frac{e}{f}$.

Identities

8. There is a unique element, $\frac{0}{1}$, such that for any $\frac{a}{b}$ in R, $\frac{a}{b} + \frac{0}{1} = \frac{0}{1} + \frac{a}{b} = \frac{a}{b}$.

9. There is a unique element, $\frac{1}{1}$, such that for any $\frac{a}{b}$ in R, $\frac{a}{b} \cdot \frac{1}{1} = \frac{1}{1} \cdot \frac{a}{b} = \frac{a}{b}$.

Additive Inverses

10. For each $\frac{a}{b}$ in R there is a unique element, $-\frac{a}{b} = \frac{-a}{b}$ in R such that $\frac{a}{b} + \left(-\frac{a}{b}\right) = \left(-\frac{a}{b}\right) + \frac{a}{b} = \frac{0}{1}$, where $\frac{0}{1}$ is the additive identity.

Multiplicative Inverses

11. For each $\frac{a}{b}$ in R, $\frac{a}{b} \neq \frac{0}{1}$, there is a unique element, $\left(\frac{a}{b}\right)^{-1} = \frac{b}{a}$ in R such that $\frac{a}{b} \cdot \left(\frac{a}{b}\right)^{-1} = \left(\frac{a}{b}\right)^{-1} \cdot \frac{a}{b} = \frac{1}{1}$, where $\frac{1}{1}$ is the multiplicative identity.

Exercise 6.10a

1. Use numerical examples to illustrate each of the laws in the preceding section.

2. State in words the meaning of each of the laws in the preceding section.

3. Is the set of rational numbers closed with respect to subtraction?

4. Use the properties of the system of integers and the definition of addition and multiplication for rational numbers to show that

(a) $\frac{a}{b} + \frac{c}{d} = \frac{c}{d} + \frac{a}{b}$

(b) $\frac{a}{b} \cdot \frac{c}{d} = \frac{c}{d} \cdot \frac{a}{b}$

(c) $\frac{a}{b} \cdot \left(\frac{c}{d} + \frac{e}{f}\right) = \frac{a}{b} \cdot \frac{c}{d} + \frac{a}{b} \cdot \frac{e}{f}$

5. Write a cancellation law for the addition of rational numbers and prove it.

6. Write a cancellation law for the multiplication of rational numbers and prove it.

7. If $a \cdot b = 0$, then $a = 0$ or $b = 0$.

Proof

If $a = 0$, then the statement is true.
If $a \neq 0$, then $1/a$ is the multiplicative inverse of a.

(a) Use this fact to show that if $a \neq 0$, then $b = 0$.
(b) Use the statement to show that if $(x-3)(x-7) = 0$, then $x = 3$ or $x = 7$.

6.10b The Integers as a Subsystem of the Rational Numbers

We have studied the system of integers as a number system apart from the system of rational numbers. The realization that each is a number system with its own interesting properties is of considerable importance. It represents the point of view of the algebraists and number theorists who are interested in the structure of mathematics. This concept is fundamental in the understanding of the solvability of equations. As we have repeatedly emphasized in our development, *the solvability of equations depends on the number system in which the equation is to be solved.*

It is desirable to interpret the system of integers as a subsystem of the system of rational numbers. Eventually we will consider the system of rational numbers as a subsystem of the system of real numbers.

We have identified $\left[\frac{m}{1}\right]$ with m and $\left[\frac{n}{1}\right]$ with n, where m and n are integers, and examined their behavior under the binary operations (see problem 9, Exercise 6.7 and problem 7, Exercise 6.8). Addition and multiplication of those rational numbers identified with the integers were consistent with addition and multiplication of the integers.

We now see that this same identification of integers with rational numbers, that is, m corresponding to $m/1$, gives us

$$m^{-1} = \frac{1}{m} \qquad \text{(see Section 6.10).}$$

This is also consistent with the definition of exponents (see Section 1.5b).

Example 1

$\frac{7}{1}+\frac{5}{1}=\frac{7\cdot 1+1\cdot 5}{1\cdot 1}=\frac{7+5}{1}=\frac{12}{1}$ Rational numbers

$7+5=12$ Integers

$12=\frac{12}{1}$ Names for the same number

$\frac{7}{1}\cdot\frac{5}{1}=\frac{7\cdot 5}{1\cdot 1}=\frac{35}{1}$ Rational numbers

$7\cdot 5=35$ Integers

$35=\frac{35}{1}$ Names for the same number

Thus the system of rational numbers actually contains a subsystem which acts just like the integers. They are not exactly the same as the integers because they are classes of ordered pairs of integers. We will take the naive point of view that because they act exactly like the integers they can be interpreted as integers. Hereafter we will use m and $\frac{m}{1}$ interchangeably as names for the same number.

6.11 ORDER IN THE RATIONAL NUMBERS

We introduce *order* in the rational numbers in much the same way that order was introduced in the integers. First we define the *positive rational numbers* and then define the order relation in terms of them.

> ***Definition 6.11a.*** The rational number m/n is *positive* if the integer $n \cdot m$ is a positive integer.

We use the notation

$$\frac{m}{n} > 0$$

to indicate that $\frac{m}{n}$ is positive. The numeral $\frac{m}{n}$ is a representative of the class $\left[\frac{m}{n}\right]$. If $\frac{m}{n}$ is positive, so also is every member of the class $\left[\frac{m}{n}\right]$, that is, the definition of positiveness is independent of the representative. We assert that this is true and the interested reader will find it easy to verify.

6.11a The Trichotomy Law and Order for Rational Numbers

If we let the single letter r represent a rational number, one and only one of the following statements is true:

1. r is positive,
2. $r = 0$, or
3. $-r$ is positive.

If r is positive, $-r$ is called a *negative* rational number.

The set of rational numbers is separated into three sets, the positive rational numbers, the negative rational numbers, and 0. (Hereafter we treat the integers as a subset of the rational numbers.)

The *positive rational numbers* are closed under addition and multiplication in the same way as the *positive integers*.

1. The sum of two positive rational numbers is positive.
2. The product of two positive rational numbers is positive.

We are now ready to define order in the rational numbers.

> ***Definition 6.11b.*** If r and s denote rational numbers, then r "is less than" s if $s - r$ is positive.

We use the same notation used earlier:

$$r < s \quad \text{if and only if} \quad s - r > 0.$$

We also write $r \leq s$ if and only if $s - r \geq 0$, and this is true if either $r < s$ or $r = s$.

We use the notation $r < x < s$ to mean that both $r < x$ and $x < s$ hold simultaneously.

We list some of the properties of order. For rational numbers p, r, and s,

1. if $r < s$, then $r+p < s+p$.
2. if $r < s$, and $p > 0$, then $rp < sp$.
3. if $0 < r < s$, then $\frac{1}{r} > \frac{1}{s}$.
4. if $r < s$, and $p < 0$, then $rp > sp$.

Exercise 6.11a

1. Verify that each of the following mathematical statements are true:

(a) $\frac{3}{5} < \frac{32}{51}$ (b) $\frac{1}{4} < \frac{1}{3}$ (c) $\frac{1}{10} < \frac{1}{7}$

(d) $\frac{1}{100} < \frac{2}{100}$ (e) $\frac{1}{1001} < \frac{1}{1000}$ (f) $\frac{2}{3} < \frac{2463}{3463}$

2. Verify that each of the following mathematical statements are true:

(a) $\frac{-2}{3} < \frac{2}{3}$ (b) $\frac{-1}{4} < \frac{1}{-5}$ (c) $\frac{1}{-7} < \frac{1}{-8}$

(d) $\frac{-1}{9} < \frac{1}{2}$ (e) $-3 < 2$ (f) $\frac{1}{5} < \frac{1}{3}$

3. If $a > 0$ and $b > 0$ and if $a < b$, how is $1/a$ related to $1/b$?

4. (a) Prove your assertion in problem 3.
(b) Given the inequality $2 < 3$, what is the effect of multiplying both sides of the inequality by 2? By -3? By -1? By $-\frac{1}{2}$?

5. (a) What is the solution set of the open sentence, $-3 < x \leq 5$, if x is an integer?
(b) List the positive integers which satisfy the open sentence $-3 < x \leq 5$.
(c) Use inequalities to specify the set consisting of $\{-3, -2, -1, 0, 1, 2, 3\}$.

6. (a) Show that if $n^2pq > q^2mn$, then $m/n < p/q$. (*Hint:* $n^2pq > q^2mn$ if $n^2pq - q^2mn > 0$. Divide by q^2n^2.)
(b) Show that if $m/n < p/q$, then $n^2pq - q^2mn > 0$. (*Hint:* $m/n < p/q$ if and only if $p/q - m/n > 0$. Use Definition 6.12a.)

7. (a) If m, n, and k are positive integers, show that $m/k < n/k$ if and only if $m < n$.
(b) Use (a) to show that if m, n, p, q, are positive integers then $m/n < p/q$ if and only if $mq < np$.

8. List some rational numbers which satisfy

(a) $\frac{1}{3} < x < \frac{1}{2}$ (b) $-\frac{1}{3} < x < \frac{1}{3}$

9. Illustrate each of the properties, (1), (2), (3), and (4), of order.

10. Locate each of the following numbers on the number line:

(a) $0, \frac{5}{7}, \frac{11}{3}, \frac{6}{8}, \frac{9}{11}$ (b) $-1, \frac{-4}{3}, \frac{-3}{5}, \frac{5}{2}, \frac{19}{7}$

11. What number lies one third of the distance from $\frac{1}{12}$ to $\frac{7}{8}$?

6.11b Absolute Value

We repeat an earlier definition in terms of the rational numbers.

Definition 6.11c. For each rational number r we define

$$|r| = \begin{cases} r \text{ if } r > 0 \\ -r \text{ if } r < 0 \\ 0 \text{ if } r = 0 \end{cases}$$

The number $|r|$ is called the *absolute value* of r.

Example 1

$$\left|\frac{-2}{3}\right| = \frac{2}{3} \quad |-6| = 6 \quad \left|\frac{4}{5}\right| = \frac{4}{5}$$

The following are some useful properties of absolute value. For rational numbers a and b.

1. if $a \neq 0$, then $|a| > 0$, and $|a| = 0$ if and only if $a = 0$.
2. $|a-b| = |b-a|$.
3. $|a+b| \leq |a| + |b|$.
4. $|a \cdot b| = |a| \cdot |b|$.
5. $||a| - |b|| \leq |a-b|$.

Exercise 6.11b

1. Find the solution set in the set of integers for each of the following open sentences:

(a) $|x| \leq 3$ (b) $|x-13| \leq 2$
(c) $|x+5| \leq 3$ (d) $|x-6| < 1$
(e) $|x+3| = 2$
(f) Use set notation and inequalities to describe the complements of the solution sets of (a) through (e)

2. Find the integer solutions.

(a) $|3x+2| = 5$ (b) $|2x-3| = 7$

3. Use numerical examples to illustrate the properties of absolute value, (1) through (5), in Section 6.11b.

4. Give a numerical example of properties (3) and (5) of absolute value in which the strict inequality holds.

5. If a is rational and $1 < a$, how is a related to a^2?

6. If a is rational and $0 < a < 1$, how is a related to a^2?

7. If a is rational and $a < -1$, how is a related to a^2?

8. If a is rational and $-1 < a < 0$, how is a related to a^2?

6.11c The Property of Denseness

By associating to each rational number the point on the *number line* whose *distance* from some fixed point (see Sections 5.12b and 5.12c) is the rational number, we define a one-to-one correspondence between the rational numbers and a subset of the points on the line. We shall see in Section 7.1 that there are points on the number line which do not correspond to rational numbers (see Figure 2).

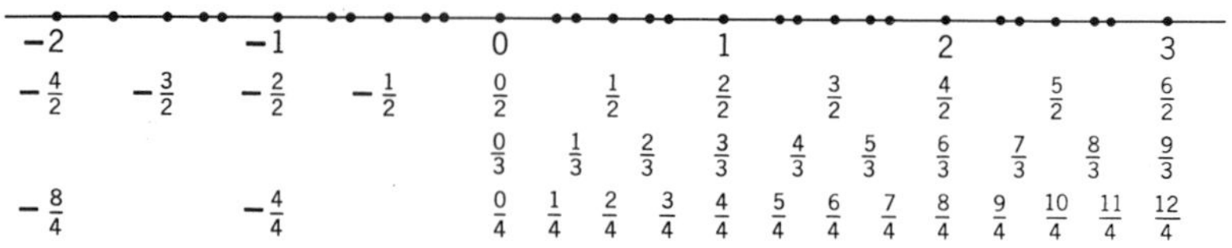

$-2 \quad -1 \quad 0 \quad 1 \quad 2 \quad 3$

$-\frac{4}{2} \quad -\frac{3}{2} \quad -\frac{2}{2} \quad -\frac{1}{2} \quad \frac{0}{2} \quad \frac{1}{2} \quad \frac{2}{2} \quad \frac{3}{2} \quad \frac{4}{2} \quad \frac{5}{2} \quad \frac{6}{2}$

$\frac{0}{3} \quad \frac{1}{3} \quad \frac{2}{3} \quad \frac{3}{3} \quad \frac{4}{3} \quad \frac{5}{3} \quad \frac{6}{3} \quad \frac{7}{3} \quad \frac{8}{3} \quad \frac{9}{3}$

$-\frac{8}{4} \quad -\frac{4}{4} \quad \frac{0}{4} \quad \frac{1}{4} \quad \frac{2}{4} \quad \frac{3}{4} \quad \frac{4}{4} \quad \frac{5}{4} \quad \frac{6}{4} \quad \frac{7}{4} \quad \frac{8}{4} \quad \frac{9}{4} \quad \frac{10}{4} \quad \frac{11}{4} \quad \frac{12}{4}$

If we let a and b be any two rational numbers (assume $a < b$), then it is always possible to find another rational number c such that c is "between" a and b; that is, given a and b with $a < b$ we can find another rational number c such that

$$a < c < b.$$

One way to do this is to take half of the sum of a and b; that is, to let

$$c = \frac{a+b}{2}.$$

Using the properties of order we can see this easily:

$a < b$; hence $\frac{a}{2} < \frac{b}{2}$, by property 2, Section 6.11

and $a = \frac{a}{2} + \frac{a}{2} < \frac{a}{2} + \frac{b}{2}$, by property 1, Section 6.11.

Also $\frac{a}{2} + \frac{b}{2} < \frac{b}{2} + \frac{b}{2}$, by the same property

Thus $a < \frac{a+b}{2} < b.$

Speaking geometrically, we note that the point $c = (a+b)/2$ is the midpoint of the line segment from a to b.

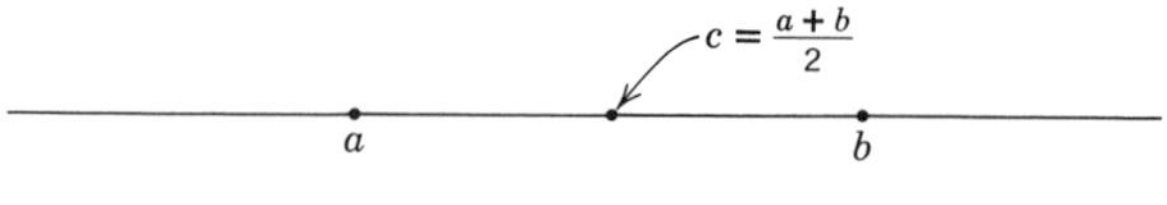

$c = \frac{a+b}{2}$

$a \qquad b$

Example 1

Find the midpoint of the line segment from $\frac{1}{2}$ to $\frac{9}{11}$ and verify that it is between $\frac{1}{2}$ and $\frac{9}{11}$.

$$\frac{\frac{1}{2}+\frac{9}{11}}{2}=\frac{\frac{11+18}{22}}{\frac{2}{1}}=\frac{29}{44}.$$

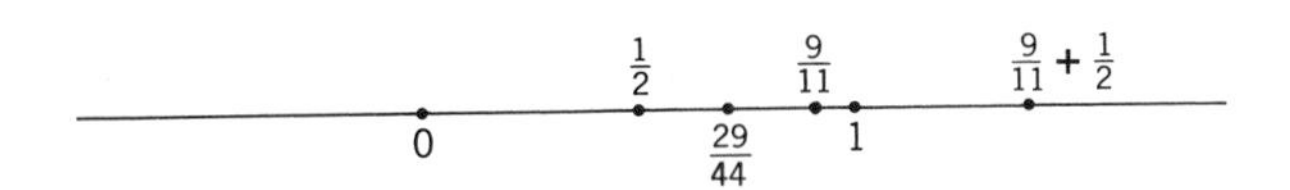

To show that $\frac{1}{2}<\frac{29}{44}<\frac{9}{11}$, we first show that $\frac{1}{2}<\frac{29}{44}$ and then show that

$\frac{29}{44}<\frac{9}{11}$. $\frac{29}{44}-\frac{1}{2}=\frac{58-44}{88}=\frac{14}{88}>0$ since $14\cdot 88>0$. Hence $\frac{1}{2}<\frac{29}{44}$.

$\frac{9}{11}-\frac{29}{44}=\frac{396-319}{484}=\frac{77}{484}>0$ since $77\cdot 484>0$. Hence $\frac{29}{44}<\frac{9}{11}$.

The fact that we can always find another rational number between any two distinct rational numbers implies that there are infinitely many rational numbers between any two distinct rational numbers. Why? No matter how "close together" two rational numbers may be, another rational number can be found which lies between them. This interesting property of the rational numbers is called *denseness*.

> ***Definition 6.11d.*** To say that the rational numbers are *dense-in-themselves* (or dense) means that between any two distinct rational numbers one can always find another rational number.

This is merely a way of saying that the rational numbers are densely distributed along the number line, that is, there is no part of the number line that contains two rational numbers without containing infinitely many rational numbers. The idea of denseness is closely related to the idea of approximations. This concept is extremely important in any situation involving measurements and indeed in much of mathematics itself. We examine it in Chapter 7.

6.11d Plotting Solution Sets

In plotting or graphing solutions sets we have adopted the convention of using parentheses to indicate that the endpoints of the interval are *not* to be included in the set. If the endpoints are to be included, square brackets are used.

Example 1

Plot the set and its complement for
$A = \{x \in R | 3 < x < 5\}$.

The graph of A:

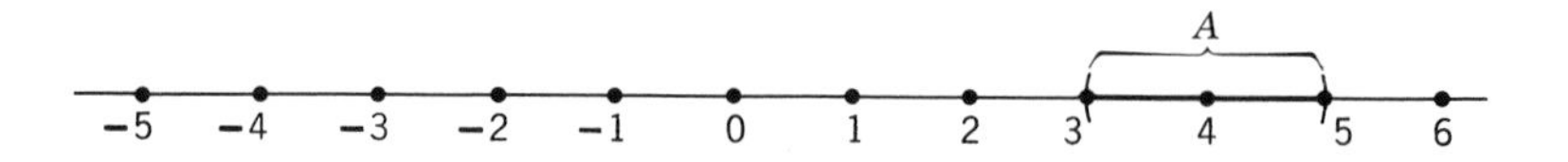

The graph of A':

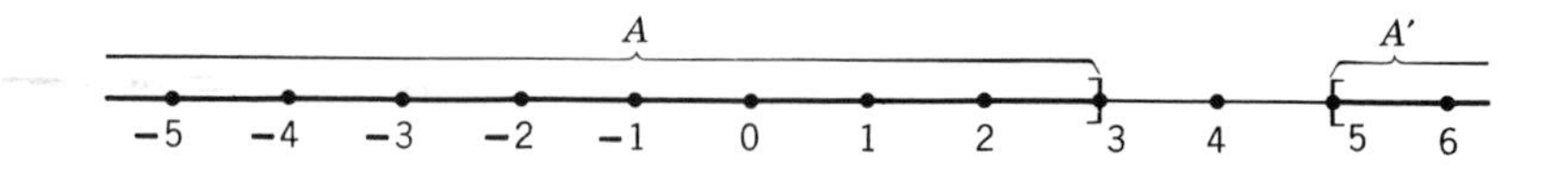

Example 2

Plot the set and its complement for
$B = \{x \in R \| x | \leqslant 2\}$.

The graph of B:

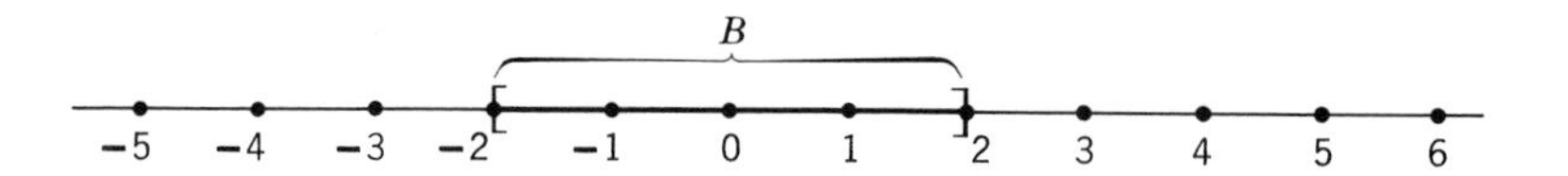

The graph of B':

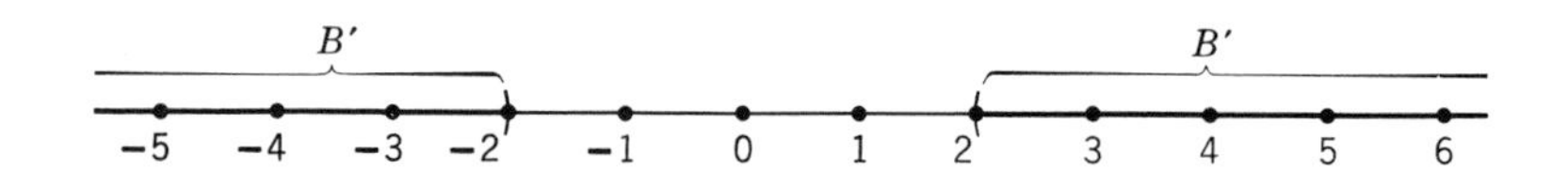

Exercise 6.11c

1. Find a rational number between 0 and 1.

2. Find 5 rational numbers between 0 and $\frac{1}{2}$.

3. Find 5 more rational numbers different from those in problem 2 which lie between 0 and $\frac{1}{2}$.

4. Is there a smallest rational number bigger than 1? Why?

5. Is there a largest rational number less than 3? Why?

6. Is there a smallest positive integer? What is it?

7. Is there a smallest positive rational number? What is it?

8. What number lies half way between 0 and $1/2^n$ for n a positive integer?

9. Is there a smallest rational number greater than $\frac{1}{3}$?

10. Is there a largest rational number smaller than $\frac{1}{3}$?

11. What is the smallest integer greater than -9?

12. What is the largest integer less than -5?

13. What is the smallest rational number x such that $x \geq \frac{3}{4}$?

14. What is the smallest rational number x such that $x \geq \frac{7}{1}$?

15. What do we mean when we say two rational numbers are close together?

16. Extend Definitions 4.13b, c, and d to the set of integers. To the set of rational numbers.

17. What is the least upper bound of the set A in each of the following (x rational)? (see Definition 4.13c.).

(a) $A = \{x||x| \leqslant 10\}$ (b) $A = \{x||x| < 1\}$
(c) $A = \{x|-3 < x \leqslant 0\}$ (d) $A = \{x|-5 < x < -2\}$

18. Plot the sets of problem 17 and their complements.

6.12 INTERPRETATIONS OF RATIONAL NUMBERS

As indicated in Section 6.2, we now turn to the other interpretations of rational numbers.

6.12a The "Division" Interpretation

We have identified the integers with particular rational numbers, that is,

$$\frac{2}{1} = 2 \quad \text{or} \quad 2 = \frac{2}{1},$$

$$\frac{3}{1} = 3 \quad \text{or} \quad 3 = \frac{3}{1},$$

$$\frac{n}{1} = n \quad \text{or} \quad n = \frac{n}{1},$$

etc. But $\frac{2}{1}$ is just one of the representatives of the class $[\frac{2}{1}]$. We could use $\frac{4}{2}, \frac{6}{3}, \frac{8}{4}$, and so forth, as representatives of this class:

$$\tfrac{4}{2} = \tfrac{6}{3} = \tfrac{8}{4} = 2.$$

To establish a connection between m/n and division, let us return to our definition of "divides" on the set of integers. Recall that $a|b$ if there is an integer k such that $b = a \cdot k$. This can also be stated, b "divided by" a is k or, with the usual symbol, $b \div a = k$.

Since $8 = 4 \cdot 2$, $4|8$. Using the "divided by" language, we have $8 \div 4 = 2$.

But from the previous statement we see that $\frac{8}{4} = 2$. Then $\frac{8}{4}$ can be interpreted as $8 \div 4$. This is the "division" interpretation of the rational number $\frac{8}{4}$.

Generalizing, we say that m/n can be interpreted as $m \div n$.

The "division" interpretation serves many purposes. We will see in a later section how it is used to obtain a very useful representation of numbers.

6.12b Division of Fractions

The usual rule associated with the division of fractions is "invert the denominator and multiply."

Example 1

$$\frac{\frac{3}{7}}{\frac{2}{11}} = \frac{3}{7} \cdot \frac{11}{2} = \frac{33}{14}$$

The explanation of this procedure is based on the fact that a/a can be used as a representative of the class $[\frac{1}{1}]$, the multiplicative identity for the rational numbers. The choice of a is that which, when multiplied by the denominator, gives $\frac{1}{1} = 1$. In this example we would choose a to be the multiplicative inverse of $\frac{2}{11}$.

Example 2

$$\left(\frac{2}{11}\right)^{-1} = \frac{11}{2}$$

$$\frac{\frac{3}{7}}{\frac{2}{11}} = \frac{\frac{3}{7}}{\frac{2}{11}} \cdot \frac{\frac{11}{2}}{\frac{11}{2}} = \frac{\frac{3}{7} \cdot \frac{11}{2}}{\frac{2}{11} \cdot \frac{11}{2}} = \frac{\frac{33}{14}}{\frac{1}{1}} = \frac{\frac{33}{14}}{1} = \frac{33}{14}.$$

Notice that we are using the fact that

$$\frac{\frac{3}{7}}{\frac{2}{11}} \cdot \frac{\frac{11}{2}}{\frac{11}{2}} = \frac{\frac{3}{7}}{\frac{2}{11}}.$$

That is, that $\frac{\frac{11}{2}}{\frac{11}{2}}$ is a form of the multiplicative identity.

Example 2 illustrates the procedure that is being presented in many of the newer arithmetic textbooks.

An alternate approach to the division of fractions would be to consider the type of problem that requires a number of the form $\frac{3}{7}/\frac{2}{11}$ as an answer. This can be interpreted as a special case of b/a, where b/a is that number which when multiplied by a gives b, where a and b are rational numbers, that is, b/a is the solution of the problem

$$a \cdot x = b.$$

The number $\frac{3}{7}/\frac{2}{11}$ can be interpreted as the solution of the equation:

$$\frac{2}{11} \cdot x = \frac{3}{7}.$$

We can solve for x by multiplying both sides of the equation by the multiplicative inverse of $\frac{2}{11}$ and making use of the commutative and associative laws.

$$\frac{2}{11} \cdot x = \frac{3}{7}$$

$$\left(\frac{2}{11}\right)^{-1} \cdot \frac{2}{11} \cdot x = \frac{3}{7} \cdot \left(\frac{2}{11}\right)^{-1}$$

$$1 \cdot x = \frac{3}{7} \cdot \left(\frac{2}{11}\right)^{-1}$$

$$x = \frac{3}{7} \cdot \frac{11}{2}.$$

It follows that

$$\frac{\frac{3}{7}}{\frac{2}{11}} = \frac{3}{7} \cdot \frac{11}{2}.$$

Exercise 6.12b

1. Carry out the indicated operations.

(a) $\frac{\frac{2}{3}}{\frac{3}{4}}$ (b) $\frac{\frac{5}{6}}{\frac{3}{6}}$ (c) $\frac{\frac{5}{2}}{\frac{3}{4}}$

2. Solve each of the following equations:

(a) $\frac{3}{4} \cdot x = \frac{5}{6}$ (b) $\frac{11}{3} \cdot x = \frac{2}{3}$ (c) $\frac{4}{7} \cdot x = \frac{7}{4}$

3. Find the multiplicative inverse of each of the following:

(a) $\frac{2}{3}$ (b) 1 (c) $\frac{1}{5}$
(d) $\left(\frac{3}{4}\right)^{-1}$ (e) 31 (f) $\left(\frac{6}{1}\right)^{-1}$

4. Find the additive inverse of each of the following:

(a) $\frac{4}{5}$ (b) $\frac{3}{4}+\frac{2}{3}$ (c) $\frac{3}{4}-\frac{2}{3}$

5. Explain in detail why the numerator is multiplied by the reciprocal of the denominator to get the quotient of two rational numbers.

6. Carry out the indicated operations.

(a) $\frac{\frac{1}{7}}{\frac{7}{2}}$ (b) $\frac{\frac{1}{2}+\frac{1}{3}}{\frac{2}{6}}$ (c) $\frac{\frac{1}{2}+\frac{2}{3}}{\frac{3}{2}-\frac{1}{3}}$

(d) $\frac{\frac{1}{2}}{\frac{1}{3}+\frac{1}{2}}$ (e) $\frac{2}{5} \cdot \left(\frac{3}{4}-\frac{1}{2}\right)$ (f) $\frac{\frac{1}{15}+\frac{1}{5}}{\frac{1}{5}}$

7. Solve each of the following equations:

(a) $\frac{3}{4} \cdot x - \frac{2}{3} = \frac{1}{2}$ (b) $\frac{4}{5} \cdot x = \frac{2}{3} - \frac{3}{7}$ (c) $\frac{5}{2} \cdot x + \frac{0}{1} = \frac{3}{3}$

8. Give an example of a problem situation in which it would be useful to write $1 \cdot 3$ instead of 3. (*Hint*: Factor $3x+3$.)

9. Give an example of a problem situation in which it would be useful to write $\frac{15}{3}$ instead of 5. (*Hint*: Add $\frac{2}{3}$ to 5.)

10. Give an example of a problem situation in which it would be useful to write 1 as $\frac{2}{2}$.

6.12c "Fraction" Interpretation

In the early development of fractions in almost any arithmetic textbook you will encounter an illustration similar to the following.

Example 1

In Figure 3, the disc is marked off in 3 equal parts. The numbers 2 and 3 can be used to tell how much of the disc is shaded.

This tells the number of equal parts. $\searrow \frac{2}{3} \nwarrow$ Two of the parts are shaded.

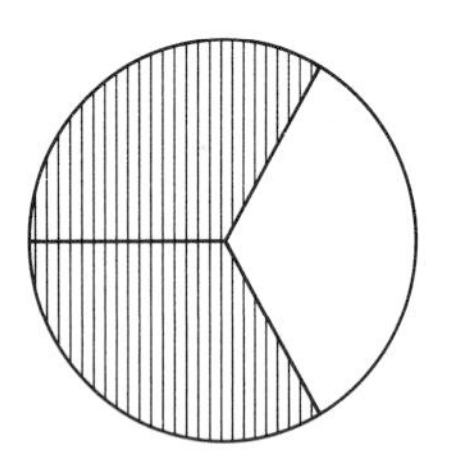

Figure 2. The rational number line.

Many other examples could be cited. This is the "fraction" interpretation or, as some authors term it, the "partition" interpretation of the rational number. This is probably the most familiar interpretation of rational numbers, so it needs but little explanation here.

In the rational number m/n, used in the "fraction" sense, the n is called the *denominator* and the m is called the *numerator*. Denominator is derived from the Latin word *denominatus*, "to call by name." It designates by name (number name) the parts into which the whole is divided. Numerator is derived from the Latin *numeratus*, "to count," and "counts" the parts under consideration. The symbol, m/n, under the "fraction" interpretation designates m of n equal parts.

$$\frac{m}{n} = m \cdot \frac{1}{n}.$$

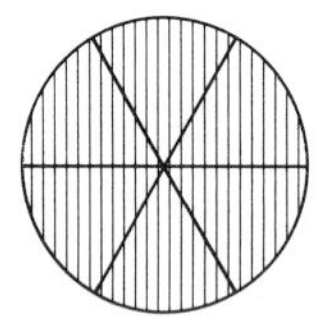
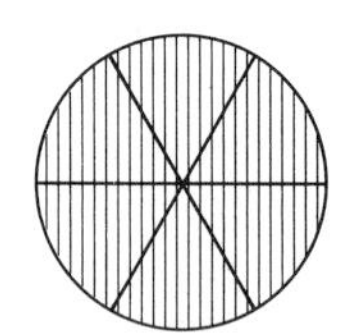
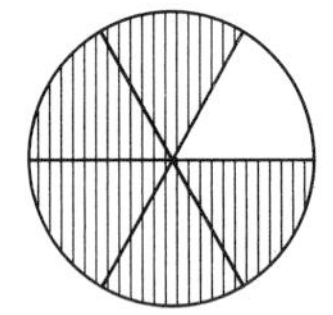

Figure 3

Example 2

In Figure 3, how do we express the shaded portion of the circles, $2\frac{5}{6}$ or $\frac{17}{6}$?

The symbol $2\frac{5}{6}$ is recognized and accepted as the name of a number. It is read, "two and five-sixths." It means $2+\frac{5}{6}$. As written, it is the sum of an integer and a rational number. We know how to add integers to integers and rational numbers to rational numbers, but we have not defined addition of integers to rational numbers.

To give the numeral $(2+\frac{5}{6})$ meaning, we must turn to the system of rational numbers. We must interpret the integer 2 as a rational number as we agreed earlier could be done. The obvious choice for a rational number to represent 2 would be $\frac{2}{1}$. By $\frac{2}{1}$ we mean $[\frac{2}{1}]$. We have considerable freedom in choosing a representative of $[\frac{2}{1}]$.

The term "five-sixths" suggests the "fraction" or "partition" interpretation of the rational number, $\frac{5}{6}$. The "5 of 6 equal parts" approach is conventional but, technically, it requires caution. To introduce the fraction $\frac{5}{6}$ as

$$\frac{5}{6}=\frac{1}{6}+\frac{1}{6}+\frac{1}{6}+\frac{1}{6}+\frac{1}{6}$$

assumes that the learner knows how to add rational numbers or to multiply an integer by a rational number.

$$\frac{1}{6}+\frac{1}{6}+\frac{1}{6}+\frac{1}{6}+\frac{1}{6}=5\cdot\frac{1}{6}=\frac{5}{6}.$$

To avoid this assumption we introduced the system of rational numbers before we discussed the various interpretations. From our previous discussions we can proceed as follows:

$$\frac{5}{6}=\frac{5}{1}\cdot\frac{1}{6}$$

But $\frac{5}{1}=5$, so we can write

$$\frac{5}{6}=\frac{5}{1}\cdot\frac{1}{6}=5\cdot\frac{1}{6}.$$

In the interpretation of $(2+\frac{5}{6})$ we can choose $\frac{12}{6}$ as the representative of $[\frac{2}{1}]$ and write this as $12\cdot\frac{1}{6}$. We then have

$$2+\frac{5}{6}=\frac{12}{1}\cdot\frac{1}{6}+\frac{5}{1}\cdot\frac{1}{6}.$$

Using the distributive law, we have

$$\begin{aligned}\frac{12}{1}\cdot\frac{1}{6}+\frac{5}{1}\cdot\frac{1}{6}&=\left(\frac{12}{1}+\frac{5}{1}\right)\cdot\frac{1}{6}\\&=(12+5)\cdot\frac{1}{6}\\&=17\cdot\frac{1}{6}\\&=\frac{17}{6}.\end{aligned}$$

This is a tedious way to show that $2\frac{5}{6}=\frac{17}{6}$, but it is based on the fundamental properties of the number systems we have discussed. The usual procedure is to write

$$2\frac{5}{6}=2+\frac{5}{6}=\frac{12}{6}+\frac{5}{6}=\frac{12+5}{6}=\frac{17}{6}.$$

The question arises as to which is the more useful name, $2\frac{5}{6}$ or $\frac{17}{6}$? The form $2\frac{5}{6}$ reflects its origin in measuring and suggests magnitude or size. The expression, "$\frac{17}{6}$ yd of material," is not a familiar way to order yard goods. On the other hand, there are times when $\frac{17}{6}$ is a more convenient form that $2\frac{5}{6}$. For example, if $2\frac{5}{6}$ yd of material is to be divided among six people, how much does each recieve?

$$2\tfrac{5}{6} \div 6 = ?$$
$$\tfrac{17}{6} \div 6 = \tfrac{17}{6} \cdot \tfrac{1}{6} = \tfrac{17}{36}.$$

Each person would recieve $\frac{17}{36}$ yd or 17 in.

To conclude this section we illustrate the use of $2\frac{5}{6}$ and $\frac{17}{6}$ as names for the same number in computation.

Example 3

$$\begin{aligned}
(2\tfrac{5}{6})(1\tfrac{1}{2}) &= (2+\tfrac{5}{6})(1+\tfrac{1}{2}) \\
&= (2+\tfrac{5}{6})(1) + (2+\tfrac{5}{6})(\tfrac{1}{2}) \\
&= 2\cdot 1 + \tfrac{5}{6}\cdot 1 + 2\cdot\tfrac{1}{2} + \tfrac{5}{6}\cdot\tfrac{1}{2} \\
&= 2 + \tfrac{5}{6} + 1 + \tfrac{5}{12} \\
&= (2+1) + (\tfrac{5}{6}+\tfrac{5}{12}) \\
&= (2+1) + (\tfrac{10}{12}+\tfrac{5}{12}) \\
&= 3 + \tfrac{15}{12} \\
&= 3 + \frac{12+3}{12} \\
&= 3 + \tfrac{12}{12} + \tfrac{3}{12} \\
&= (3+1) + \tfrac{1}{4} \\
&= 4 + \tfrac{1}{4} = 4\tfrac{1}{4}
\end{aligned}$$

$$\begin{aligned}
(2\tfrac{5}{6})(1\tfrac{1}{2}) &= (\tfrac{17}{6})(\tfrac{3}{2}) \\
&= \frac{17\cdot 3}{6\cdot 2} \\
&= \frac{17}{2\cdot 2} \\
&= \tfrac{17}{4} \\
&= \frac{16+1}{4} \\
&= \tfrac{16}{4} + \tfrac{1}{4} \\
&= 4 + \tfrac{1}{4} \\
&= 4\tfrac{1}{4}
\end{aligned}$$

Example 4

$$\begin{aligned}
2\tfrac{5}{6} + 1\tfrac{1}{2} &= 2 + \tfrac{5}{6} + 1 + \tfrac{1}{2} \\
&= (2+1) + (\tfrac{5}{6}+\tfrac{1}{2}) \\
&= 3 + (\tfrac{5}{6}+\tfrac{3}{6}) \\
&= 3 + \tfrac{8}{6} \\
&= 3 + \frac{6+2}{6} \\
&= 3 + \tfrac{6}{6} + \tfrac{2}{6} \\
&= (3+1) + \tfrac{1}{3} \\
&= 4\tfrac{1}{3}
\end{aligned}$$

$$\begin{aligned}
2\tfrac{5}{6} + 1\tfrac{1}{2} &= \tfrac{17}{6} + \tfrac{3}{2} \\
&= \tfrac{17}{6} + \tfrac{9}{6} \\
&= \frac{17+9}{6} \\
&= \tfrac{26}{6} \\
&= \frac{24+2}{6} \\
&= \tfrac{24}{6} + \tfrac{2}{6} \\
&= 4 + \tfrac{1}{3} \\
&= 4\tfrac{1}{3}
\end{aligned}$$

Exercise 6.12c

1. (a) Add $2\frac{1}{6}$ to $3\frac{2}{3}$ as you were taught in arithmetic.
(b) Add the same numbers as in Example 4, Section 6.12c, and justify each step.

2. Carry out the indicated operations and write your answer as a rational number and as an integer plus a rational number.

(a) $\dfrac{2+\frac{5}{6}}{2}$ (b) $\dfrac{3\frac{4}{7}}{2\frac{1}{3}}$

(c) $\dfrac{(5+\frac{1}{4})+2}{3+\frac{1}{3}}$ (d) $(2\frac{1}{3})(3\frac{1}{2})$

3. Give an example of a situation in which each of the following numerals is in a convenient form.

(a) $4\frac{1}{6}$ (b) $\frac{33}{11}$ (c) $\frac{9}{6}$
(d) $\frac{5}{5}$ (e) $9\frac{2}{6}$ (f) $4+\frac{1}{7}$

4. The following diagram represents the product $(2\frac{1}{3})(3\frac{1}{2})$. Label each part properly.

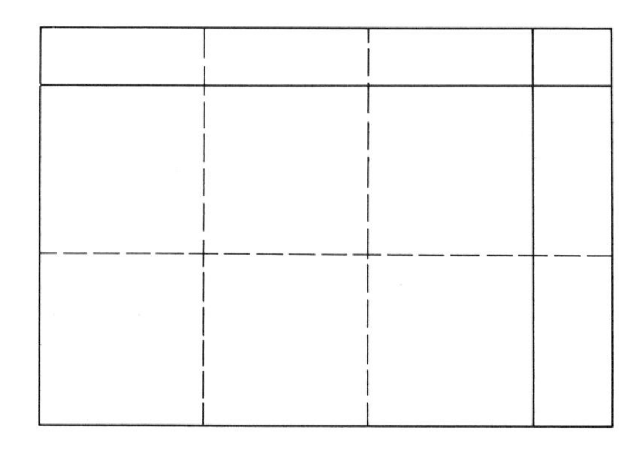

5. Carry out the computation and explain each step in the following:

$$\frac{17\frac{2}{5}}{3\frac{5}{9}}$$

6. Carry out the indicated operations.

(a) $2\frac{3}{7} \div 4\frac{3}{5}$ (b) $1 \div \dfrac{1}{10}$

(c) $\left(5+\dfrac{2}{3}\right) \div \left(6+\dfrac{1}{2}\right)$ (d) $\dfrac{4\frac{1}{3}}{2}$

(e) $2 \div 5\frac{3}{4}$ (f) $\dfrac{5+\frac{5}{7}}{7}$

7. Solve for x: $(3\frac{1}{2})x = 4\frac{2}{9}$.

8. Multiply each of the following by direct application of the distributive law.

(a) $(5+\frac{4}{5})(6+\frac{3}{10})$ (b) $(6+\frac{1}{9})(9+\frac{1}{6})$
(c) Represent each product as area and label each part.
(d) Change the numerals of (a) and (b) to rational form and carry out the multiplication.

9. Carry out the indicated operations and simplify.

(a) $\frac{3+\frac{1}{3}}{2}$ (b) $\frac{3^{-1}+3}{3}$

(c) $\frac{5-\frac{1}{3}}{\frac{1}{3}+1}$ (d) $2(\frac{2}{3}-\frac{1}{6})$

(e) $\frac{1}{2}\cdot(3+\frac{5}{8})$ (f) $5\frac{2}{3}-4\frac{7}{8}$

10. How much is one half of

(a) 0.00000000012? (b) 10^{23}?

(c) $5+\frac{3}{7}$?

6.12d "Rate Pair" Interpretation (Ratio)

We have had occasion to consider whole numbers in situations in which the only properties used were their distinctiveness and their inexhaustible supply. In these situations the numbers are used as names of objects or places, for example, telephone numbers, positions on a baseball team, etc. Addition and multiplication in these situations have no meaning. The numbers are used as names and not as numbers.

Similarly, many situations call for the use of ordered pairs of whole numbers; for example, "Mary bought 3 pencils for 10 cents." The pair of whole numbers (3, 10), is an ordered pair. "Mary bought 10 pencils for 3 cents" describes a completely different situation. (One looks like a bargain.) Another way of describing this situation is to say that Mary bought pencils which cost "3 for 10 cents." The sign in the window of the store probably read, "pencils, 3/10¢."

The ordered pair of whole numbers is being used here to describe a *many-to-many correspondence*, in this instance, a 3 to 10 correspondence. The statement, "The odds are 8 to 5 on the Yankees to win," describes an 8 to 5 correspondence. As stated before, these situations involve the ordered pairs of natural numbers but not as numbers in a number system. What would it mean to add or multiply the ordered pairs in these situations? The number pairs are used as *rate pairs*. As such, they are often written in the same way as ordinary rational numbers. As *rate pairs* they should be read "3 for 10," or "3 to 10," or "8 to 5," and not as three-tenths or eight-fifths.

In many situations involving these ordered pairs, different pairs describe the same situation. Thus 3 pencils for 10¢ describes the same situation as 6 pencils for 20¢ or 12 pencils for 40¢. Similarly, 20 miles to 1 gal is equivalent to 40 miles to 2 gal, or 100 miles to 5 gal. The criterion for determining when two rate pairs describe the same many-to-many correspondence is the same as determining when two rational numbers are *equivalent*. Thus the rate pair, m/n, will be *equivalent* to the rate pair, r/s, if and only if $m \cdot s = n \cdot r$ as integers. Let us use the symbol $=$ to indicate that two rate pairs describe the same many-to-many correspondence.

Example 1

$\frac{3}{10} = \frac{6}{20}$ because $3 \cdot 20 = 10 \cdot 6$.
$\frac{3}{10} = \frac{9}{30}$ because $3 \cdot 30 = 10 \cdot 9$.
$\frac{8}{5} = \frac{16}{10}$ because $8 \cdot 10 = 5 \cdot 16$.

This relation between rate pairs is an equivalence relation and partitions all rate pairs into equivalence classes. The essential difference between working with the ordered pairs as *rate pairs* as opposed to working with ordered pairs as *rational numbers* is that as rational numbers they are added, multiplied, subtracted, and divided. These operations associate with ordered pairs of two classes, an ordered pair of a third class. On the other hand, rate pairs involve essentially working with one class at a time.

The usual problem situation involving the rate pairs is built around the following simple idea. *Three* of the four components of two equivalent rate pairs are known. The problem is to find the *fourth*.

Example 2

If 6 apples cost 25¢, how much do 30 apples cost?

The equivalence class to which the rate pair $\frac{6}{25}$ belongs is

$\frac{6}{25}, \frac{12}{50}, \frac{18}{75}, \frac{24}{100}, \frac{30}{125}, \frac{36}{150}, \ldots .$

Now the answer to the question is quite obvious. The rate pair with 30 as its first component is $\frac{30}{125}$, which is to be interpreted as 30 apples for $1.25. This is usually shortened as follows:

Let N be the cost of the 30 apples; then

$$\frac{6}{25} = \frac{30}{N} \quad \text{if and only if} \quad 6 \cdot N = 25 \cdot 30.$$

Then

$$N = \frac{25 \cdot 30}{6},$$

and

$$N = 125.$$

Exercise 6.12d

1. Give five examples of the use of numbers as names only.

2. What is the meaning of the baseball expression, "Out, 6 to 3"?

3. What is the meaning of the numbers on the doors in a building?

4. A certain canned food sells at 3 cans for 38¢. How many cans can you get for $1.90?

5. A bank charges $3 for every $500 of the amount of a certified check it issues. If the charges on a certain check was $12, what was the value of the check?

6. On a map 1 in. represents 16 miles. What distance is represented by $5\frac{1}{2}$ in.?

7. A tree 60 ft high casts a shadow 45 ft long. What is the height of a tree that casts a shadow 30 ft long at the same time of day?

8. A man in an automobile made a trip of 125 miles in $2\frac{1}{2}$ hours. At the same rate how long would it take him to make a trip of 500 miles?

9. The scale in an architect's drawing is 1 ft to $\frac{1}{4}$ in. A distance of 10 in. on the drawing represents how many feet in the structure?

10. Every 3 gal of radiator fluid contains 2 qt of pure antifreeze. How many quarts of antifreeze are there in 45 gal of this mixture?

6.12e "Rate Pair" Interpretation (Percent)

The rate pair is also used in making relative comparisons. Thus a person invests $3000 in a particular stock and sells it for $3450. He earns $450 profit. Another person invests $450 in some stock which he sells for $900. He also shows a profit of $450. The total amount earned was the same, but the rate of return on the original investment is markedly different. The rate pairs, $\frac{450}{3000}$ and $\frac{450}{450}$, describe quite different aspects of this investment situation. It is possible to make a comparison of these two rate pairs; however, rate pairs with the same denominator are more conveniently compared. The usual practice is to use 100 as the common denominator. Rate pairs whose common denominator is 100 are called *percents*. Percents are rate pairs in which the second place number is 100, and when this is understood, they are usually written with the symbol % or "percent" in place of the 100 in the denominator.

From these examples we have

$$\tfrac{450}{3000} = \tfrac{15}{100} = 15\%$$

$$\tfrac{450}{450} = \tfrac{100}{100} = 100\%$$

Thus 15% really means 15 per 100. In terms of percent, the investment situation can be compared by inspection.

The problem of changing from percent to decimal form and vice versa simply involves remembering what percent means:

$$15\% = \frac{15}{100} = 0.15$$

$$0.02 = \frac{2}{100} = 2\%$$

Most rational numbers can be expressed only approximately in decimal form. The same is true of rate pairs and percents. The rate pair $\frac{1}{3}$ can only be approximated by a percent. For example,

$$\tfrac{1}{3} \cong \tfrac{33}{100} = 33\%.$$

Convention has given meaning to the following rate pairs:

$33\frac{1}{3}\%$ $\quad$ $66\frac{2}{3}\%$

The meaning here is clear. Strangely enough, we seldom see such rate pairs as

$14\frac{2}{7}\%$ $\quad$ $81\frac{9}{11}\%$ $\quad$ $74\frac{74}{99}\%$

These are usually expressed as "about 14%," "about 82%," and "about 75%." We simply mention this as a convention and do not attempt to clarify it as a mathematical idea.

The traditional approach has been to present percent problems, or problems involving percent, as three different types or cases. With the use of rate pairs the percent problems are essentially of one type—finding the fourth component of two equivalent rate pairs when the other three components are known.

Let P symbolize percent. Then, for example, 32% would give us $P = 32$ and would be expressed as the rate pair $\frac{32}{100}$. When using P to represent the percent, $P/100$ is the rate pair. For the other equivalent rate pair in the percent problem we shall use A/B, where A is called the "amount" and B the "base." Now the "three" types of percent problems can be expressed in terms of these two rate pairs.

Find 32% of 400. $\quad \dfrac{32}{100} = \dfrac{A}{400}, \quad A = 128.$

What percent of 400 is 128? $\quad \dfrac{P}{100} = \dfrac{128}{400}, \quad P = 32.$

128 is 32% of what number? $\quad \dfrac{32}{100} = \dfrac{128}{B}, \quad B = 400.$

Exercise 6.12e

1. Express the following rate pairs as percents:

(a) $\frac{32}{100}$ (b) $\frac{2}{100}$ (c) $\frac{3}{4}$ (d) $\frac{13}{20}$
(e) $\frac{4}{25}$ (f) $\frac{18}{50}$ (g) $\frac{18}{75}$ (h) $\frac{96}{400}$

2. Express the following percents as rate pairs:

(a) 25% (b) 34% (c) 40% (d) 85%
(e) 125% (f) 250% (g) 400% (h) 12.5%

3. Use the rate pair idea to answer the following questions:

(a) What is 15% of 80?
(b) What is 45% of 2300?
(c) What percent of 248 is 31?
(d) What percent of $2000 is $160?
(e) 65% of a number is 260. What is the number?
(f) 34% of a number is 170. What is the number?

6.13 DECIMAL FRACTIONS

Some of the problems in earlier sections suggest that some other representation of the rational numbers is needed; a representation which is amenable to arithmetic computation and which gives the user a readily understandable idea of magnitude. For instance, which of the two rational numbers listed below is the larger and what is their sum?

$$\frac{3{,}692{,}846}{15{,}496{,}321} \quad \frac{29{,}487{,}692}{58{,}836{,}857}$$

Without some other representation, computations using numbers of this form would have to be, and have been, performed with not much more than brute force and perseverance. The necessary innovations in our system of numeration have occurred in relatively recent times. The use of exponents to express extremely small and extremely large numbers originated in the seventeenth century. The idea of the decimal fraction was introduced the century before. We turn our attention to the latter.

Historians report that Al-Kashi was using *decimal fractions* systematically in the fifteenth century (see "Historically Speaking," *The Mathematics Teacher*, April 1964). Before this report, more available records showed that during the middle of the sixteenth century, Simon Stevin, a Belgian, introduced the idea of *decimal fractions*. By combining the idea of using fractions whose denominators are integral powers of the base with the idea of place value, the arithmetician is freed of computations of the foregoing type.

Definition 6.13. A *decimal fraction* is a rational number whose denominator is an integral power of 10.

Example 1

$$\frac{25}{100}, \quad \frac{874}{10{,}000}, \quad \frac{21}{100{,}000}, \quad \frac{337{,}601}{10{,}000}.$$

In place of writing the power of 10 in the denominator, a point called the decimal point is inserted between two digits of the numerator so that *the number of places to the right of this point tells the power of the base in the denominator.* The point serves as a "separatrix." Digits to the left of the point form the whole or integer part of the number and digits to the right of the point form the numerator of the fraction whose denominator is the power of 10 with exponent equal to the number of digits to the right of the decimal point.

Example 2

$$\frac{25}{100} = \frac{25}{10^2} = 0.25.$$

$$\frac{874}{10{,}000} = \frac{874}{10^4} = 0.0874.$$

$$\frac{337{,}601}{10{,}000} = \frac{337{,}601}{10^4} = 33.7601.$$

The number of digits to the right of the decimal point is called the number of *decimal places* in the numeral.

Example 3

1. The numeral 3.1416 is given to four decimal places.

$$\frac{31{,}416}{10{,}000} = 3.1416.$$

2. The numeral 0.000163 is given to six decimal places.

$$\frac{163}{1{,}000{,}000} = 0.000163.$$

3. The numeral 0.001 is given to three decimal places.

$$\frac{1}{1000} = 0.001.$$

6.13a The Number Line

If we consider the number line again, it is a straight line along which the *integers* are evenly marked in both directions from the origin which we labeled 0.

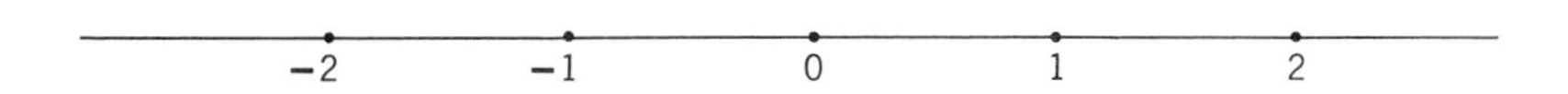

The decimal fractions which we call tenths separate the line segment into equal segments so that ten such segments fit between any consecutive pair of integers.

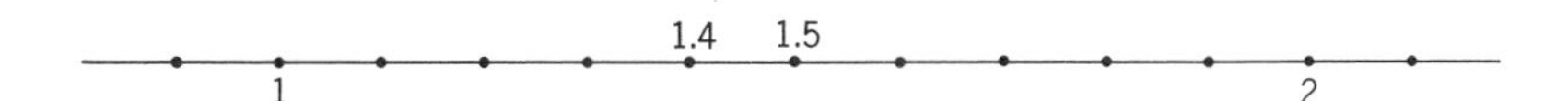

The hundredths divide each of the tenths into ten equal parts. The thousandths divide each of the hundredths segments into ten equal parts. The ten thousandths, hundred thousandths, millionths, and so on are scattered along the number line in the same way as just indicated. If

the line is divided into millionths, there would be one million such segments between any two consecutive integers. This suggests that the decimal fractions are also *dense* on the number line. This is actually the case although the fact requires proof.

One can imagine plotting the numbers 0.3, 0.33, 0.333. . . indefinitely, approaching in the limit the rational number $\frac{1}{3}$. Conversely, it is plausible that if we let the letter r denote the point on the number line corresponding to the number r, by taking successive subdivisions, tenths, hundredths, thousandths, and so forth, choosing at each successive step the nearest mark to the point r, we will either arrive at the point corresponding to r or we will approximate the number r by a decimal fraction. By continuing indefinitely, the number r can then be expressed as an infinite decimal.

Exercise 6.13a

1. Plot the points 0.1, 0.14, and 0.142.

2. Find decimal approximations of $\frac{57}{40}$. Compare these to table values of $\sqrt{2}$.

3. Find the decimal approximations $\frac{6449}{4560}$ and compare these to table values of $\sqrt{2}$.

4. Find decimal approximations of $\frac{22}{7}$ and compare them to the value of π given by the mnemonics in Section 8.13c.

5. Find decimal approximations of $\frac{3927}{1250}$, $\frac{754}{240}$, $\frac{355}{113}$ and compare these to π.

6. The decimal fractions are dense in the real numbers.

(a) Interpret this statement, using $\sqrt{2}$ as an illustration.
(b) Interpret this statement, using π as an illustration.

6.13b Computations with Decimal Fractions

There is nothing essentially new in computations involving decimal fractions. Decimal fractions are, after all, special rational numbers written in "decimal form." At most, "placement of the decimal point" in the resultant of a binary operation is the only new procedure not previously discussed. We consider this problem only briefly, for it can be explained as a simple consequence of the behavior of exponents and the properties of the system of rational numbers.

6.13c Addition of Decimal Fractions

The usual procedure in adding two or more decimal fractions is to add the numbers by column after lining up the decimal points. As in the case of addition by column of integers, the decimal points are "lined up" so that the distributive law can be applied.

Example 1

Find the sum of 3.92, 406.7273, and 0.076.

$$\begin{array}{r} 3.92 \\ 406.7273 \\ 0.076 \\ \hline 410.7233 \end{array}$$

These three numbers can also be added as rational numbers, remembering that the number of decimal places indicates the power of the base in the denominator.

$$3.92+406.7273+0.076=\frac{392}{10^2}+\frac{4{,}067{,}273}{10^4}+\frac{76}{10^3}.$$

Writing these rational numbers with a common denominator, we have

$$\frac{39{,}200}{10{,}000}+\frac{4{,}067{,}273}{10{,}000}+\frac{760}{10{,}000} \qquad \text{judicious choice of representatives of the classes}$$

$$=\frac{39{,}200+4{,}067{,}273+760}{10{,}000} \qquad \text{by the distributive law.}$$

The indicated addition in the numerator of the last expression can be carried out by "column addition" (the distributive law again).

$$\begin{array}{r} 39{,}200 \\ 4{,}067{,}273 \\ 760 \\ \hline 4{,}107{,}233 \end{array}$$

Notice that the digits are lined up in the same columns as they are in the decimal form. We have then

$$\frac{4{,}107{,}233}{10^4}=410.7233.$$

The reason for the procedure of lining up the decimal points in the decimal numbers to be added in column addition may be more apparent if the numbers are written in the expanded form. A numeral written in a place-value system of numeration is a convenient expression for a *sum of multiples of powers of the base.* For instance, the integer 1066 can be written in "expanded form" as follows:

$$1066=1\cdot 10^3+0\cdot 10^2+6\cdot 10^1+6\cdot 10^0.$$

Using negative exponents, decimal numerals can be written in the expanded form.

Example 2

1. $2033.3906=2\cdot 10^3+0\cdot 10^2+3\cdot 10^1+3\cdot 10^0+3\cdot 10^{-1}$
   $+9\cdot 10^{-2}+0\cdot 10^{-3}+6\cdot 10^{-4}$
2. $0.027=0\cdot 10^{-1}+2\cdot 10^{-2}+7\cdot 10^{-3}.$

The addition algorithm can be extended to decimal numerals directly, and the distributive law applied to the "like" powers justifies the column addition as before.

6.13d Multiplication of Decimal Fractions

Multiplication is quite literally a binary operation. Indicated products of more than two numbers can be expressed because of the associative law, but the actual computation is strictly binary. The placement of the decimal point in the product of two decimal numerals is again no problem. We illustrate with a numerical example.

Example 1

$$\begin{array}{r} 33.9 \\ \underline{4.27} \\ 2373 \\ 6780 \\ \underline{135600} \\ 144.753 \end{array}$$

To place the decimal point in the product, add the number of places in the multiplier and the multiplicand. The number of places in the product is this sum. This procedure is an immediate consequence of the behavior of exponents.

$$(33.9)(4.27) = \left(\frac{339}{10^1}\right) \cdot \left(\frac{427}{10^2}\right) = \frac{(339)(427)}{10^3} = \frac{144{,}753}{10^3} = 144.753$$

$$(0.0339)(0.427) = \left(\frac{339}{10^4}\right) \cdot \left(\frac{427}{10^3}\right) = \frac{144{,}753}{10^7} = 0.0144753.$$

6.13e Division of Decimal Fractions

The usual procedure in placing the decimal point in the quotient of two decimal numbers is as follows. Move the decimal point in the dividend and the divisor enough places to the right to produce whole numbers, adding zeros where needed. Carry out the division as whole numbers with the decimal point in the quotient directly above the decimal point in the new position in the dividend.

Example 1

$60.69 \div 0.017 = ?$

$0.017.\overline{)60.690.}$

$$\begin{array}{rr} 17\overline{)60690.} & 3000 \\ \underline{51000} & \\ 9690 & 500 \\ \underline{8500} & \\ 1190 & 70 \\ \underline{1190} & \\ & \overline{3570.} \text{ quotient} \end{array}$$

Moving the decimal place in both the divisor and the dividend to produce division of integers, as we have indicated, is actually accomplished by multiplying the divisor and dividend by a high enough power of 10 to produce integers. If the indicated division is expressed in rational form, then multiplying the dividend and the divisor by the same power of 10 is the same as multiplying by the multiplicative identity.

Example 2

$$\frac{60.69}{0.017} = \left(\frac{60.69}{0.017}\right) \cdot \left(\frac{10^3}{10^3}\right) = \frac{(60.69)(10^3)}{(0.017)(10^3)} = \frac{60{,}690}{17} = 3570.$$

An alternative procedure, which is quite common, is to perform the division as a division of rational numbers. In using this procedure, the decimal points are moved just enough places to the right to produce a whole number divisor, adding zeros where necessary. Place the decimal point in the quotient directly above the decimal point in the dividend in its new position.

Example 3

$$\frac{6.069}{0.17} = \frac{\frac{6069}{10^3}}{\frac{17}{10^2}} = \left(\frac{6069}{10^3}\right) \cdot \left(\frac{10^2}{17}\right) = \left(\frac{6069}{10}\right) \cdot \left(\frac{1}{17}\right) = \frac{606.9}{17}$$

$$0.17.\overline{)6.06.9} \quad \text{quotient } 35.7$$

Exercise 6.13e

1. Add the following rational numbers:

(a) $\frac{3}{10} + \frac{2}{100}$ (b) $\frac{2}{10} + \frac{6}{1000}$

(c) $\frac{7}{100} + \frac{9}{1000}$ (d) $\frac{2}{10} + \frac{3}{100} + \frac{7}{1000}$

2. Find the solution set of the open sentence:

(a) $0.0065 \times 10^{26} = 6.5 \times 10^n$

(b) $740 \times 10^{-23} = 7.4 \times 10^n$

(c) $10^n \cdot 10^n = 10^{30}$

3. A number is written in expanded form when it is written multiplicatively as a sum of powers of the base 10; for example,

$73.25 = 7 \cdot 10^1 + 3 \cdot 10^0 + 2 \cdot 10^{-1} + 5 \cdot 10^{-2}.$

Write each of the following in expanded form:

(a) 700.125 (b) 333.3333 (c) 10,000.0001

(d) 27.272727 (e) 3.1416

4. Carry out the indicated operations.

(a) $(3.1416) \cdot (17.74)$
(b) $17.74 \div 3.1416$
(c) $3.1416 \div 17.74$
(d) $(293.004)(7.46)(2.917)$
(e) $1001.0102 \div 0.005$

5. Carry out the indicated operations.

(a) $(2.99776 \cdot 10^{10})(1.673 \cdot 10^{-24})$
(b) $(3.5 \cdot 10^{4})(7.69)$
(c) $(6.0228 \cdot 10^{23})(1.673 \cdot 10^{-24})$
(d) $(8.015 \cdot 10^{5}) \div (0.005)$
(e) $(6.45 \cdot 10^{-3}) \div (2.4 \cdot 10^{5})$

6. If we take the diameter of a silver dollar to be the unit length of a number line, how many of these units must the dollar roll in order for it to start with the head upright, turn over once, and end upright?

7. If two George Washington quarters are placed next to each other with the heads on both coins upright and one coin is rolled over the top of the other, what will be the position of the head on the rolled coin when it comes to rest on the side opposite from which it started? Try this with coins.

8. Find the smallest n such that $10^{-n} < 0.00005$.

9. Find the largest n such that $1/96000000 < 10^{-n}$.

REVIEW EXERCISES

1. Define the binary operation $\odot$ on $W \times N$ as follows:

$$((m, n), (p, q)) \xrightarrow{\odot} (m, n) \odot (p, q) = (mp, nq)$$

(a) Is $W \times N$ closed with respect to $\odot$?
(b) Does $\odot$ obey the commutative law?
(c) Does $\odot$ obey the associative law?
(d) Is there an identity element for $\odot$? If so, what is it?

2. The even numbers are numbers of the form $2k$, where k is in W. The odd numbers are numbers of the form $(2k-1)$ or $(2k+1)$, where k is in W.

(a) Show that the product of two even numbers is even and justify each step.
(b) Show that the sum of two even numbers is even and justify each step.
(c) Show that the product of two odd numbers is odd and justify each step.

Give precise definitions of each of the following:

3. A proper subset of a set.

4. The intersection of two sets.

5. The universal set.

6. The "inclusion" relation for sets.

7. An equivalence relation.

8. Domain of a relation.

9. The "divides" relation for natural numbers.

10. A one-to-one correspondence.

11. A system of numeration.

12. The order relation, $\leq$, for the integers.

13. Denseness of the rational numbers.

14. Absolute value of n, for n an integer.

List the properties of each of the following relations (a and b are integers):

15. a Ⓡ b if and only if $|a-b| = k > 0$.

16. a Ⓡ b if and only if $|a-b| = 0$.

17. a Ⓡ b if and only if $(a-b) \geqq 0$.

18. a Ⓡ b if and only if $(a-b) > 0$.

Give a formal definition of each of the following:

19. The system of integers.

20. The system of rational numbers.

21. What does it mean to say that a set is finite?

22. What does it mean to say that a set is infinite?

REFERENCE

"Historically Speaking", The Mathematics Teacher, April 1969.

CHAPTER 7

The System of Real Numbers

7.1 INTRODUCTION TO IRRATIONAL NUMBERS

Some numbers are rational numbers, some are not; for example, $\frac{3}{4}$ is a rational number, $\frac{5}{1}$ is a rational number. Because we also interpret m/n as a division, we may consider 5 as a rational number; 0 is also a rational number; $1, -1, \frac{17}{4}, \frac{2}{3}, -\frac{77}{13}$ are rational numbers.

Numbers that are not rational numbers are called *irrational numbers.* Examples of irrational numbers are $\sqrt{2}$, $\sqrt[3]{5}$, $\sqrt[5]{11}$, $\sqrt{13}$, π, ... It is not uncommon to find people who see $\sqrt{2}$ or $\sqrt{87}$ as some complicated arithmetic operation. (We remind the reader again that the addition algorithm, the multiplication algorithm, and now the square root algorithm are merely arithmetic procedures for determining another name of a number.) We repeat: $2+5$ is the name of a number and we recognize it as the number 7. So also is $3,746,297+894,299$. A student may feel that he has to know *how* to add before he can say that it is *a number*. Still it is a perfectly good number as it is written. In this spirit, $\sqrt{2}$ is a number. *It is that positive number which when multiplied by itself gives the number 2.*

$$\sqrt{2} \cdot \sqrt{2} = 2.$$

This number is not a rational number. It is called an irrational number and occurs quite naturally. There is a well-known theorem in geometry which was known to the Babylonians almost 2000 years B.C. but whose discovery is frequently credited to the Greeks. It is called the Pythagorean Theorem. The theorem states that the sum of the squares on the legs of a right triangle is equal to the square on the hypotenuse (see Chapter 8). If we label the leg of the triangle a and b and the hypotenuse c, then the theorem states that $a^2+b^2=c^2$ (see Figure 1).

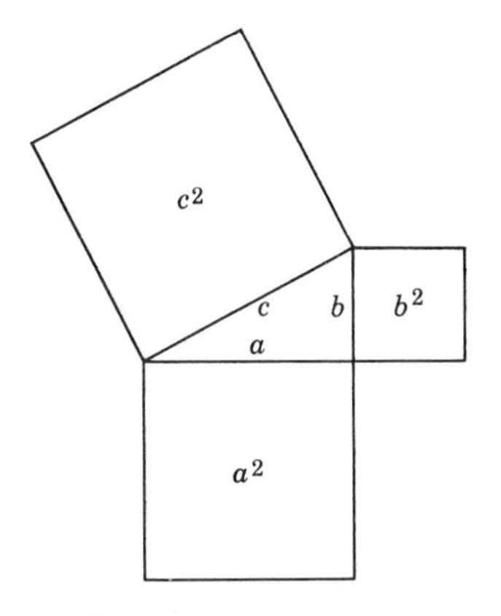

Figure 1. $a^2+b^2=c^2$

Let us consider the triangle formed by the diagonal and two sides of a square whose sides are of unit length. Then we have

$$1^2+1^2=2.$$

This means the hypotenuse c must have a length such that $c^2=2$. The Greeks referred to the hypotenuse of such a triangle as incommensurable. We will see that the set of rational numbers developed for the purpose of measuring was found to be inadequate for the simplest kind of measurement.

Example 1

The side of a square whose area is 2 is $\sqrt{2}$ and $\sqrt{2}$ is an irrational number.

The side of a square whose area is 3 is $\sqrt{3}$ and $\sqrt{3}$ is an irrational number.

The side of a cube whose volume is 2 is $\sqrt[3]{2}$ and $\sqrt[3]{2}$ is an irrational number.

The radius of a circle whose area is 1 is $\sqrt{\pi}/\pi$ and $\sqrt{\pi}/\pi$ is an irrational number.

These simple examples suggest that there are many numbers which are irrational numbers (not rational). There are, in fact, many more irrational numbers than rational numbers. They may not be as familiar as the rational numbers, but such numbers exist. For example, π is an irrational number, but it was a long time before it was shown to be irrational. The number whose symbol is e is familiar to the student of calculus. It is useful as a base for a system of logarithms.

At this point we demonstrate that $\sqrt{2}$ is not a rational number to strengthen the assertion that irrational numbers do, in fact, exist.

7.1a The Irrationality of $\sqrt{2}$

The argument to show that there is no rational number x such that $x^2=2$ is an example of the method of indirect proof. The procedure is as follows. We *suppose* that there is a rational number p/q whose square is 2. By using correct mathematical operations we arrive at a contradictory situation. Since we arrive at this contradiction by mathematically correct steps, the only conclusion left is that our original assumption is false.

Statement to be Proved. There is no rational number p/q such that $(p/q)^2=2$.

Assumption. Suppose the statement is false; that is, suppose there does exist a rational number p/q such that $p^2/q^2 = 2$.

1. We assume that p/q is in reduced form; that is, p and q have no common factor except $+1$ or -1. We may do this without loss of generality. For if $p^2/q^2 = 2$ where p/q is not a rational number in simplest form, then

$$\frac{p}{q} = \frac{dp'}{dq'} = \frac{p'}{q'} \qquad \text{where} \quad \text{g.c.d. } (p', q') = 1.$$

That is, p'/q' is in reduced form.

2. $p^2/q^2 = 2$, so $p^2 = 2q^2$.

3. $p^2 = 2q^2$ implies p^2 is even. But if p^2 is even, so also is p. For if p is odd, p^2 is odd. Hence p is even.

4. p is even so we can write p in the form $p = 2m$, where m is an integer.

5. Hence $p^2 = 2m \cdot 2m = 4m^2$, that is, $p^2 = 4m^2 = 2q^2$.

6. $4m^2 = 2q^2$, so $2m^2 = q^2$.

7. Hence q^2 is even. But this means q is even. So p and q must have the common factor of 2. This cannot happen if their only common factor is $+1$ or -1.

8. Hence our assumption must be false.

9. Therefore $\sqrt{2}$ is irrational.

7.2 THE NUMBER LINE

In Sections 6.11 and 6.13 we discussed the disttribution of the rational numbers on the number line. We noted then that the rational numbers were *dense.* Intuitively this means that no matter which point we choose on the number line, there are infinitely many *rational numbers* arbitrarily close to it. The "point on the number line" of the last sentence may *not* be a rational number. In Section 7.1 we indicated that there are numbers such as $\sqrt{2}$ and π which are not rational. We now show informally that there are points on the number line which correspond to these irrational numbers and that there are rational numbers arbitarily close to them.

Let a square with sides of length 1 have the line segment from 0 to 1 as a base (see Figure 2). Consider the diagonal with one end at 0. The length of this diagonal is $\sqrt{2}$. If we rotate the diagonal clockwise about the point 0 until it lies on the line, the free end of the diagonal marks the point

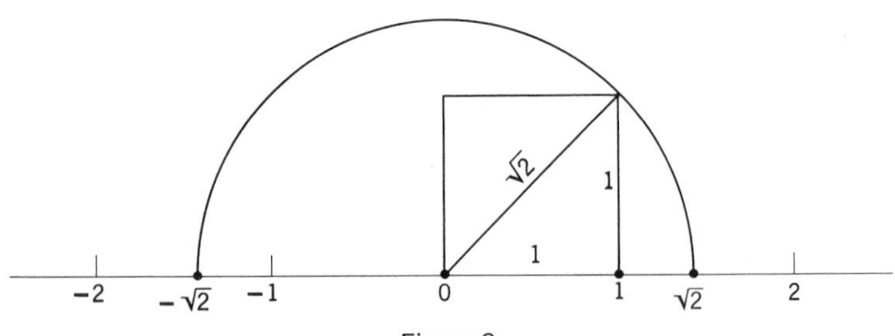

Figure 2

whose distance from 0 is $\sqrt{2}$. We label this point $\sqrt{2}$. If we rotate the diagonal counterclockwise until it lies on the number line, the free end marks the point which we label $-\sqrt{2}$.

The irrational number $\sqrt{2}$ lies between 1 and 2. In fact, $\sqrt{2}$ lies between 1 and $\frac{3}{2}$, since $1^2 = 1$ and $(\frac{3}{2})^2 = \frac{9}{4} = 2\frac{1}{4}$. Any rational number between 1 and $\frac{3}{2}$ will be reasonably close to $\sqrt{2}$, for example, the midpoint of the line segment from 1 to $\frac{3}{2}$. (This midpoint is also called the *arithmetic mean* or *average* of the numbers 1 and $\frac{3}{2}$.)

$$\frac{1+\frac{3}{2}}{2} = \frac{5}{4} \qquad \text{first approximation}$$

$$\left(\frac{5}{4}\right)^2 = \frac{25}{16} < \frac{32}{16} = 2.$$

We see that $\frac{3}{2}$ is greater than $\sqrt{2}$ and $\frac{5}{4}$ is less than $\sqrt{2}$. (These rational numbers which are close to $\sqrt{2}$ will be called *estimates* or *approximations* of $\sqrt{2}$.)

In order to obtain a closer approximation to $\sqrt{2}$, we divide 2 by the first estimate, $\frac{5}{4}$, and take the *average* of this quotient and $\frac{5}{4}$.

$$\frac{2}{\frac{5}{4}} = \frac{8}{5} \qquad \text{quotient}$$

$$\frac{\frac{5}{4}+\frac{8}{5}}{2} = \frac{57}{40} \qquad \text{second approximation}$$

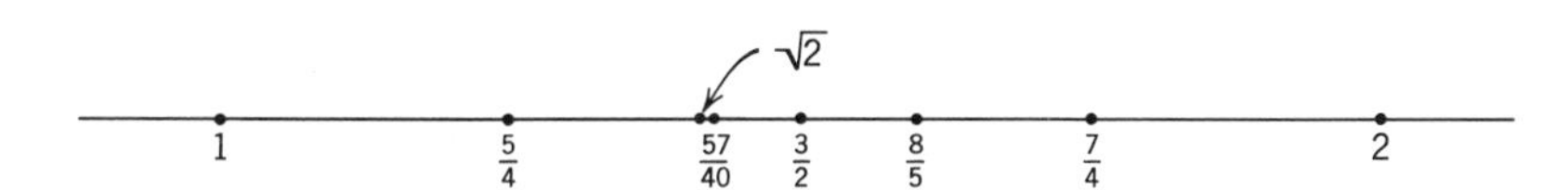

One might wonder why we went through this unexpected procedure to obtain our second approximation. Actually it is a very reasonable procedure. The symbol $\sqrt{2}$ is the name of the number which when multiplied by itself gives 2:

$$\sqrt{2} \cdot \sqrt{2} = 2.$$

We are looking for a number, call it x, such that

$$x \cdot x = 2.$$

In our first attempt, we try $\frac{5}{4}$. We have the following relation to consider:

$$\tfrac{5}{4} \cdot x = 2.$$

If we solve this relation for x, we get $x = \frac{8}{5}$. Since we are using $\frac{5}{4}$ as an estimate of $\sqrt{2}$, it is reasonable to expect $\frac{8}{5}$ also to be an approximation of $\sqrt{2}$. Furthermore, if one of these estimates is too small, the other will be too large. This is the reason we take the *average* of these two "estimates" to obtain $\frac{57}{40}$ as a better approximation. This process of dividing the number by the last approximation and taking the average can be continued indefinitely. In this way we get a sequence of rational numbers approaching $\sqrt{2}$.

Any of the rational numbers $\frac{5}{4}$, $\frac{57}{40}$, . . . can be used as *rational* approximations to $\sqrt{2}$. We saw earlier that there is no rational number whose square is 2, but the fact that the rational numbers are *dense* means that there are rational numbers whose squares differ from 2 by less than any preassigned amount.

Exercise 7.2

1. Divide $\frac{57}{40}$ into 2 and take the average of this quotient and $\frac{57}{40}$ as a new approximation to $\sqrt{2}$. Compare this approximation to the last approximation by looking at the difference of 2 and the squares of the approximation.

$$\left|2-\left(\frac{57}{40}\right)^2\right| = ?$$

$$\left|2-\left(\begin{matrix}\text{third}\\ \text{estimate}\end{matrix}\right)^2\right| = ?$$

2. Find the fourth approximation to $\sqrt{30}$, using 5 as your first approximation. Check your result by squaring.

$$\frac{30}{5} = 6$$

$$\frac{5+6}{2} = \frac{11}{2} \qquad \text{second approximation}$$

3. Find the third approximation to $\sqrt{300}$, using 17 as your first estimate.

4. If a circle of diameter 1 which touches the number line at 0 is rolled along the number line, the point on the circle which was touching 0 will touch the line again at the point corresponding to the irrational number π.

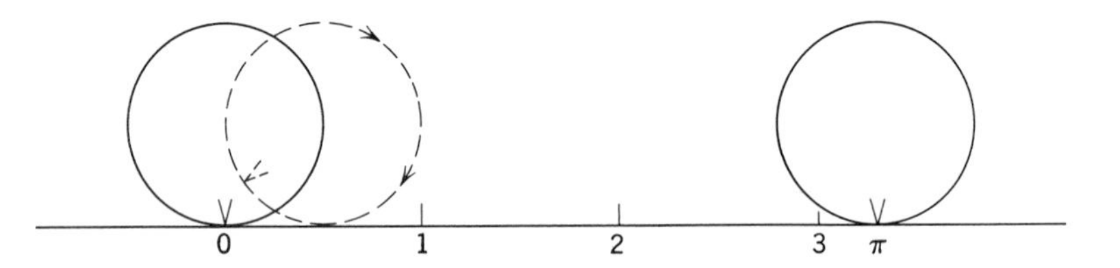

What are some rational numbers close to π?

5. Find the third approximation to $\sqrt{2}$, using $\frac{7}{5}$ as your first estimate. Compare this with the third approximation obtained in problem 1.

6. Find the second approximation to $\sqrt{30}$, using $\frac{27}{5}$ as the first approximation, and compare this with the third approximation obtained in problem 2.

7.2a More Irrational Numbers on the Number Line

We plotted the irrational numbers $-\sqrt{2}$ and $\sqrt{2}$ on the number line by placing the lower left-hand corner of the unit square on the point 0. If we place this corner of the square at the point $\frac{1}{2}$ and again rotate the diagonal, the free end of the diagonal will determine new numbers. This translation of the unit square corresponds to the operation of adding $\frac{1}{2}$ to each of the numbers $-\sqrt{2}$ and $\sqrt{2}$.

The binary operation addition assigns to the pair of numbers $\frac{1}{2}$ and $\sqrt{2}$ a number which we write as $\frac{1}{2}+\sqrt{2}$. Similarly, for the pair $(\frac{1}{2}, -\sqrt{2})$ we write $\frac{1}{2}+(-\sqrt{2})$ or $\frac{1}{2}-\sqrt{2}$. When we add $\frac{1}{2}$ and $\frac{1}{4}$ we write this $\frac{1}{2}+\frac{1}{4}$, but then we find a new name for this number, $\frac{3}{4}$. For $\frac{1}{2}$ and $\sqrt{2}$, the numeral $\frac{1}{2}+\sqrt{2}$ *is* a name of the number. It is not the only name; another is $(1+2\sqrt{2})/2$.

These new numbers $\frac{1}{2}-\sqrt{2}$ and $\frac{1}{2}+\sqrt{2}$ are again irrational numbers (see Figure 3). The following argument can be used to prove this. Suppose $\frac{1}{2}+\sqrt{2}=p/q$, where p/q is a rational number; then $-\frac{1}{2}+\frac{1}{2}+\sqrt{2}=-\frac{1}{2}+p/q$; this is, $\sqrt{2}=-\frac{1}{2}+p/q$. But $-\frac{1}{2}+p/q$ is the sum of two rational numbers. By the closure law for addition of rational numbers we infer $\sqrt{2}$ is rational, but this is not so. Hence $\frac{1}{2}+\sqrt{2}$ cannot be rational.

These statements are true if we use *any* rational number n/m in place of $\frac{1}{2}$; that is, for each rational number n/m, the numbers $n/m-\sqrt{2}$ and $n/m+\sqrt{2}$ are irrational numbers. Thus it is seen that there are as many irrational numbers of the form $n/m+\sqrt{2}$ as there are rational numbers.

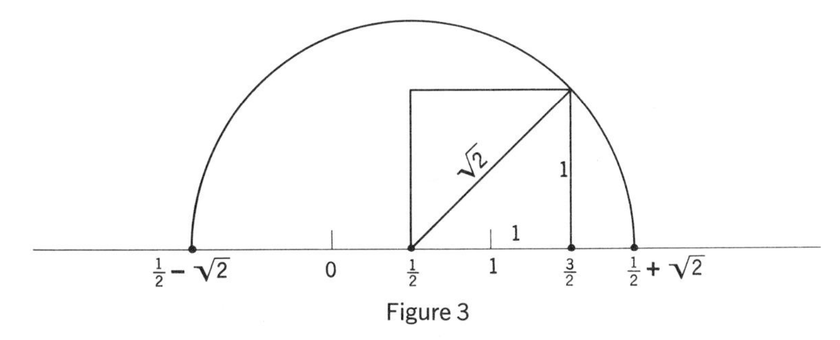

Figure 3

Also, there are as many irrational numbers of the form $n/m-\sqrt{2}$ as there are rational numbers; the same is true for $n/m+\pi$, $n/m-\pi$, $n/m+\sqrt{3}$, and many more. Therefore it is reasonable to believe that there are more irrational numbers than rational numbers. Such is actually the case.

These arguments indicate that the *rational number line* is full of "holes." These "holes" in the line correspond to the irrational numbers. The set of all numbers which correspond to the points on the line is called the *set of real numbers*. Hereafter we shall refer to the line to which the real numbers correspond as the *real line*. The irrational numbers which fill the "holes" in the line make the set of real numbers *complete*. This fact is extremely important in mathematics and will be discussed further in the next section.

Exercise 7.2a

1. Show that $5+\sqrt{2}$ is an irrational number.

2. Show that $-\frac{2}{3}+\sqrt{2}$ is an irrational number.

3. What is the largest *integer* less than 3? Less than 0?

4. Is there a largest rational number less than 5?

5. List several upper bounds of each of the following sets:

(a) $A=\{x|x \leq 17\}$
(b) $B=\{x|x=1-1/n \quad \text{for } n \text{ a positive integer}\}$
(c) $C=\{x|x^2 \leq 2\}$
(d) $D=\{x|x^2 \leq 25\}$

6. State informally the meaning of the statement that the rationals are dense in the real numbers.

7. If we lived in a world in which there were no irrational numbers, it would be possible for a circle to pass through the number line without "cutting" the line.

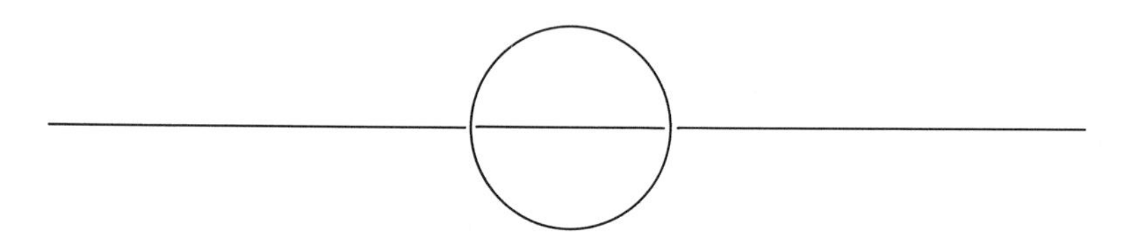

Give an example of such a circle.

7.3 THE SET OF REAL NUMBERS

The set of real numbers was introduced loosely as consisting of the rational and irrational numbers which correspond to the points on the real line. Without proving it, we have indicated that there is a one-to-one correspondence between the points on the real line and the set of real numbers. This is a very useful idea. It is the link between arithmetic and geometry. If we think of the points on the line as being indexed or addressed by the real numbers, it allows us to locate the points very conveniently. The plane can then be thought of as the Cartesian product of the real line with itself. The points in the plane will then be in one-to-one correspondence with the ordered pairs of real numbers (see Figure 4).

Rather than pursue this geometric line of thought we return to the arithmetic of the real numbers.

We note that we have not fully attempted to answer the question, "What is a real number?" Answering this question involves "limiting processes" and operations involving infinite sets. We will discuss this informally in the section on decimal fractions later. For the present we say that the *set of real*

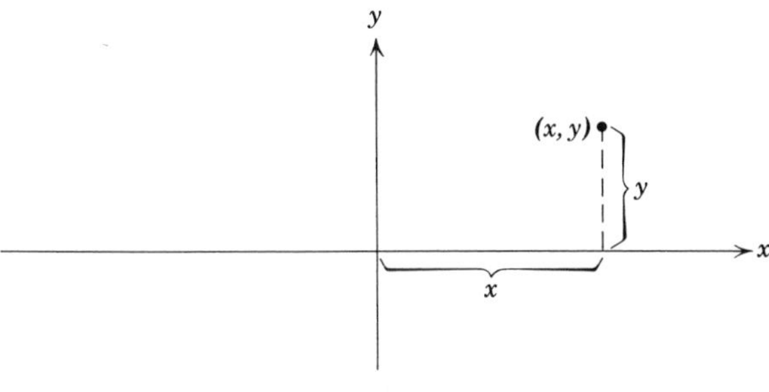

Figure 4

numbers consists of the rational numbers (which include the integers) and the irrational numbers, as introduced earlier.

We indicate briefly a connection between rational numbers and irrational numbers, using $\sqrt{2}$ as an example. Recall that in Section 7.2 we showed how we could find infinitely many rational numbers m/n such that $(m/n)^2$ is close to 2. Among these there are those which we write as p/q such that $(p/q)^2 < 2$ and those which we write as r/s such that $(r/s)^2 > 2$; that is, some are greater than $\sqrt{2}$ and some are less than $\sqrt{2}$. We denote the two sets of rational numbers as follows:

$$A = \left\{\frac{p}{q}\middle| p^2 < 2q^2\right\}$$

$$B = \left\{\frac{r}{s}\middle| r^2 > 2s^2 \text{ and } \frac{r}{s} > 0\right\}.$$

The number $\sqrt{2}$ is an *upper bound* of the set A and a *lower bound* of the set B. It is, in fact, the smallest or *least upper bound* of the set A. It is also the largest or *greatest lower bound* of the set B. Similarly, *every* real number may be thought of as the *least upper bound* of a set of rational numbers as is the case here with $\sqrt{2}$. For instance, 3 can be thought of as the least upper bound of the set of all rational numbers of the form $3 - 1/n$, where n is a positive integer. When $n = 1$, $3 - \frac{1}{1} = 2$; when $n = 2$, $3 - \frac{1}{2} = \frac{5}{2}$; when $n = 3$, $3 - \frac{1}{3} = \frac{8}{3}$; and when $n = 4$, $3 - \frac{1}{4} = \frac{11}{4}$. This set looks like this:

$$\{2, \tfrac{5}{2}, \tfrac{8}{3}, \tfrac{11}{4}, \tfrac{14}{5}, \ldots\}.$$

This is the concept involved in the statement that the set of real numbers is *complete*. We state this more precisely.

> ***Definition 7.3a.*** A set S of real numbers is *bounded* if there is a positive number b such that $|s| \le b$ for all s in S.

> ***Definition 7.3b.*** The statement that the set of real numbers is *complete* means that every nonempty bounded set of real numbers has a least upper bound.

Once we admit irrational numbers, such as $\sqrt{2}$, $\sqrt{30}$, $5 + \sqrt{7}$, $\pi, \ldots$, the following questions arise, "How are such numbers used in arithmetic computation? How are such numbers used to describe the quantitative aspects of our environment?"

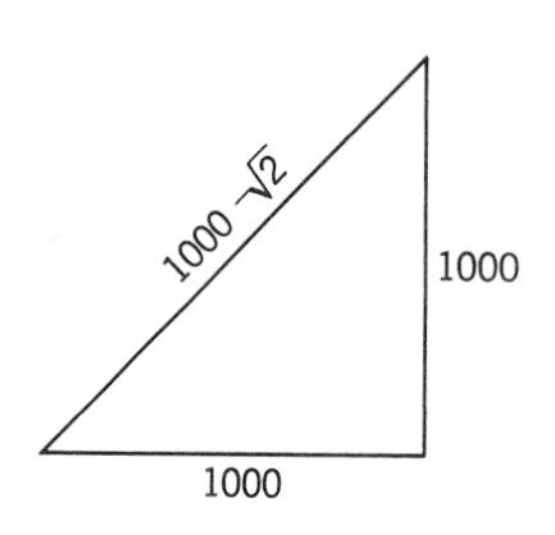

Example 1

How many feet of fencing must be purchased to enclose a field in the shape of a right triangle whose legs are 1000 feet long? We see that

length $= 1000 + 1000 + 1000\sqrt{2}$ feet, or
length $= 2000 + 1000\sqrt{2}$ feet.

Example 2

What is the circumference of a circle whose radius is $\sqrt{2}$? It is $2\pi\sqrt{2}$.

7.4 ORDER RELATIONS IN THE REALS

The order relations $<$ and $\leqslant$ in the real numbers are defined in the same way as in the system of rational numbers and in the system of integers. In each instance we specified *positive* elements and defined the order relations in terms of the positive elements. We do the same now. We indicate that the real number r is positive by the notation $r > 0$ or $0 < r$. We indicate that it is nonnegative by the notation $r \geq 0$ or $0 \leq r$.

How we determine whether a real number is positive or negative depends on the way the real numbers are introduced.

If we think of the real numbers as numbers corresponding to the points on the real line, then the real number is positive if it is associated with a point to the right of the 0 point (the origin). If r and s are any two numbers, $r < s$ or $r \leqq s$ if $s - r > 0$ or $s - r \geqq 0$, respectively.

If we think of the real numbers as infinite decimals, the decimal fraction approximation obtained by neglecting all the decimal places after a particular decimal place is a rational number. A positive real number has a positive decimal approximation. A negative real number has a negative decimal approximation. Thus the positive elements can again be determined and the order relations defined.

7.5 THE SYSTEM OF REAL NUMBERS

There are other ways of defining the real numbers. They can be defined as equivalence classes of Cantor sequences (Cauchy sequences). The real numbers can also be defined as "Dedekind Cuts." Our informal treatment below of the real numbers as infinite decimals on the one hand and as the least upper bounds of sets of rational numbers on the other introduces the student to both of the formal approaches previously indicated in a way which makes the nature and properties of the real numbers more accessible. Each of the number systems previously studied in this book is a subsystem of the system of real numbers. The natural numbers are real numbers. The integers are real numbers. The rational numbers are real numbers. Without formally listing the properties, as was done with the other number systems, we define the *system of real numbers* as a number system which satisfies the same laws as the system of rational numbers and which has an order relation with respect to which the set is *complete.* A mathematical system which satisfies the same laws as the system of rational numbers is called a *field.* In this language, the real numbers are specified as a *complete ordered field.* Rather than pursue the discussion in this direction further, we make a few comments regarding computations with real numbers. Arithmetic at the elementary level is primarily concerned with the arithmetic processes: addition, subtraction, multiplication, division, comparing, and estimating square roots. Arithmetic computations are limited to these operations involving rational numbers except in a few isolated instances, such as $\sqrt{7} \cdot \sqrt{7} = 7$.

Computations using irrational numbers involve infinite processes; hence, in actual numerical calculations, the irrational numbers are replaced by decimal fractions.

Awareness of this situation is important and has been stressed in this book.

7.6 THE REAL NUMBERS AS INFINITE DECIMALS

The decimal 0.333 . . . is called an *infinite decimal.* We indicated that $\frac{1}{3}$ can be expressed as an infinite decimal. On the other hand, decimal fractions of the form 0.5, 0.25, etc., are called finite decimals or terminating decimals. However, if we adjoin zeros to 0.5 in an unending sequence as

0.50000. . . ,

we can think of the decimal fraction representation for $\frac{1}{2}$ as an infinite decimal. We must add a note of caution. The number 1 can be written as an infinite decimal as 1.00000 . . . , but at the same time the infinite decimal 0.99 . . . is also the whole number 1 in the same sense that $\frac{1}{3} = 0.333\ldots$. If we identify 0.999. . . with 1 as well as 3.279999. . . with 3.28, and so on, then we can express any rational number uniquely as an *infinite decimal.* This is also true of any irrational number, although it is not always easy to find the decimal representation of an irrational number. In fact, the *square root algorithm* is simply the process of finding the decimal representation of the square root of a number. We can now define the real numbers.

> ***Definition 7.6.*** The real numbers are the numbers named by the infinite decimals.

The *set of real numbers* was first introduced as those numbers corresponding to points on the number line. We defined the real numbers as named by the infinite decimals. We used both approaches to the real numbers in order to have a broader understanding of the nature of the real numbers.

These two approaches are quite different but in no way inconsistent. The definition of the real numbers in terms of the infinite decimals is a unifying concept not inconsistent with the other approach.

7.6a Infinite Radix Fractions

A change of bases in our system of numeration leads to infinite radix fractions. Following the procedure described in Chapter 6 we find that

$$0.2_{\text{ten}} = 0.001100110011\ldots_{\text{two}}$$

$$0.333\ldots_{\text{ten}} = 0.01010101\ldots_{\text{two}}$$

Any finite binary radix fraction can be expressed as a finite decimal fraction since $\frac{1}{2} = 0.5$ and $(\frac{1}{2})^n = (0.5)^n$. The converse is not true, however, as illustrated above. The finite decimal fraction 0.2 is represented by an infinite binary fraction.

7.7 REPEATING DECIMALS

We were careful to distinguish between rational numbers and irrational numbers, rational numbers being of the form m/n, where m and n are integers and $n \neq 0$. Irrational numbers cannot be expressed in this form. We now ask, "Can these numbers be distinguished when expressed in decimal form?" The answer is "Yes." Some rational numbers are finite or terminating decimals; the other rational numbers turn out to be those infinite decimals which "repeat," that is, a certain block of consecutive digits will be repeated over and over in an unending sequence.

Example 1

0.825 0.75	Finite or terminating decimals.
$0.27272\overline{27}\ldots$ $0.46314631\overline{4631}\ldots$ $29.3785485\overline{4854}\ldots$	Infinite repeating or non-terminating repeating decimals. The bars indicates the repeating "blocks."

That such is the case can be easily verified by using the "division" interpretation on a few rational numbers, such as $\frac{5}{7}$, $\frac{2}{9}$, $\frac{4}{11}$. We are interested more in *why* the rational numbers have representations as infinite repeating decimals.

In examining the division process which yields the decimal representation of a rational, such as $\frac{5}{7}$, we see that we are actually dividing repeatedly by the same number 7.

$$\begin{array}{r} 0.7 \\ 7\overline{)5.0} \\ 4\,9 \\ 1 \end{array} \qquad 50 = 7 \cdot 7 + 1$$

The next step is to divide 7 into 10.

$$\begin{array}{r} 0.71 \\ 7\overline{)5.00} \\ \underline{490} \\ 10 \\ \underline{7} \\ 3 \end{array} \qquad 10 = 7 \cdot 1 + 3$$

The next step is to divide 7 into 30.

$$\begin{array}{r} 0.714 \\ 7\overline{)5.000} \\ \underline{4900} \\ 100 \\ \underline{70} \\ 30 \\ \underline{28} \\ 2 \end{array} \qquad 30 = 7 \cdot 4 + 2$$

Notice that at each step we are dividing 7 into the remainder from the previous division times a power of 10. But, how many different remainders can we have when we divide all possible numbers by 7? We saw in Section 5.8 that the only possible remainders when dividing by 7 are 0, 1, 2, 3, 4, 5, 6. Now let us continue the division:

$$\begin{array}{l} \quad 0.714285\overline{714285} \\ 7\overline{)5.000000000000} \\ \quad \underline{49} \\ \quad\ \ 10 \\ \quad\ \ \ \underline{7} \\ \quad\ \ \ 30 \\ \quad\ \ \ \underline{28} \\ \quad\ \ \ \ \ 20 \\ \quad\ \ \ \ \ \underline{14} \\ \quad\ \ \ \ \ \ \ 60 \\ \quad\ \ \ \ \ \ \ \underline{56} \\ \quad\ \ \ \ \ \ \ \ \ 40 \\ \quad\ \ \ \ \ \ \ \ \ \underline{35} \\ \quad\ \ \ \ \ \ \ \ \ \ \ 50 \end{array}$$

It is quite obvious now that "714285" will repeat endlessly. Can you look at the above division and tell what the decimal expansion of $\frac{1}{7}$, $\frac{2}{7}$, $\frac{3}{7}$, etc., will be?

The reason that rational numbers are infinite repeating decimals is actually embodied in what we have previously called the *division algorithm.* To reiterate, if m and n are any two positive integers, then there are positive integers q and r such that $m = n \cdot q + r$ and $0 \leq r < n$.

Since there are only n distinct integral values that the remainder r can assume, the argument implies that the decimal numeral for m/n will be repeating. This is not to be interpreted as saying that the number of digits in the repeating sequence will be equal to the divisor n. It does say that it can be less than *or* equal to the divisor but never greater.

It can be shown that, conversely, every repeating decimal is the decimal representation of a rational number; that is, given a repeating decimal, we can find a rational number whose repeating decimal is the given one. We indicate how this is done. Let us use the letter N to denote an infinite repeating decimal. Suppose

$$N = 0.273273273\overline{273}\ldots.$$

We multiply by that power of 10 whose exponent is equal to the number of digits in the repeating block and subtract N from this product. Thus

$$\begin{aligned} 10^3N = 1000N &= 273.273273\overline{273}\ldots \\ N = \quad 1N &= \quad 0.273273\overline{273}\ldots \\ \hline 999N &= 273 \\ N &= \frac{273}{999}. \end{aligned}$$

Modifications of this procedure must be made for those infinite repeating decimals which do not start repeating immediately, for example,

$$N = 32.49631631631\overline{631}\ldots.$$

We want the repeating blocks to "match up" so that subtraction will eliminate the decimal fraction. Hence we multiply first by 10^5 and then by 10^2.

$$\begin{aligned} 10^5N = 100{,}000N &= 3{,}249{,}631.631631\overline{631}\ldots \\ 10^2N = \quad 100N &= \quad 3{,}249.631631\overline{631}\ldots \\ \hline 99{,}900N &= 3{,}246{,}382 \\ N &= \frac{3{,}246{,}382}{99{,}900}. \end{aligned}$$

The previous argument indicates that every rational number is a repeating decimal and the foregoing examples suggest that every infinite repeating decimal is a rational number. This leaves but one conclusion. *The irrational numbers are the infinite nonrepeating decimals.* If one reflects about this fact for a moment, it may seem reasonable that there are many more irrationals than rationals.

Exercise 7.7

1. (a) How many possible remainders are there when 11 is the divisor? List them.

(b) Write $\frac{1}{11}$ as an infinite repeating decimal.
(c) Write $\frac{2}{11}$ as an infinite repeating decimal.
(d) Write $\frac{3}{11}$ as an infinite repeating decimal.

2. (a) How many possible remainders are there when 12 is the divisor?
(b) Write $\frac{1}{12}$ as an infinite repeating decimal.
(c) Write $\frac{2}{12}$ as an infinite repeating decimal.
(d) Write $\frac{3}{12}$ as an infinite repeating decimal.

3. (a) How many possible remainders are there when 13 is the divisor?
(b) Write $\frac{1}{13}$ as an infinite repeating decimal.
(c) Write $\frac{2}{13}$ as an infinite repeating decimal.

4. Each of the infinite repeating decimals represents a rational number. Find the rational numbers.

(a) $0.17721772\overline{1772}\ldots$

(b) $0.314314314\overline{314}\ldots$

(c) $0.2935353\overline{535}\ldots$

5. Find a decimal fraction which approximates each of the rational numbers of problem 4 to within an error of one part in 10,000.

6. Find the rational number represented by each of the following infinite repeating decimals:

(a) $4.99999\overline{9}\ldots$

(b) $0.100100100\overline{100}\ldots$

(c) $0.009009009\overline{009}\ldots$

7.8 APPROXIMATIONS

Returning to the definition of a decimal fraction as a rational number whose denominator is a power of 10, we find that very few rational numbers can be written exactly as decimal fractions. The prime factors of 10 are 2 and 5, so that any nontrivial positive power of 10 is a product of the corresponding powers of 2 and 5.

$$10 = 2 \cdot 5$$
$$10^2 = 10 \cdot 10 = (2 \cdot 5)(2 \cdot 5) = 2^2 \cdot 5^2$$
$$10^3 = 10 \cdot 10 \cdot 10 = 2^3 \cdot 5^3$$
$$\vdots$$
$$10^n = 2^n \cdot 5^n$$

This means that in order for a rational number in lowest terms to be written exactly as a decimal fraction, its denominator can include only powers of 2 and/or powers of 5 as factors.

Example 1

1. $\frac{1}{2} = \frac{1}{2} \cdot \frac{5}{5} = \frac{5}{10} = 0.5$

2. $\frac{1}{4} = \frac{1}{4} \cdot \frac{25}{25} = \frac{25}{100} = 0.25$

3. $\frac{1}{5} = \frac{1}{5} \cdot \frac{2}{2} = \frac{2}{10} = 0.2$

4. $\frac{1}{20} = \frac{1}{2^2 \cdot 5} = \frac{1}{2^2} \cdot \frac{5}{5^2} = \frac{5}{(2 \cdot 5)^2} = \frac{5}{10^2} = 0.05$

5. $\frac{1}{50} = \frac{1}{2 \cdot 5^2} = \frac{2}{2^2} \cdot \frac{1}{5^2} = \frac{2}{(2 \cdot 5)^2} = \frac{2}{10^2} = 0.02$

Thus rational numbers of the form $\frac{1}{3}$, $\frac{13}{11}$, . . . which have primes other than 2 and 5 as factors of the denominator cannot be written exactly as decimal fractions. This undoubtedly prevented the predecessors of Simon Stevin from discovering this powerful innovation in arithmetic. The *idea* essential to the discovery of decimal fractions was one of the problems of antiquity. It is embodied in such familiar problems as the frog which at each hop leaps half of the distance from itself to the end of the log. Simon Stevin's inspiration must have come from his realization that most numerical situations do not demand *exactness* as much as *accuracy to within an allowable error*. This is a very important idea. It is essentially the idea of approximations which we used earlier in this section.

We have referred to approximation as an *idea*. It is, in fact, a very important one. It is the idea of using a simpler, more workable, acceptable representation in lieu of something inaccessible, unknown, or inconvenient. As a mathematical idea it is related to the concept of denseness. We found that the rational numbers are densely distributed along the number line. This means that such numbers as $\sqrt{2}$, π, and other irrational numbers which lie on the number line must have many rational numbers very close to them. The rational number $\frac{22}{7}$ is larger than the irrational number π but is close enough to π so that for many problem situations involving circles, the error involved in using $\frac{22}{7}$ instead of π is allowable. We found several rational numbers "close" to $\sqrt{2}$. Some of these may be used instead of $\sqrt{2}$ in some situations.

Exercise 7.8

1. List ten rational numbers that can be written exactly as terminating decimal fractions. Write these numbers in decimal form.

2. List five rational numbers that can be written exactly as terminating decimal fractions. Write them as rational numbers with powers of 10 in the denominator, and in decimal form.

3. A rectangular field is 1276.32 ft long and 789.44 ft wide.

(a) Compute the area of the field rounding given measurements to the nearest foot.
(b) Compute the area to two decimal places, using the numbers as given.
(c) What is the difference in these two results?

4. How would you find the approximate area of a field shaped in the following way?

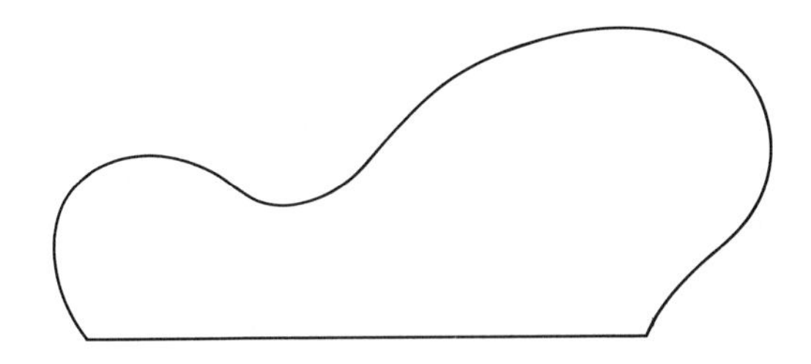

5. How would you obtain a more accurate measure of the area? Could you justify that this would be more accurate?

6. What are some other *rational numbers* which are very close to the number π? What decimal fractions are close to π?

7. What are some rational numbers close to $\sqrt{10}$? Use the divide and average process to find five of them.

8. What are some rational numbers close to $\sqrt{5}$? Use the divide and average process to find five of them.

7.9 DECIMAL APPROXIMATION OF RATIONAL NUMBERS

Section 7.8 showed that we cannot express all rational numbers exactly as decimal fractions. However, we can find decimal fractions which differ from a particular rational number by an amount which we are willing to neglect, the particular amount depending on the problem situation. We find this *decimal approximation* by the simple process of division; that is, we find decimal approximations of $\frac{1}{3}$ by dividing 1 by 3 to obtain 0.3, 0.33, 0.333, and so on.

We have been careful to call the decimal fractions 0.3, 0.33, and 0.333 decimal approximations of $\frac{1}{3}$. The numerals 0.3, 0.33, and 0.333 are convenient numerals which we use instead of $\frac{3}{10}$, $\frac{33}{100}$, and $\frac{333}{1000}$, respectively. From this form it is easy to compute the error involved in using these decimal fractions in place of the rational number $\frac{1}{3}$.

$$\frac{1}{3} - \frac{3}{10} = \frac{1}{30}$$
$$\frac{1}{3} - \frac{33}{100} = \frac{1}{300}$$
$$\frac{1}{3} - \frac{333}{1000} = \frac{1}{3000}$$

These computations indicate that $\frac{1}{3}$ is larger than any of the decimal fractions 0.3, 0.33, 0.333, etc. If we were to continue the division process indefinitely, we would have infinitely many decimal fractions of the form 0.3333, 0.33333, ... Again, $\frac{1}{3}$ is an *upper bound* of the decimal fractions of this type and, indeed, $\frac{1}{3}$ is the *least upper bound* of all such decimal fractions. This is the same idea used in regarding $\sqrt{2}$ as the least upper bound of the rational numbers of the form p/q, where $p^2 < 2q^2$. For this reason we use the convention that the rational number $\frac{1}{3}$ *is the same* as the decimal number 0.333 . . . , where the dots indicate that the 3's continue without end; that is, we write

$$\tfrac{1}{3} = 0.3333\ldots.$$

Any one of the numerals 0.3, 0.33, 0.333, . . . is a decimal approximation involving an error which can be computed. This error can be made smaller by using a decimal fraction with more decimal places. At some point one must decide when the convenience of using decimal fractions outweighs the error or loss of accuracy and what is the maximum error that one is willing to accept.

Exercise 7.9

1. How much would it cost to accept 0.3 of a \$100 gift instead of $\frac{1}{3}$ of \$100?

2. How many places in a decimal approximation of $\frac{1}{3}$ would you use in order to have an error of less than one-millionth?

3. How many places in a decimal approximation of $\frac{1}{7}$ would you use in order to have an error of less than one-millionth?

4. What is the error involved in using a three-place decimal approximation of $\frac{1}{7}$? A four-place decimal approximation?

5. Find a ten-place decimal approximation of $\frac{1}{7}$ and of $\frac{1}{11}$.

7.10 ROUNDING OFF DECIMAL APPROXIMATIONS

Since it is physically impossible to write down an infinite decimal, the decimals after a certain finite number of places are dropped. The part that is retained is a decimal fraction, and the part that is dropped constitutes the error involved in using the approximation:

$$\tfrac{5}{7} = 0.714285\overline{714285}\ldots.$$

$$\tfrac{1}{3} = 0.333\overline{3}\ldots.$$

$$\tfrac{5}{7} = \underbrace{0.7142}_{\text{decimal fraction approximation}} + \underbrace{0.000085\overline{714285}\ldots.}_{\text{error}}$$

$$\tfrac{1}{3} = \underbrace{0.333}_{\text{decimal fraction approximation}} + \underbrace{0.00033333\overline{3}\ldots.}_{\text{error}}$$

The usual purpose of "rounding off" a decimal fraction is to reduce the error involved in the approximation. There are several conventions used in rounding off numbers. The procedure we shall use is to look at the first digit dropped. If this digit is 0, 1, 2, 3, or 4, the last digit in the approximation is retained. If the first digit dropped is 5, 6, 7, 8, or 9, the last digit of the approximation is increased by 1. Thus we would use 0.333 as a three-place decimal approximation of $\frac{1}{3}$ and 0.7143 as a four-place decimal approximation of $\frac{5}{7}$.

Decimal numerals, in general, are often rounded off to a certain number of significant digits. A digit of a numeral naming an approximate number is *significant* unless its only function is to help place the decimal point. Whenever digits to the right of the decimal point are dropped, they must never be replaced by zeros in keeping with the meaning of *significant digits*. All nonzero digits in a number are significant. All zeros between significant digits are significant. A zero following a nonzero digit may or may not be significant.

Example 1

673,924	has six significant digits.
674,000	as a rounding of 673,924 has three significant digits. As an independent numeral it has three or more, possibly six, significant digits.
0.07003	has four significant digits.
0.07	has one significant digit.
1.2370	has five significant digits.
200,001	has six significant digits.

Example 2

The population of Montana in the 1970 census was listed as 680,000. This is to be interpreted as a number of two significant digits. The population at the time of the census might have been any number, such as 682,924, or any other number between 675,000 and 685,000, but because the population is constantly changing as a result of people moving into or out of the state for one reason or another, it is meaningless to list 682,924 exactly. For most purposes the approximate figure 680,000 is accurate enough.

7.11 DECIMAL APPROXIMATIONS OF IRRATIONAL NUMBERS

The full significance of characterizing the rational numbers as terminating decimals or infinite repeating decimals, and irrational numbers as infinite nonrepeating decimals may not have been realized. Finding the decimal representation of a rational number involves only a simple division process, and when the repeating block of digits has been determined, the division process does not have to be continued; that is, it is always possible to determine the exact digit in any decimal place in an infinite repeating decimal once the repeating block has been determined.

Example 1

What digit is in the 105th decimal place of the infinite repeating decimal of the rational number $\frac{1}{7}$? Since there are six digits in the repeating block of digits and since $105 \equiv 3 \pmod 6$, the digit in the 105th place is the same as the digit in the third place of the repeating block.

It is also possible to compute the *exact* error involved in using a decimal fraction as an approximation to a rational number. Such is not the case when decimal fractions are used to estimate irrational numbers. Without actually computing the decimal, there is no known way of predicting the digit in the fifth decimal place of π, or the digit in the seventh decimal place of $\sqrt{3}$, or the digit in any decimal place of any irrational number. This means we cannot compute the *exact* error involved in using decimal fractions in place of irrational numbers. It is possible to give *bounds* on the error. We can do this using only the properties of the place-value system and what we mean by rounding off a number. We illustrate with some examples.

Example 2

What does it mean to say 1.4 is a one-place decimal approximation of $\sqrt{2}$? This means that $\sqrt{2}$ is a number which is between 1.35 and 1.45.

$1.35 \leq \sqrt{2} \leq 1.45.$

The error involved in using 1.4 as an approximation to $\sqrt{2}$ is less in absolute value than 0.05; that is, 0.05 is a *bound* on the error. By extracting the square root of 2 to two decimal places we get 1.41. The error in using 1.41 as an approximation to $\sqrt{2}$ is less in absolute value than 0.005. Similarly, the error in using 3.1416 in place of π is less in absolute value than 0.00005 or five parts in one hundred thousand.

Exercise 7.11

1. In what sense is 3.14 an approximation of π? Is 3.142 an approximation of π? Is 3 an approximation of π?

2. Which is the better approximation of π, 3.142 or $\frac{22}{7}$?

3. What is the error in using 0.6666667 for $\frac{2}{3}$?

7.12 SQUARE ROOTS

Before considering the problem of finding decimal approximations of the square roots of numbers (square root algorithms), we shall discuss more fully the meaning of the square root of a number. It was carefully pointed out earlier in this chapter that the symbol "$\sqrt{2}$" is a numeral, that is, a name for that number which when multiplied by itself gives the number 2. It was also pointed out that many people do not distinguish $\sqrt{2}$ from some complicated arithmetic operation. Indeed, many people do not think of $\sqrt{2}$ as a number at all. Unfortunately, there are many teachers in the elementary and secondary schools who think of $\sqrt{2}$ not as a number but as something to do. This is partly because of the existence of certain natural numbers which are perfect squares. The perfect squares are the squares of the natural numbers.

$$1, 4, 9, 16, 25, 36, 49, \ldots$$

These numbers are part of the real number system, so there is a natural answer to the question, "What number multiplied by itself is 4, or 9, or 16, etc.?" This is indicated by

$$\sqrt{4} = 2$$
$$\sqrt{9} = 3$$
$$\vdots$$
$$\sqrt{144} = 12$$
$$\sqrt{625} = 25$$
$$\vdots$$
$$\sqrt{2401} = ?$$
$$\vdots$$
$$\sqrt{82369} = ?$$

When the number N^2 becomes unfamiliarly large, such as 2401 or 82,369, the question naturally arises whether these numbers are perfect squares, and if they are how does one find their square roots? One way, of course, is to write down the squares of the natural numbers to see if these numbers occur among the squares. That there is actually an algorithm which can be used to answer the question is part of the reason for the confusion between the number and the algorithm.

The square root of a number a is actually a solution of the equation

$$x^2 = a.$$

We noted repeatedly that the solvability of equations, in general, depends on the number system in which the coefficients lie, that is, if the letter "a" denotes a perfect square and we require the solution to be a *natural number*, there is only *one* solution. If we allow *integer* solutions, there are *two* solutions; for example, if

$$x^2 = 25,$$

then

$$x = 5 \qquad \text{or} \qquad x = -5.$$

If we allow a to be any positive rational number and require *rational* solutions, we may have no solution or two solutions, depending on the nature of a; that is, *if* a is the square of a rational number, we will have two solutions. If a is not the square of a rational number, we will have no rational solutions. The equation

$$x^2 = \tfrac{25}{64}$$

has two rational solutions,

$$x = \tfrac{5}{8} \qquad \text{and} \qquad x = -\tfrac{5}{8},$$

whereas

$$x^2 = 2$$

has no rational solutions, since there is no rational number whose square is the number 2.

If we allow a to be a nonnegative real number and allow the solutions of $x^2 = a$ to be real numbers, there will be exactly two solutions when $a \neq 0$. We indicate the two square roots of the number a by the symbols $\sqrt{a}$ and $-\sqrt{a}$. The square roots of 3 will be written $\sqrt{3}$ and $-\sqrt{3}$. The $\sqrt{3}$ is the positive number which when multiplied by itself gives the number 3, whereas $-\sqrt{3}$ is the additive inverse of $\sqrt{3}$ and is the negative number which when multiplied by itself gives 3. In order to avoid any confusion about the square roots of those numbers written as squares, we use the absolute value to define $\sqrt{a^2}$.

Definition 7.12. For every real number a, $\sqrt{a^2} = |a|$ and $-\sqrt{a^2} = -|a|$.

Example 1

$$\sqrt{(-3)^2} = |-3| = 3.$$

There is one situation that we have not yet discussed and that is the case when a is allowed to be a negative number in the equation $x^2 = a$ that is, do the equations

$$x^2 = -1,$$
$$x^2 = -5,$$
$$\vdots$$

have solutions? These equations do not have solutions which are *real* numbers. In the same way that we introduced the negative integers as solutions of the equation

$$a + x = b,$$

and the rational numbers as solutions of the equation

$$ax = b,$$

and the irrational numbers as solutions of the equation

$$x^2 = a \qquad \text{when} \qquad a > 0,$$

we now introduce the number i as the solution of the equation

$$x^2 + 1 = 0$$

or

$$x^2 = -1.$$

Hence

$$i^2 = -1.$$

It follows that

$$i^3 = i^2 \cdot i = -i,$$

and

$$i^4 = i^2 \cdot i^2 = 1.$$

In general, numbers of the form $2+3i$, $1+i$, $7i$, $\frac{1}{2}+(\sqrt{2/3})i$, $\pi - i$ are called *complex numbers*. Numbers of the form $7i$, $2i$, $\sqrt{3}i$ are called *pure imaginary* numbers. The terms *complex* and *pure imaginary* are used here simply as the names of sets of numbers. The reader is cautioned about attaching any literal meaning to these terms. The system of complex numbers is a number system which is very much a part of the scientific world, but will not be discussed further in this book.

Exercise 7.12

1. Is $3+\sqrt{5}$ a number?

2. What are the sum, difference, and product of $3-\sqrt{5}$ and $3+\sqrt{5}$?

3. What are the sum, difference, and product of $\sqrt{a}+\sqrt{b}$ and $\sqrt{a}-\sqrt{b}$?

4. $\sqrt{10}$ is approximately 3.162. Give a decimal approximation for $1/\sqrt{10}$.

5. $\sqrt{2}$ is approximately 1.4142. Give a decimal approximation for $1/\sqrt{2}$.

6. Give a decimal approximation for $1/(2+\sqrt{2})$.

7. Give a decimal approximation for $1/(10-\sqrt{2})$.

8. Give a decimal approximation for $\sqrt{10/10}$, $\sqrt{2/2}$, $(2+\sqrt{2})/2$, $(\sqrt{10}+\sqrt{2})/8$.

9. Give a decimal approximation for $2/\sqrt{2}$, $10/\sqrt{10}$, $(2+\sqrt{2})/(2-\sqrt{2})$.

10. Express the following, using the number i (for example, $\sqrt{-5} = i\sqrt{5}$, since $(i\sqrt{5})^2 = i^2(\sqrt{5})^2 = -5$.

$\sqrt{-4}$, $\sqrt{-9}$, $\sqrt{-7}$, $\sqrt{-1}$, $\sqrt{-50}+\sqrt{-32}$.

7.12a The Square Root Algorithm

A square root algorithm is an arithmetic process of finding or approximating the square root of a number. We review a standard process.

Example 1

Find $\sqrt{82369}$.

Step 1. Mark off the digits in pairs to the left, starting from the decimal point. Each pair determines a *place* in the square root.

$$\begin{array}{r} * \;\; * \;\; * \,. \\ \sqrt{8'23'69.} \end{array}$$

Step 2. Find the largest whole number whose square is less than or equal to the number in the left-hand pair or single digit:

$$\begin{array}{r|l} & 2 \;\; * \;\; * \,. \\ \sqrt{} & 8'23'69. \\ & 4 \\ \hline & 4 \end{array}$$

$2^2 = 4 < 8 < 9 = 3^2.$

Write 2 above this "pair." Square 2 and subtract from 8. Bring down the two digits of the next pair.

Step 3. Double the number 2 in the root and place an asterisk as indicated to the right of the 4.

$$\begin{array}{r|l} & 2 \;\; * \;\; * \,. \\ \sqrt{} & 8'23'69. \\ & 4 \\ \hline 4(*) & 4\ 23 \end{array}$$

Step 4. Estimate how many times 40 divides 423. We try 9. Replace the asterisk with 9 and multiply by 9. 9 is too large. We try 8. Replace the asterisk with 8 and multiply by 8. Place 8 above the pair brought down.

$$\begin{array}{r|l} & 2 \;\; 9 \;\; * \,. \\ \sqrt{} & 8'23'69. \\ & 4 \\ \hline 4(9) & 4\ 23 \\ 9 & 4\ 41 \\ \hline \end{array}$$

Step 5. Repeat steps 3 and 4.

$$\begin{array}{r|l} & 2 \;\; 8 \;\; 7 \,. \\ \sqrt{} & 8'23'69. \\ & 4 \\ \hline 4(8) & 4\ 23 \\ 8 & 3\ 84 \\ \hline 56(7) & 39\ 69 \\ 7 & 39\ 69 \\ \hline \end{array}$$

$\sqrt{82369} = 287.$

We now consider what takes place at each step. Recall that multiplying a number by 100 moves the decimal point *two* places to the right. But multiplying a number by 100 only multiplies the square root of the number by 10. This follows from the fact that the square root of a product is equal to the product of the square roots.

$$\sqrt{M \cdot N} = \sqrt{M} \cdot \sqrt{N}.$$

$$\sqrt{36} = \sqrt{4 \cdot 9} = \sqrt{4} \cdot \sqrt{9} = 2 \cdot 3 = 6.$$

$$\sqrt{100 \cdot N} = \sqrt{100} \cdot \sqrt{N} = 10\sqrt{N}.$$

This is the reason that the digits in the number are marked off in pairs. The number of pairs determines the number of digits in the square root.

At each step of the algorithm we are interested in finding the largest whole number whose square is less than or equal to the number determined by those pairs directly concerned. In step 2, for instance, we estimate the largest whole number whose square is less than or equal to 8. In the next step we seek the largest whole number whose square is less than or equal to 823. At this point we are working with two pairs, so the square root of 823 is a two-digit number of which the first digit is 2. We can write this as $20+x$. We want to estimate the largest integer x so that the square of the number $20+x$ is less than or equal to 823; that is, we would like the largest x which satisfies

$$(20+x)^2 \leq 823.$$

Squaring $(20+x)$, we get

$$400+2 \cdot 20x+x^2 \leq 823.$$

Subtracting 400 from both sides of this inequality, we see that

$$2 \cdot 20x+x^2 \leq 423.$$

This step corresponds to subtracting 4 from 8 and bringing down the next pair of digits (note this step in the example).

$$(2 \cdot 20+x)x \leq 423.$$

$$(40+x)x \leq 423.$$

This accounts for doubling the digit in the root and leaving a place for a digit as indicated by the * in the example. We divided 423 by 40 to get an estimate of x. This procedure is based on the observation that 40 is much larger than the digit x so that dividing by 40 is almost like dividing by $40+x$. This is the reason that we often overestimate x. We saw in the example that 9 was too large so 8 is the number we sought.

The next step is essentially a repetition of the previous step. We want the largest whole number whose square is less than or equal to 82,369. This is a three-digit number whose first two digits are known. That is, we seek the largest whole number x such that

$$(280+x)^2 \leq 82{,}369$$

$$78{,}400+2 \cdot 280x+x^2 \leq 82{,}369.$$

$$560x+x^2 \leq 3969$$

$$(560+x)x \leq 3969$$

$$x = 7$$

$$280+7 = 287$$

We extracted the square root of the square of a natural number. The algorithm is valid for any positive real number and the process may be continued beyond the decimal point. We must add zeros to make as many pairs to the right of the decimal point as we want decimal places in the square root. We illustrate with the following:

Example 2

Find $\sqrt{3.4}$ to three decimal places.

$$
\begin{array}{r|l}
 & \ \ 1.\ 8\ \ 4\ \ 3 \\
\sqrt{} & 3.40'00'00 \\
 & 1 \\
\hline
28 & 2\ 40 \\
8 & 2\ 24 \\
\hline
364 & 16\ 00 \\
4 & 14\ 56 \\
\hline
3683 & 1\ 44\ 00 \\
3 & 1\ 10\ 49 \\
\hline
\end{array}
$$

If we carry the work out one place further, we would find that $\sqrt{3.4}$ is closer to 1.844 than to 1.843. This extra step is less troublesome than worrying about "least absolute remainders." We point out instead that we can greatly improve the approximation by a simple division process which is discussed in the next section.

Exercise 7.12a

1. Find $\sqrt{502.3}$ rounded to 3 decimal places.
2. Find $\sqrt{50.23}$ rounded to 3 decimal places.
3. Find $\sqrt{1000.}$ rounded to 4 decimal places.
4. Find $\sqrt{5.023}$ rounded to 4 decimal places.
5. Find $\sqrt{0.5023}$ rounded to 4 decimal places.

7.12b Newton's Method of Approximating Square Roots

The method of approximating the square root of a number by the "divide and average" process discussed earlier is actually a very special case of *Newton's method of approximation.* (Sir Isaac Newton, 1642–1727, an Englishman, was one of the great scientists of all time. Newton made many contributions to mathematics and physics, many of which still bear his name. At about the same time as Gottfried Wilhelm Leibnitz, 1646–1716, but independently, he originated the calculus.)

Newton's method is based on the calculus and is a powerful method for approximating roots of very general equations. In the particular case of approximating square roots, the method reduces to a very simple *iterative* or *repetitive* process. The repetitive nature of the process, "divide and take the average, divide and take the average" makes this method well suited to our modern high-speed electronic computers. Another important feature of Newton's method is the fact that the first "guess" or approximation does not have to be very accurate, that is, making a "poor guess" the first time does not always mean the work has to be abandoned and the process begun all over again. It may mean that another one or two repetitions of the process will be needed to yield an accurate approximation. We illustrate the method by finding a decimal approximation to $\sqrt{300}$. We label our first "guess" A_1 and successive approximations $A_2, A_3, \ldots$.

Example 1

Find $\sqrt{300}$

Step 1. Since $17^2 = 289 < 300 < 324 = 18^2$, we choose 17 as a reasonable first guess.

$A_1 = 17$ first approximation

Step 2. Divide the number N by A_1, that is, 300 by 17, and carry out the division to twice as many digits as in the previous approximation and round off the last digit to an even number. To get the second approximation A_2, take the average of this quotient and the previous approximation.

$$\frac{300}{17} = 17.64.$$

Rounding to the nearest even digit we get 17.64 to twice as many digits as in A_1.

$$A_2 = \frac{17.64 + 17.00}{2} = 17.32 \qquad \text{second approximation}$$

Notice that A_2 *verifies* our first guess as being a good guess because when we round off 17.32 to a two-digit number we get $A_1 = 17$.

Repeating the process, we have

$$A_3 = \frac{\frac{300}{17.32} + 17.320000}{2} = \frac{17.321016 + 17.320000}{2}$$
$$= 17.320508 \qquad \text{third approximation}$$

The fact that A_3 verifies A_2 implies A_2 is accurate to two decimal places. A_3 is accurate to within 1 in the last place.

To show that a poor guess yields the same decimal fraction after a few more repetitions of the process, suppose we let

$$A_1 = 10$$

$$A_2 = \frac{\frac{300}{10} + 10}{2} = 20.$$

Notice that A_2 does not verify A_1 because A_2 rounded to two digits is not equal to A_1. We carry out the division to one less than twice as many digits in the previous approximation when the verification fails.

$$A_3 = \frac{\frac{300}{20} + 20.0}{2} = 17.5$$

$$A_4 = \frac{\frac{300}{17.5} + 17.5000}{2} = \frac{17.142 + 17.500}{2} = 17.321$$

$$A_5 = 17.320508.$$

Newton's method can be used to approximate cube roots as well as square roots.

The various corrective measures in those cases where the approximation fails to verify the previous approximation were presented as "rules to follow." The reasons for the validity of these measures as well as the method itself have their roots in the calculus. The method was presented here as another way of estimating square roots, a method adaptable to electronic computers.

Exercise 7.12b

1. Find $\sqrt{3}$ rounded to 7 decimal places.

2. Find $\sqrt{30}$ rounded to 7 decimal places.

3. Find $\sqrt{3000}$ rounded to 6 decimal places.

4. Find $\sqrt{5}$ rounded to 6 decimal places.

5. Find $\sqrt{50}$ rounded to 7 decimal places.

6. Find $\sqrt{500}$ rounded to 6 decimal places.

7. Find $\sqrt{3.4}$ rounded to twice as many digits as in Example 2 of the previous section.

REVIEW EXERCISES

1. With 4 as the first estimate of $\sqrt{15}$, use the divide and average method with rational numbers to find the third estimate.

2. Show that $5 + \sqrt{2}$ is an irrational number.

3. Consider the set $\{x | x = 1 - 1/n,\ n \text{ a positive integer}\}$.

(a) List the first five elements of this set.
(b) Give an upper bound for this set.
(c) What is the least upper bound for this set?

4. Write 3.1416 in expanded form.

5. How many places in the decimal approximation for the rational number $\frac{1}{3}$ would you need in order to have an error less than one one-millionth?

6. (a) Express $\frac{2}{11}$ as an infinite decimal.

(b) Express $0.17\overline{17}\ldots$ as a rational number.

7. Find $\sqrt{5.023}$ rounded to two decimal places.

8. If $a = 1.83$ and $b = 0.5$, find

(a) $a+b$ (b) $a-b$
(c) $a \cdot b$ (d) a/b

9. If $\sqrt{2}$ is approximately 1.414, then

(a) $1/\sqrt{2} = ?$ Give the answer as a three-place decimal fraction.
(b) $1/(1+\sqrt{2}) = ?$

10. (a) Infinite repeating decimals are called —.
(b) Infinite nonrepeating decimals are called —.

CHAPTER 8

Topics from Geometry

8.1 INTRODUCTION

Until now we have been concerned with numbers and their properties. However, numbers are not the only things in mathematics that should interest the elementary school teacher. Points, lines, planes, and space are also discussed in current mathematics programs for the elementary school.

It is our objective in this chapter to present *some* of the basic concepts of geometry without the formality of an axiomatic development. In many instances we shall rely on the conventional interpretation of terms rather than formalize definitions.

8.2 POINTS, LINES, AND SPACE

We are inclined to associate physical things with the ideas of geometry. It is true, however, that abstractions form the basis for the study of geometry. It is impossible to make a precise definition of "point" in the geometric sense. It is also impossible to find a precise physical example. Certainly we use the word "point" in a sense comparable to the abstract geometric concept, such as pen point, pencil point, and point of a pin or needle. But these are not precisely the interpretation we wish to give the word in a geometric sense. Current texts for the elementary school use the notion of "fixed location in space" as descriptive, but emphasize that the concept of geometric point is *undefined*. Similarly, the concept of *space* is undefined but may be thought of as the *set of all points*. Geometric space and geometric points are, like numbers, abstractions existing in the mind. The concept of *line* is also undefined, but we may think of it as a set of points and hence, a subset of space.

Just as with numbers, these abstractions may be symbolized, but, as with numbers, we must bear in mind that we are symbolizing for communication purposes and that the symbol is not the concept itself. We symbolize points with pencil dots and chalk dots, space with our everyday three-dimensional environment, and draw "paths" on paper or on the blackboard to symbolize lines. It is well to remember that these are but representative symbols for communication purposes and not to be confused with the abstract concepts.

8.3 NOTATION FOR BASIC CONCEPTS

In writing of these abstract concepts it is customary to use capital letters to represent points in space. We may refer to a particular point as "the point A," or "the point B." We may generalize, as we do in algebra, and say, "Consider any point C in space."

If we have two distinct points A and B in space we often consider the set of points that could be *represented* by an endless, tightly drawn, fine cord through the points A and B. This set of points is symbolized $\overleftrightarrow{AB}$, where the doubleheaded arrow is used to indicate that a line is endless (see Figure 1).

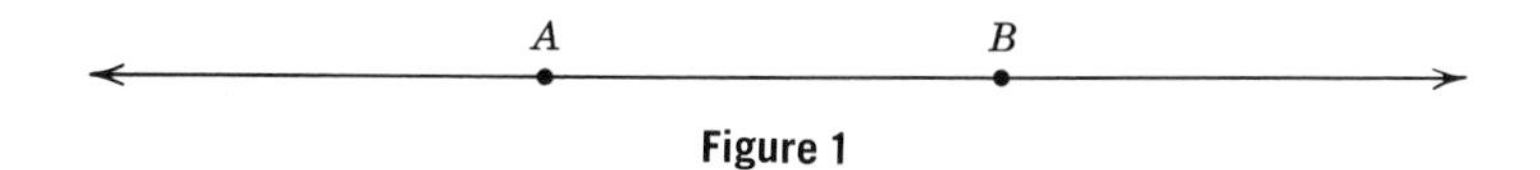

Figure 1

Here we are assuming "straight" line. Other "paths" through A and B are referred to as *curves*. The *straight line* (or simply *line*) is a special element of the set of all curves through the given points, that is, the straight line is regarded as a special kind of curve.

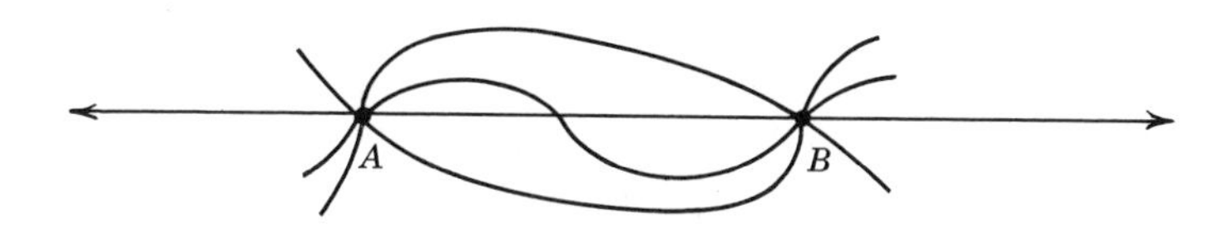

There are infinitely many lines through a single point A, but given two distinct points A and B, there is only one line through *both A and B*. We say that *two distinct points in space determine a line.*

When it is not necessary to associate lines with particular points, lowercase letters are used to symbolize them, such as "the line l" or "the line m."

A point of a line (we use the word "of" because a point is an element *of* the set of points comprising the line) separates the line into three disjoint subsets, the singleton set consisting of the point itself, and the two disjoint subsets called *half-lines*. Each of these half-lines together with the point of separation is called a *ray* (see Figure 2).

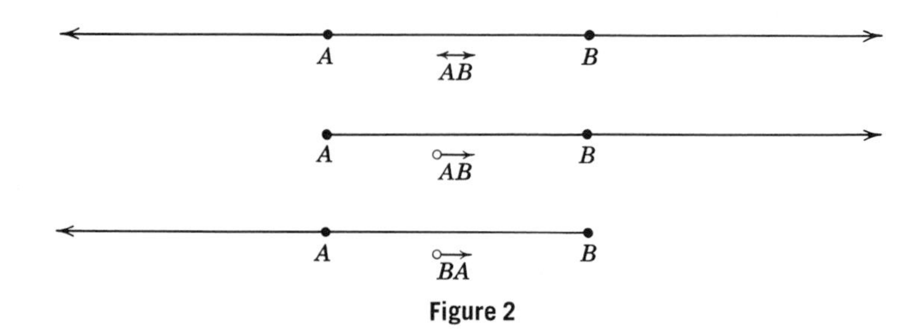

Figure 2

A ray through distinct points A and B with *endpoint* A is symbolized $\overset{\circ\!\rightarrow}{AB}$. From the line of Figure 2 we can obtain the rays $\overset{\circ\!\rightarrow}{AB}$ and $\overset{\circ\!\rightarrow}{BA}$. Note that the ray $\overset{\circ\!\rightarrow}{AB}$ is not the same as the ray $\overset{\circ\!\rightarrow}{BA}$. Since lines and rays are sets of points we can use the familiar set notation and write:

$$\overset{\circ\!\rightarrow}{AB} \cup \overset{\circ\!\rightarrow}{BA} = \overleftrightarrow{AB}.$$

The intersection of these two rays is also a familiar and interesting set that we refer to as a *line segment*. We symbolize this $\overset{\circ\!-\!\circ}{AB}$. We have

$$\overset{\circ\!\rightarrow}{AB} \cap \overset{\circ\!\rightarrow}{BA} = \overset{\circ\!-\!\circ}{AB}.$$

The points A and B are called the *endpoints* of the segment. We say a point C is *between* points A and B if C is distinct from A and B and $C \in \overset{\circ\!-\!\circ}{AB}$.

8.3a Open and Closed Segments

It is interesting at this point to reconsider the real number line. Without the real numbers labeling the points, the line is just the *geometric* line but with the numbers attached we say that the line is *coordinatized.*

If we consider a portion of the line coordinatized with the real numbers we can speak of special kinds of line segments.

The line segment $\overset{\circ\!-\!\circ}{AB}$ above is said to be *closed.* This is just another way of saying that the line segment includes the endpoints. We use the convention of using the dots to indicate this. For example, we might consider the segment from 3 to 7 on the number line which includes the point labeled 3 and the point labeled 7. This segment is a closed line segment.

Now let us delete the two endpoints. The line segment without the endpoints is called an *open* line segment. We use the convention of putting parentheses in place of the points to indicate that the endpoints are missing. *Note that the line segment now has no point at either end.*

To understand this phenomenon, remember that the real numbers are dense on the number line. This means that between any two distinct real numbers we can find a third number. If, after we delete the point labeled 7, there were a "next" point on the end of the line segment, it would have a real number different from 7 associated with it. Now, using the denseness of the reals, we can find a distinct number between it and 7. There would be a point associated with it which would lie between the so-called "next" point and the point labeled by the number 7. This argument gives credibility to the idea of the open line segment.

A ray, as discussed above, is a closed half line. We can also speak of an open half line as meaning the ray without the endpoint.

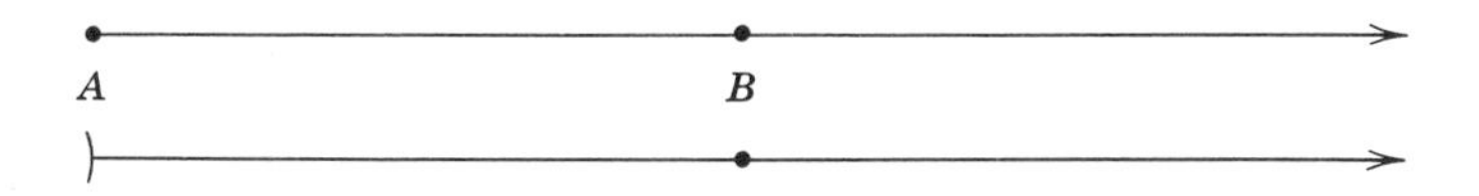

Exercise 8.3a

1. Plot the solution sets of each of the following compound open sentences where the variable x denotes a real number.

(a) $0 < x < 1$ (b) $0 \leq x \leq 1$
(c) $|x| < 2$ (d) $|x| \leq 2$

2. In problem 1, indicate which of the solution sets constitutes an open line segment and which are closed line segments.

3. On the coordinatized line, let the points A, B, and C be associated with the points labeled 7, 10, and 12, respectively:

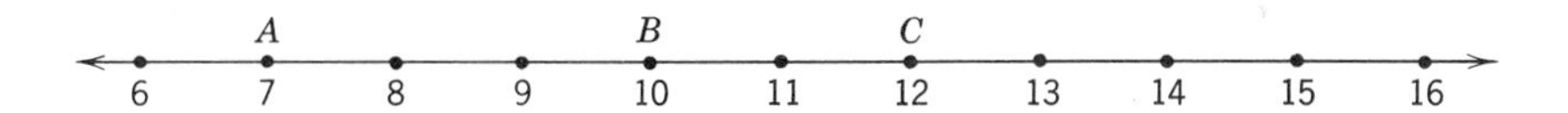

Use open compound statements to describe the following:

(a) $\overset{\circ\!-\!\circ}{AB}$ (b) $\overset{\circ\!-\!\circ}{AC}$
(c) $\overset{\circ\!-\!\circ}{AB} \cup \overset{\circ\!-\!\circ}{BC}$ (d) $\overset{\circ\!-\!\circ}{AB} \cap \overset{\circ\!-\!\circ}{BC}$

8.4 LENGTHS OF LINE SEGMENTS

The length of a straight-line segment has been discussed indirectly in Section 5.12. There we spoke of "distance" between two points on the number line where the integers were used to label the points. The *distance* from a point labeled "a" to a point labeled "b" is defined to be $|b-a|$. This can be defined as the *length of the line segment* from the point labeled "a" to the point labeled "b." Note the distinction between "a point labeled a" and "the point A." In the first instance "a" represents a real number and in the second "A" is a name for a point. The length of the straight-line segment from point A to point B is usually symbolized

$m(\overset{\circ\!-\!\circ}{AB})$. If, in the correspondence of real numbers to points on the line $\overleftrightarrow{AB}$, point A is associated with the real number a, and point B is associated with the real number b, then the distance from A to B, or the length of the line segment $\overset{\circ\!-\!\circ}{AB}$ is

$$|b-a| = m(\overset{\circ\!-\!\circ}{AB}).$$

$m(\overset{\circ\!-\!\circ}{AB})$, then, is a real number and properties of the real number system apply to such measures. Moreover, the properties of distance, as listed in Section 5.12c, apply. It should be noted that the length of a closed line segment is the same as the length of an open line segment.

8.5 PLANES AND HALF-PLANES

As with points, lines, and space, the concept of geometric *plane* is undefined. It is a particular set of points, hence a subset of space, and can be characterized by thinking of it in relation to a flat table top extending indefinitely in all directions. Any flat surface, such as a wall of a room, a floor, or a flat sheet of cardboard, is suggestive of a plane. A geometric plane can best be characterized by properties. Given any two distinct points in space there are infinitely many planes that contain these two points. Given any three, distinct, noncollinear (not on the same line) points in space, there is one and only one plane that contains all three points. We say *any three distinct noncollinear points determine a plane.* If two distinct points of a line are in a plane, then all points of the line are in the plane. Since two intersecting lines have at least three noncollinear distinct points as subsets, we can say that two distinct intersecting lines determine a unique plane.

Two distinct lines in a plane either intersect (have exactly one point in common) or they are parallel (have no points in common).

If we have a line $\overleftrightarrow{BC}$ that is a subset of a plane π it partitions the set of points of π into three disjoint subsets, the line $\overleftrightarrow{BC}$ and two open *half-planes*, denoted by π_1 and π_2. We have

$$\overleftrightarrow{BC} \cup \pi_1 \cup \pi_2 = \pi.$$

If there is a point A such that $A \in \pi_1$, $A \notin \overleftrightarrow{BC}$, we speak of π_1, in relation to $\overleftrightarrow{BC}$ and π, as the A-side of $\overleftrightarrow{BC}$ (see Figure 3).

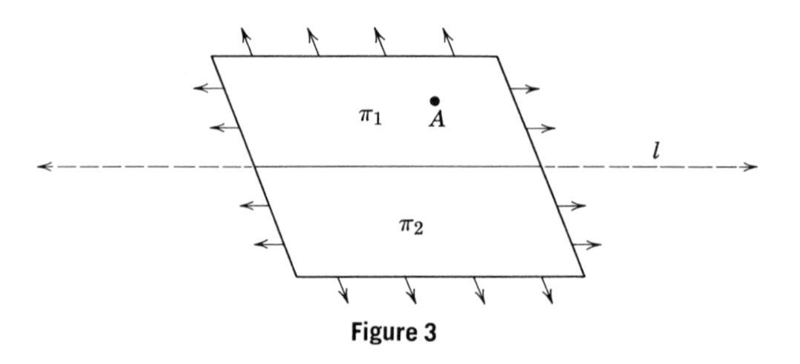

Figure 3

8.6 ANGLES AND THEIR MEASURE

An *angle* can be defined as the union of two rays with a common endpoint (see Figure 4). We symbolize this $\angle BAC$. The common endpoint A is called the *vertex* of the angle, and when three letters are used in the symbol for an angle, the letter associated with the vertex occurs between the other two. The order of the other two is immaterial. If no ambiguity results, the single letter associated with the vertex is used to designate the angle.

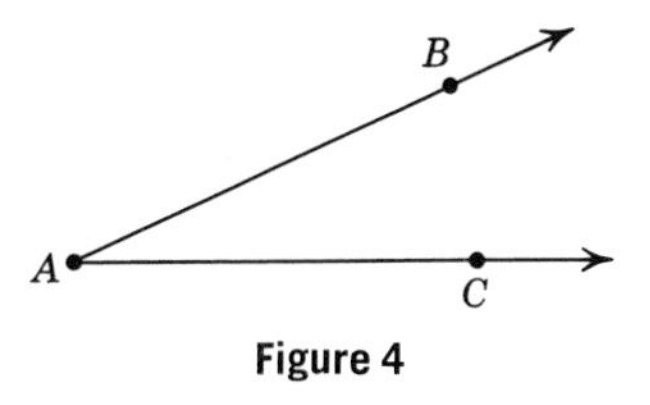

Figure 4

$$\angle BAC = \angle CAB = \angle A.$$

An angle is a set of points made up of the union of two distinct rays with a common endpoint.

$$\angle BAC = \overset{\circ\!\!\rightarrow}{AB} \cup \overset{\circ\!\!\rightarrow}{AC}.$$

If rays $\overset{\circ\!\!\rightarrow}{AB}$ and $\overset{\circ\!\!\rightarrow}{AC}$ are such that $\overset{\circ\!\!\rightarrow}{AB} \cup \overset{\circ\!\!\rightarrow}{AC} = \overleftrightarrow{CB}$ (see Figure 5) then the angle thus formed is called a *straight angle*.

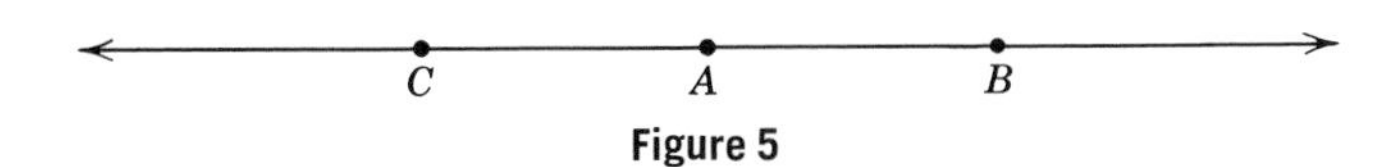

Figure 5

The intersection of the C-side of $\overleftrightarrow{AB}$ and the B-side of $\overleftrightarrow{AC}$ is called the *interior* of the angle (see Figure 6). The set consisting of points in the plane that are not in $\angle BAC$ and not in its interior is referred to as the *exterior* of $\angle BAC$.

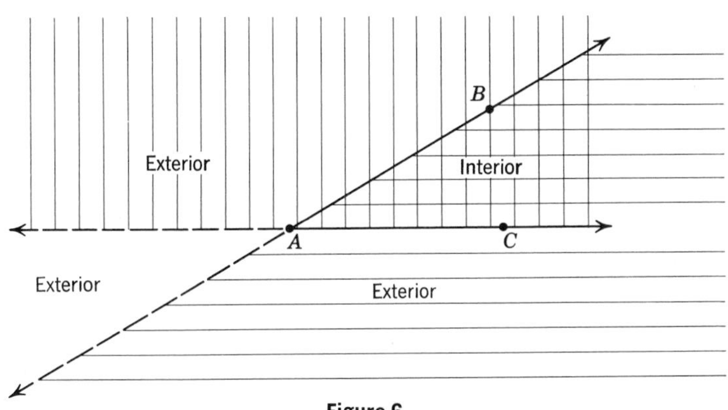

Figure 6

Associated with angles, as with line segments, is a one-to-one correspondence between angles and real numbers, or more commonly, a subset of the real numbers. The usual measure associated with angles is the *degree* measure. The instrument used to establish this correspondence is called a *protractor*. The unit of measure is called the *degree* and is a real number between 0 and 180. We symbolize this as, for example,

$$m\angle BAC = 45.$$

Notice that we do *not* say, "The measure of the angle is 45 degrees." However, it is common practice to say, "Angle BAC is a 45-degree angle." We will not dwell on the difference.

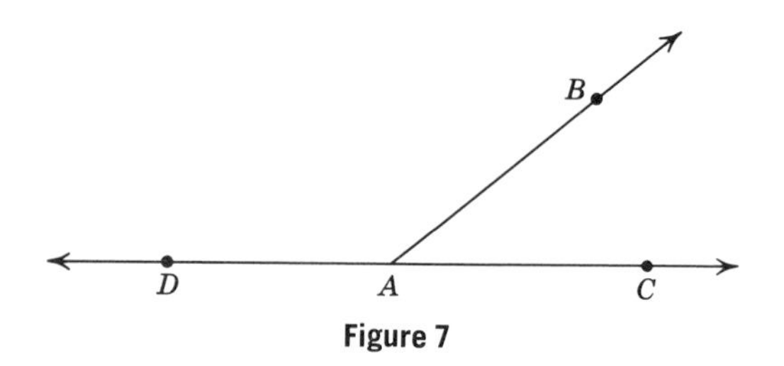

Figure 7

An angle formed by two identical rays is said to have measure 0 or to be a 0° angle (zero-degree angle). A straight angle has measure 180 or is said to be a 180° angle. Two angles, $\angle BAC$ and $\angle BAD$, where A, B, C, and D are distinct points and where A is *between* D and C, are said to be *supplementary* (see Figure 7). If two supplementary angles have the same measure, they are called *right angles* and each has measure 90. If $m\angle BAC <$ 90, $\angle BAC$ is called an *acute angle*. If $90 < m\angle BAC < 180$, $\angle BAC$ is called an *obtuse angle*.

8.7 PLANE FIGURES

A *closed curve* in a plane can be thought of as a "path" which, if one follows with a trace of some sort, leads back to the point of origin (see Figure 8).

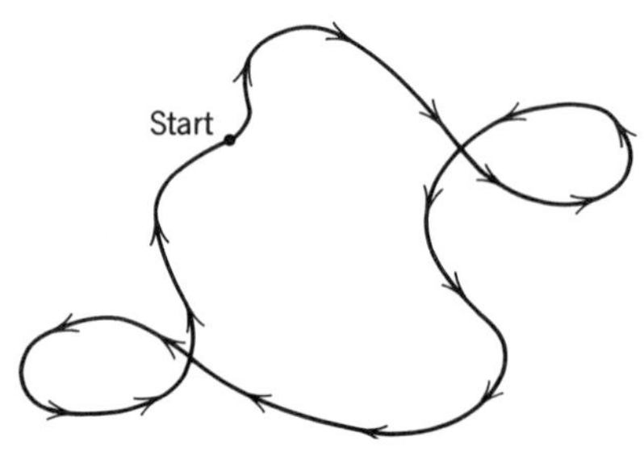

Figure 8

A *simple closed curve* in a plane is a closed curve which does not "intersect itself" (see Figure 9).

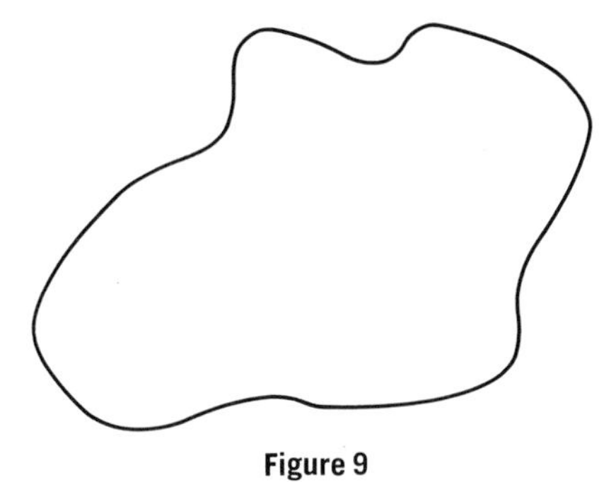

Figure 9

A circle is a simple closed curve. The portion of the plane bounded by a circle is called a *disc*. (An abuse of language should be noted when we speak of the area of a circle. We really mean the area of a circular region or the disc.)

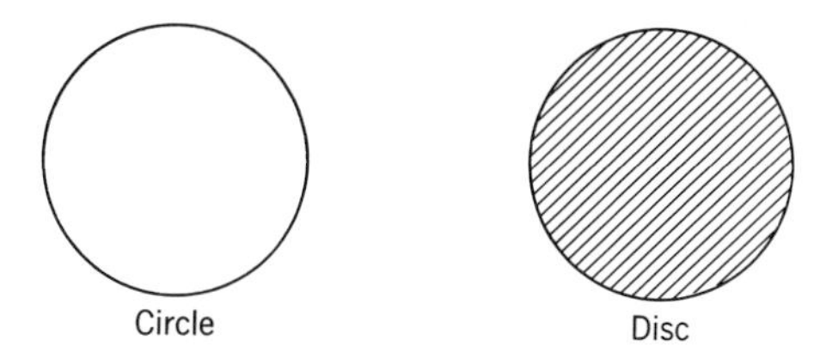

The disc consists of the circle and the portion of the plane bounded by the circle. It is also called the *closed disc*. The term "closed" is used in two distinct ways. On the one hand, a *closed* curve is a curve which comes back to itself. On the other hand, a closed disc is a disc which includes the circle which bounds it. The expression "open curve" is not used nor does it have any specific meaning. On the other hand, "open disc" is commonly used and refers to the disc with the bounding circle deleted.

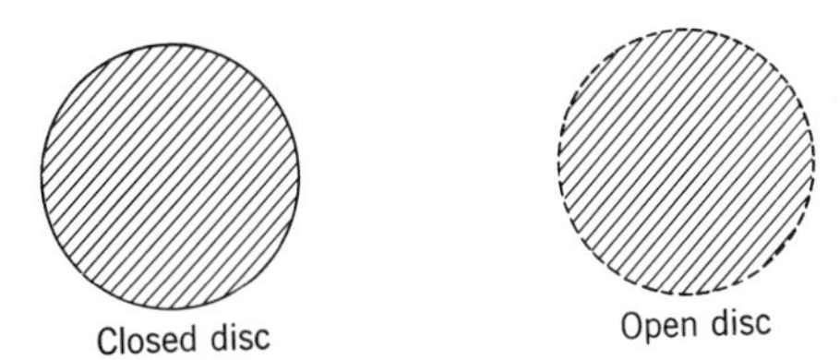

The open disc has a most interesting property. For any point p in the open disc, we can always find another open disc which contains the point p and which is itself contained in the open disc.

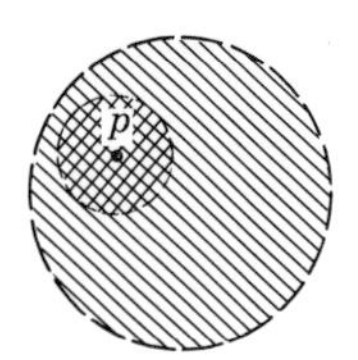

Students in middle school find these ideas extremely interesting.

The open line segment discussed in Section 8.3a has a similar property. That is, for any point p in the open interval, we can always find another open interval which contains the point p and is itself contained in the open interval.

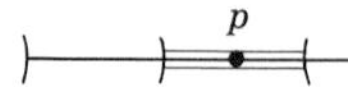

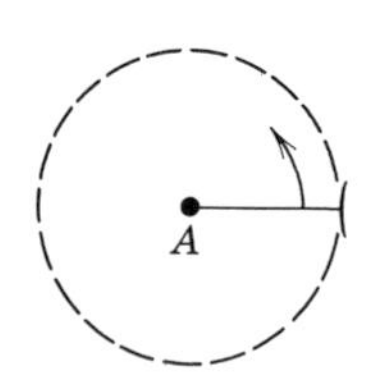

Another way to think of the open disc is to consider a closed line segment $\overset{\circ\!-\!\circ}{AB}$. Now delete the point B and rotate the segment about the point A. The region "swept out" is the open disc.

Several plane figures are made up of the union of line segments. A *triangle* may be defined as the union of three line segments which, by pairs, have a common endpoint but are otherwise distinct. For example, consider three noncollinear points A, B, and C. Then the triangle associated with these points and symbolized $\triangle ABC$ is

$$\triangle ABC = \overset{\circ\!-\!\circ}{AB} \cup \overset{\circ\!-\!\circ}{AC} \cup \overset{\circ\!-\!\circ}{BC},$$

where $\overset{\circ\!-\!\circ}{AB}$, $\overset{\circ\!-\!\circ}{AC}$, and $\overset{\circ\!-\!\circ}{BC}$ are called the *sides* of the triangle. Associated with a triangle are three "interior" angles. The angles are not subsets of the triangle since angles are formed of *rays* and not line segments; therefore we say the angles are "associated" with the triangle (see Figure 10).

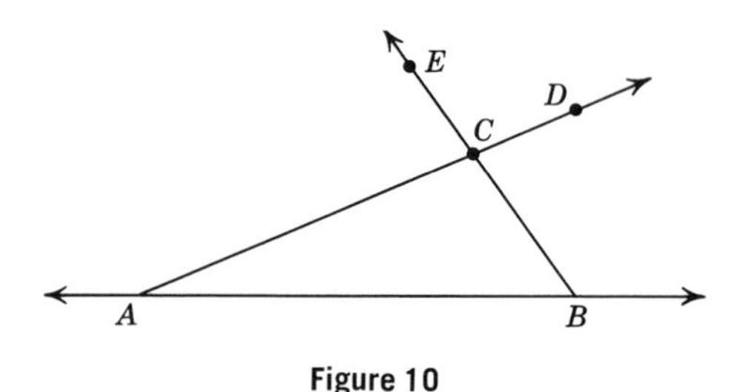

Figure 10

The portion of the plane consisting of the triangle and the part bounded by the triangle is called a *closed* triangular region. If the triangle is deleted, the portion remaining is an *open* triangular region. The open triangular region has the same interesting property as the open disc.

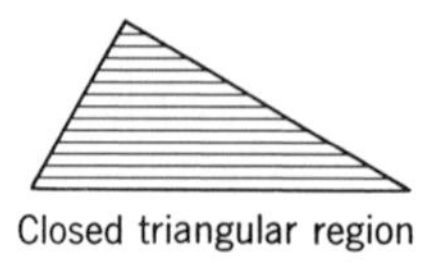
Closed triangular region

Open triangular region

A triangle is said to be a *right triangle* if one of its associated angles is a right angle. (Although incorrect, the expression, "one of its angles" is often used.) If two of its sides are equal in measure, the triangle is called *isosceles* and it can be shown that the angles opposite the equal sides must also be equal in measure. If all three sides are equal in measure, the triangle is said to be *equilateral.* An equilateral triangle is also *equiangular.*

Other plane figures may be defined in a similar way. Once we are familiar with operations on sets and definitions of rays, line segments, etc., as sets of points, we find these definitions very straightforward and easy to understand. Rather than give this further treatment we choose to turn to other geometry-related topics of interest.

Exercise 8.7

1. How many different lines may contain:

(a) one specific point?
(b) a specific pair of distinct points?

2. How many different planes may contain:

(a) one specific point?
(b) a specific pair of distinct points?
(c) a set of three distinct noncollinear points?

3. (a) If two distinct lines intersect, how many points are there in the intersection?
(b) If two distinct planes intersect, how many lines are there in their intersection? How many points?

4. Mark three points A, B, and C in that order from left to right on the same line. Indicate $\overset{\circ\!\!\rightarrow}{AB}$, $\overset{\circ\!\!\rightarrow}{BA}$, $\overset{\circ\!\!\rightarrow}{BC}$, $\overset{\circ\!\!\rightarrow}{CA}$, $\overset{\circ\!\!\rightarrow}{CB}$. Which of these are names for the same ray?

5. Mark a point P on a sheet of paper. Draw three distinct rays with endpoint P. How many rays are there with endpoint P?

6. Draw a ray $\overset{\circ\!\!\rightarrow}{AB}$. How many rays are there with endpoint A which contain point B? Is $\overset{\circ\!\!\rightarrow}{AB}$ contained in $\overset{\circ\!\!\rightarrow}{AB}$? How many line segments are there which are subsets of $\overset{\circ\!\!\rightarrow}{AB}$ and which have A as endpoint?

7. Label two distinct points on a sheet of paper P and Q.

(a) Draw a ray with endpoint P.
(b) Draw a ray with endpoint Q.
(c) Draw the ray $\overset{\circ\!\!\rightarrow}{PQ}$.
(d) Draw a ray $\overset{\circ\!\!\rightarrow}{QP}$.
(e) Make a heavier line to indicate the segment $\overset{\circ\!-\!\circ}{PQ}$.

8. How many endpoints has

(a) a line?

(b) a ray?
(c) a closed line segment?
(d) an open line segment?

9. How many lines can be drawn through four points, a pair of them at a time, if the points lie:

(a) in the same plane? Draw a figure to illustrate.
(b) not in the same plane? Draw a figure to illustrate.

10. Consider

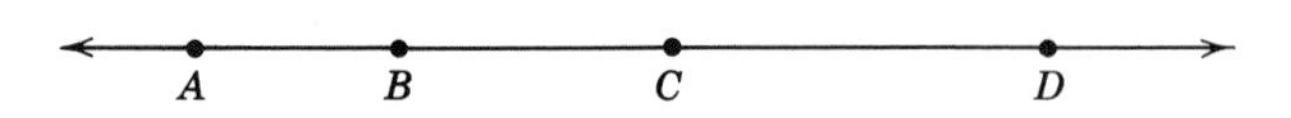

(a) $\overset{\circ\!-\!\circ}{AB} \cup \overset{\circ\!\rightarrow}{BC} = ?$
(b) $\overset{\circ\!-\!\circ}{AB} \cup \overset{\circ\!\rightarrow}{CB} = ?$
(c) $\overset{\circ\!-\!\circ}{AC} \cup \overset{\circ\!-\!\circ}{CD} = ?$
(d) $\overset{\circ\!\rightarrow}{BC} \cup \overset{\circ\!\rightarrow}{BA} = ?$

11. Draw examples of each of the following:

(a) curves that are neither closed nor simple.
(b) curves that are closed but not simple.
(c) curves that are both closed and simple.

12. Name the angle, the vertex, and sides of each of the following angles:

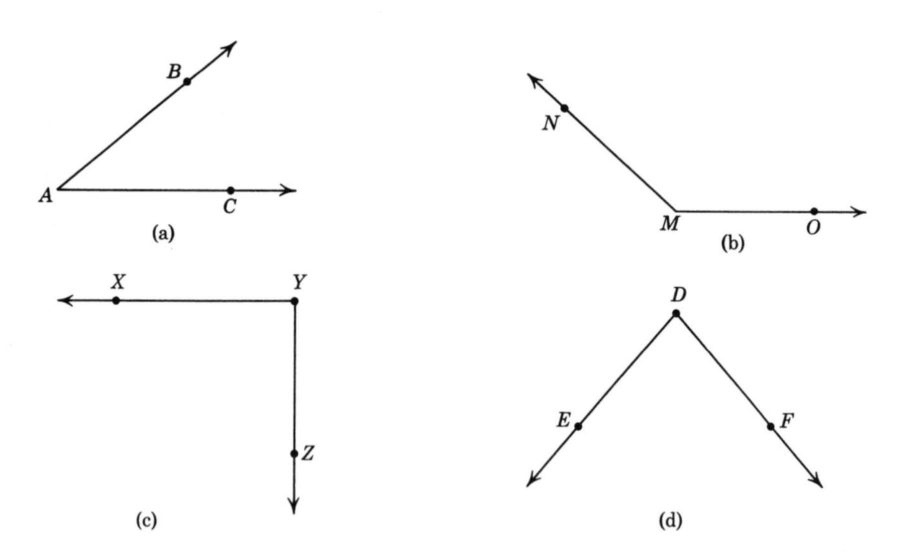

13. Consider the accompanying figure:

(a) Is the point C in the interior of $\angle BAD$?
(b) Is $\overset{\circ\!\rightarrow}{AC} \cap \{A\}'$ in the interior of $\angle BAD$?

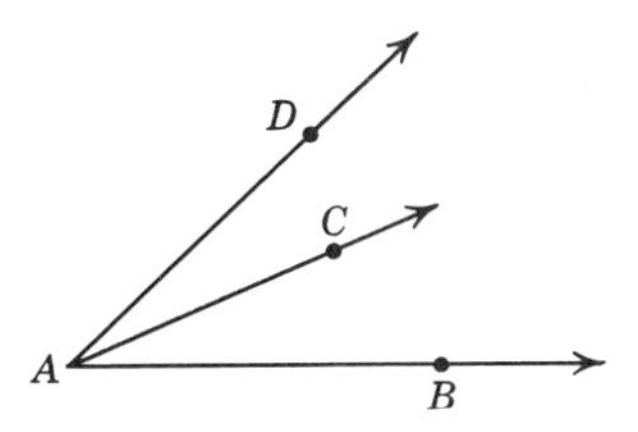

14. Consider the accompanying figure and answer the questions in problem 13.

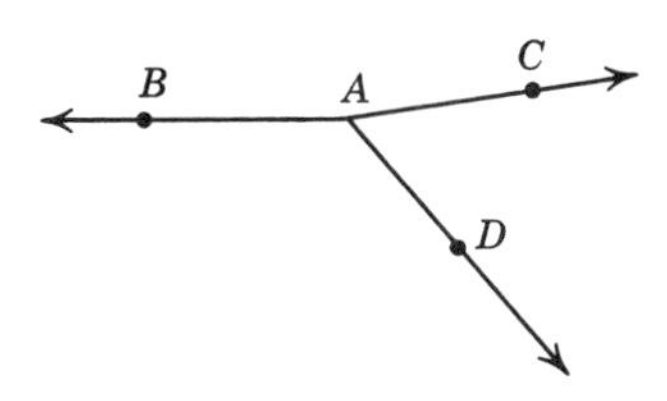

15. An angle is a 37 degree angle.
  (a) The unit of measure is —.
  (b) The measure of the angle is —.
  (c) The angle is referred to as a — angle.

8.7a Convexity

Some plane figures are convex and others are not. The following figures are convex:

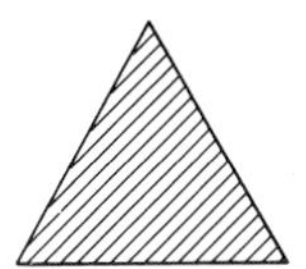
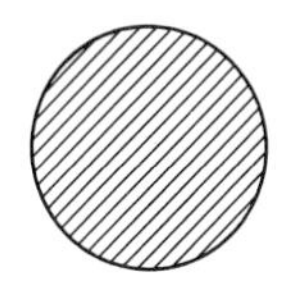
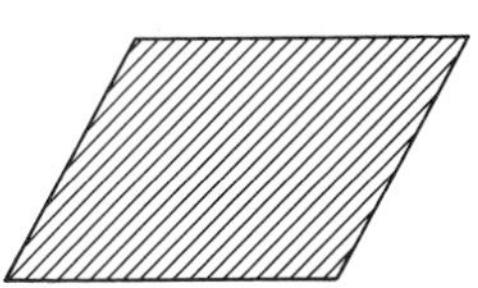
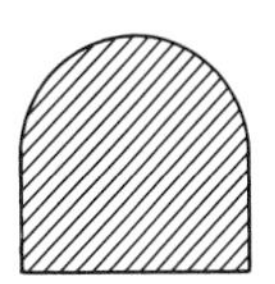

The following figures are not convex:

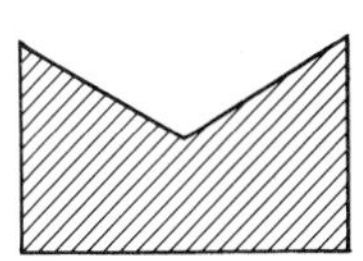
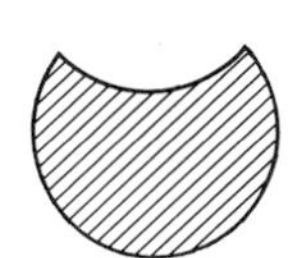
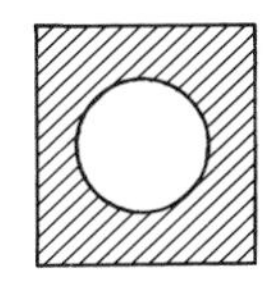

Comparing with the figures shown above, classify each of the following figures as convex or not convex.

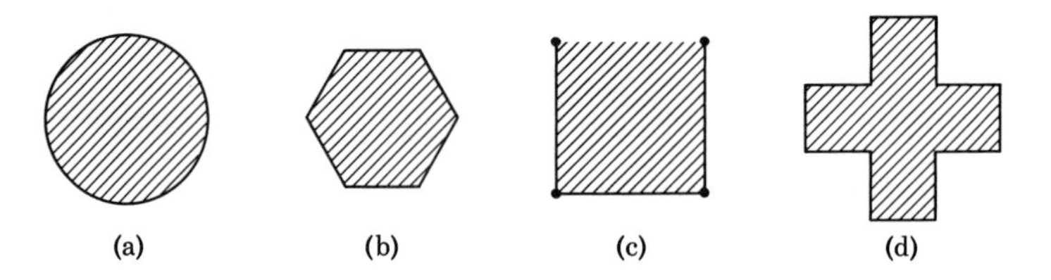

Figures (a) and (b) are convex and (d) is not.

Definition 8.7a. A plane figure is convex if for any two points in the figure the line segment joining the two points also lies in the figure.

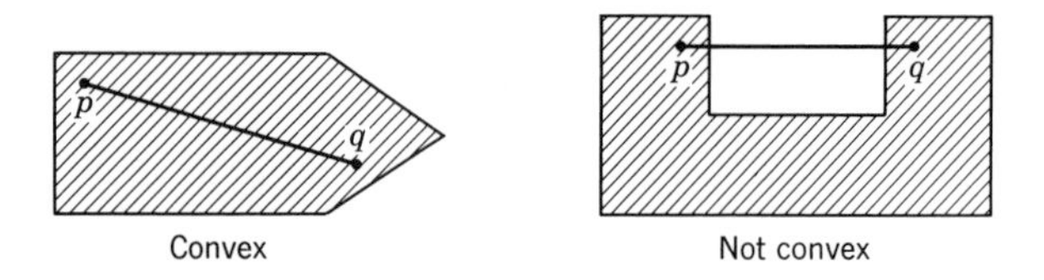

The figure labeled (c) above is not convex.

Another way of describing a convex figure is as follows: Every point in the set can "see" every other point.

Just as open sets and closed sets are used in higher mathematics, so also is the concept of convexity. Here is a simple way it arises in application.

Take a piece of string and tie it into a loop. Place the loop on a flat surface.

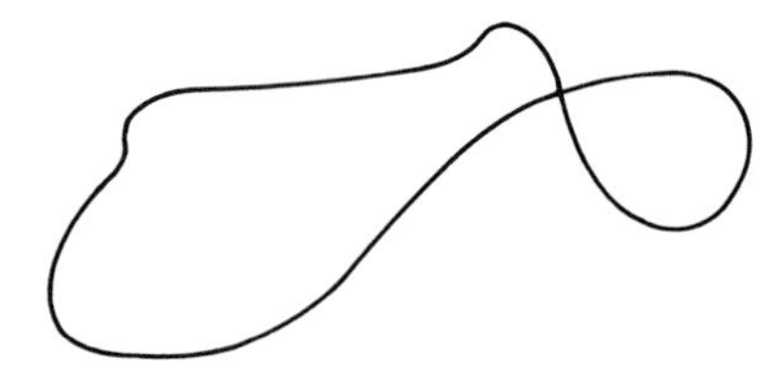

Rearrange the string so that it encloses the region having the largest area. You may recall that this is the situation which confronted Dido, the legendary lady who founded the city of Carthage. She received from a king as much land as could be encompassed by an ox hide. She cut the hide into very thin strips and made a long cord to encompass enough land to build the city of Carthage. Mathematically, this is a well-known problem called the *isoperimetric* problem. A simple argument shows that the region bounded by the cord must be *convex*. Assume the string is laid out so that the region enclosed is not convex. Suppose the string is laid out as shown below.

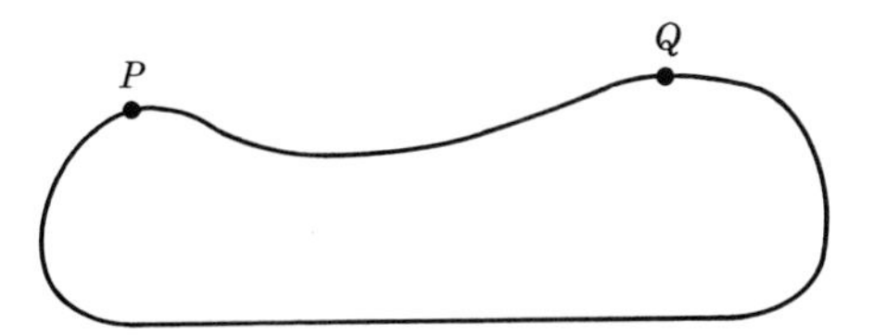

Since the region is not convex, there are points P and Q such that the line segment joining them is not in the region.

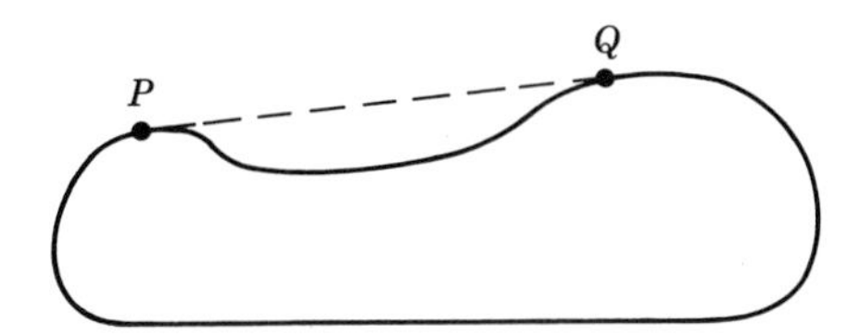

Using the idea of symmetry, flip the string below the line to the other side of the line as shown below.

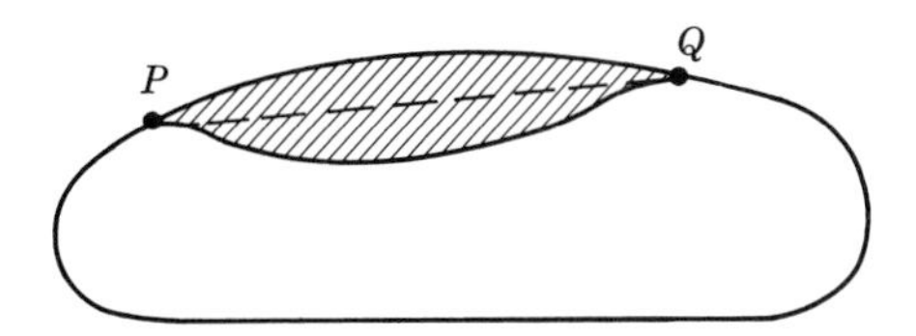

The shaded region increases the area bounded by the same length of string. Continuing in this way wherever the region is not convex has the effect of "blowing out" the string to bound the maximum area. Other arguments show, in fact, that the region should be circular.

Special problem: Use the definition of convex sets and the intersection of sets to show that the intersection of convex sets is convex.

8.8 THE LENGTH OF A SEGMENT OF A CURVE

The length of a segment of a curve is not easily defined; when it *is* defined, the task of determining it generally requires the use of calculus. Consider the special case of the arc of a circle. With the help of a little imagination and by proceeding carefully step by step, we will arrive at a suitable definition of the length of an arc of a circle.

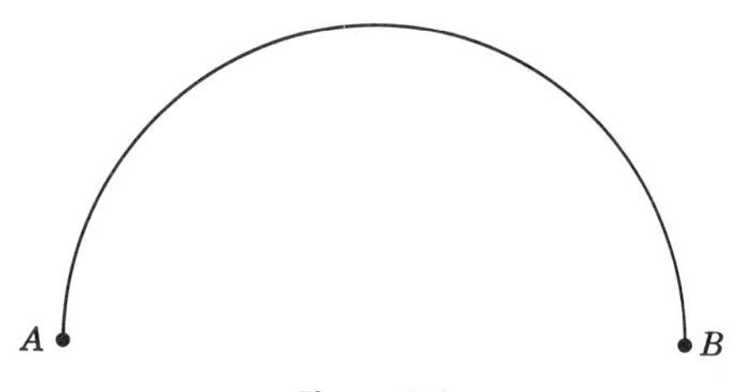

Figure 11

Consider the circular arc, which we denote $\widehat{AB}$, as in Figure 11. We wish to define the length of AB. We use the length of straight-line segments, the triangular inequality, and the concepts of least upper bound and greatest lower bound to arrive at the definition.

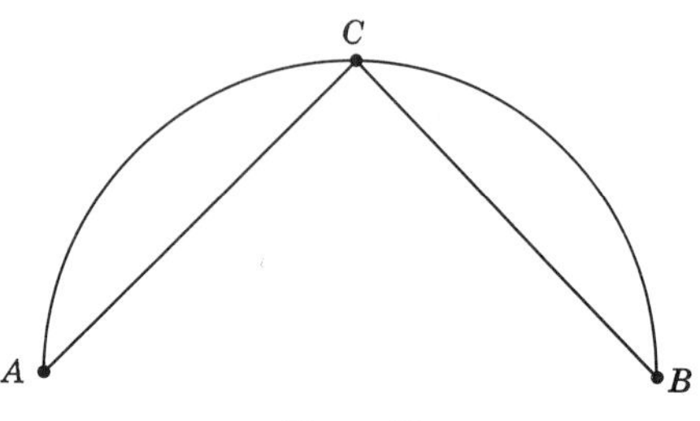

Figure 12

Let C be the point on the arc (Figure 12) midway between A and B (C can be found by construction). Then, considering the straight-line segments $\overline{AC}$ and $\overline{CB}$, we can regard the sum of the lengths of the straight-line segments as an approximation to the length of $\widehat{AB}$. Let $m(\overline{AC})$ denote the length of the straight-line segment $\overline{AC}$. (Note that $m(\overline{AC})$ is a real number.) Then our first approximation to the length of $\widehat{AB}$, denoted by l_2 (the "2" for two segments), is

$$l_2 = m(\overline{AC}) + m(\overline{CB})$$

For a second approximation, choose points D and E such that D is on the arc midway between A and C, and E is on the arc midway between C and B (see Figure 13).

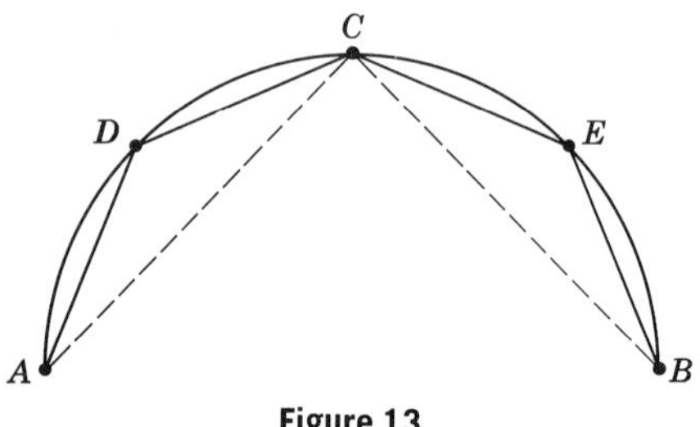

Figure 13

Then a second approximation, l_4, to the length of $\widehat{AB}$ is given by

$$l_4 = m(\overleftrightarrow{AD}) + m(\overleftrightarrow{DC}) + m(\overleftrightarrow{CE}) + m(\overleftrightarrow{EB})$$

Notice that $l_2 \leq l_4$ by the triangular inequality (see Section 5.12c, 3), that is,

$$m(\overleftrightarrow{AC}) \leq m(\overleftrightarrow{AD}) + m(\overleftrightarrow{DC})$$

and $$m(\overleftrightarrow{CB}) \leq m(\overleftrightarrow{CE}) + m(\overleftrightarrow{EB})$$

so $$l_2 = m(\overleftrightarrow{AC}) + m(\overleftrightarrow{CB}) \leq m(\overleftrightarrow{AD}) + m(\overleftrightarrow{DC}) + m(\overleftrightarrow{CE}) + m(\overleftrightarrow{EB}) = l_4$$

Continuing in this manner we can obtain *inscribed polygonal paths* from A to B of 2, 4, 8, 16, etc., sides. Let us designate the approximation obtained by using a path of 2^n sides by l_{2^n}.

Associated with each inscribed polygonal path is a circumscribed polygonal path obtained by constructing the tangents at the endpoints of the arc and at each point selected. These tangents meet to form polygonal paths as indicated in Figure 14.

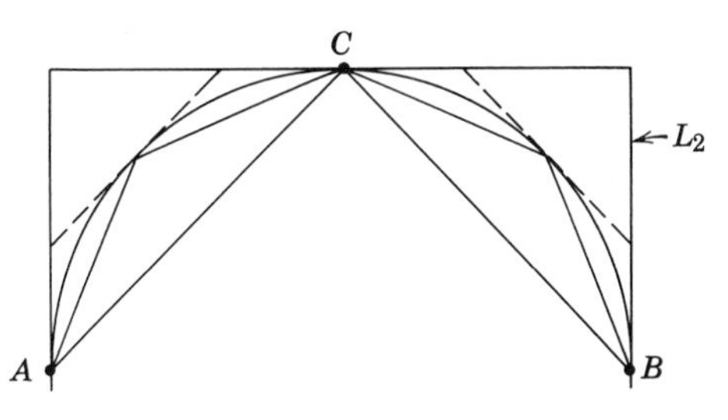

Figure 14

Let L_2 designate the length of the circumscribed path associated with the length of the inscribed path l_2. Using the *triangular inequality*, it can be shown that

$$l_{2^n} < L_{2^n} \quad \text{for every } n,$$

and $$l_{2^n} < l_{2^{n+1}} \quad \text{for every } n.$$

The sequence of numbers $l_2, l_4, l_8, \ldots l_{2^n}, \ldots$ is an increasing sequence of real numbers *bounded* above by the number L_2. By the property of *completeness* (see Section 7.3b) this sequence has a *least upper bound*. Similarly, the lengths of the circumscribed paths form a decreasing sequence of numbers *bounded* below by the length of the chord $\overleftrightarrow{AB}$. Again by the property of *completeness* this sequence has a *greatest lower bound*. If the least upper bound of the increasing sequence is equal to the greatest lower bound of the decreasing sequence, the common value is defined to be the *length* of the circular arc $\widehat{AB}$.

In the case of the circle, the length is called the *circumference*. In a circle of radius r it is given by the formula $c = 2\pi r$.

Example 1

To illustrate this procedure let us use successive approximations as outlined previously to estimate the length of a semicircular arc of a circle of diameter 2, which we know has length π.

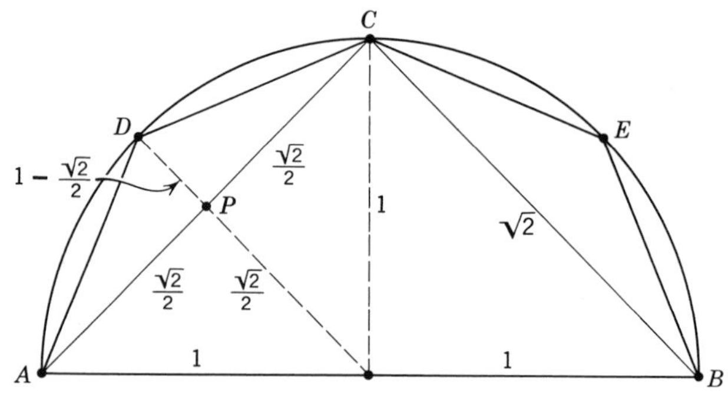

Figure 15

$m(\overline{AB}) = 2 \qquad \text{Length of } \widehat{AB} = \pi \cong 3.14$

$l_2 = m(\overline{AC}) + m(\overline{CB}) = \sqrt{2} + \sqrt{2} \cong 2(1.4142) \cong 2.83$

$l_4 = m(\overline{AD}) + m(\overline{DC}) + m(\overline{CE}) + m(\overline{EB})$, but since $\overline{AD}$, $\overline{DC}$, $\overline{CE}$, and $\overline{EB}$ are equal by construction,

$l_4 = 4m(\overline{AD})$, where $m(\overline{AD})$ can be computed from ΔAPD, Figure 15.

$$m(\overline{AD}) = \sqrt{\left(1 - \frac{\sqrt{2}}{2}\right)^2 + \left(\frac{\sqrt{2}}{2}\right)^2} = \sqrt{1 - \sqrt{2} + \tfrac{1}{2} + \tfrac{1}{2}}$$

$$= \sqrt{2 - \sqrt{2}} \cong \sqrt{2 - 1.41421356} \cong 0.76537$$

$l_4 = 4m(\overline{AD}) \cong 4(0.76537) \cong 3.06$

$l_8 = m(\overline{AF}) + m(\overline{FD}) + m(\overline{DG}) + m(\overline{GC}) + m(\overline{CH}) + m(\overline{HE}) + m(\overline{EI}) + m(\overline{IB})$, but, by construction, these are all equal, so

$l_8 = 8m(\overline{AF})$, where $m(\overline{AF})$ can be computed from ΔAQF, Figure 16.

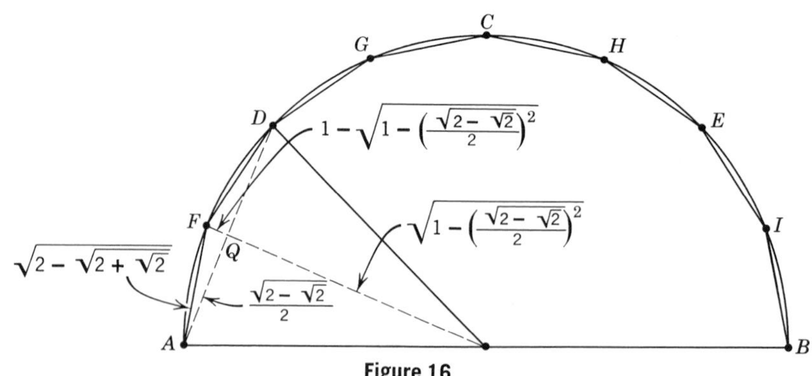

Figure 16

$$m(\overset{\circ\!-\!\circ}{AF}) = \sqrt{\left(\frac{\sqrt{2-\sqrt{2}}}{2}\right)^2 + \left[1-\sqrt{1-\left(\frac{\sqrt{2-\sqrt{2}}}{2}\right)^2}\right]^2}$$

$$= \sqrt{2-\sqrt{2+\sqrt{2}}}$$

$$\cong \sqrt{2-\sqrt{3.41421356}} \cong \sqrt{2-1.8477}$$

$$\cong \sqrt{0.1523} \cong 0.39$$

$$l_8 \cong 8(0.39) = 3.12$$

$$2.83 < 3.06 < 3.12 < 3.14$$

$$l_2 < l_4 < l_8 < \pi.$$

Notice that the third approximation yields accuracy to two significant figures.

Exercise 8.8

1. Show that $l_2 \leqq l_4$, using Figure 13.

2. Show that $l_{2^n} \leqq l_{2^{n+1}}$.

3. Find l_4 for a semicircular arc of diameter 3.

4. Estimate the length of the following curved path consisting of two semi-circles by

(a) using a ruler to estimate the measure of each segment of polygonal paths on a full-size drawing.
(b) using the calculations from problem 3.

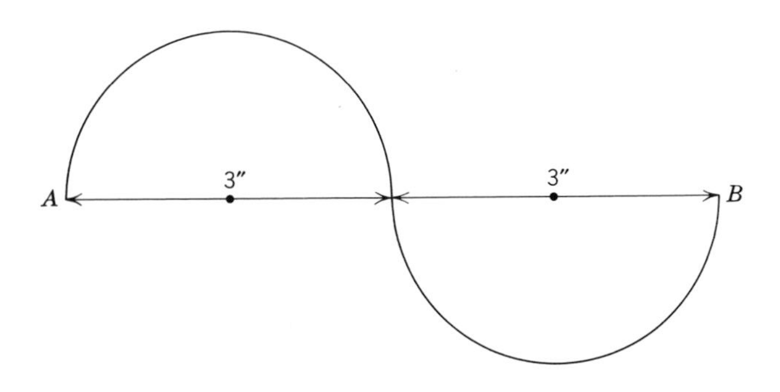

5. In Section 8.8 we used the triangular inequality to show that $m(\overset{\circ\!-\!\circ}{AC}) \leqq m(\overset{\circ\!-\!\circ}{AD}) + m(\overset{\circ\!-\!\circ}{DC})$. Sketch the triangle involved and verify the validity of our argument.

6. What is the exact length of the arc of problem 4?

7. Sketch the triangles referred to in Figure 14 and verify that $l_2 < L_2$.

8. Label the figure properly and verify that $l_4 < L_2$.

8.9 AREAS

In this section we present derivations of the standard formulas for the areas of a few selected plane figures.

As with the length of a line segment, the *area* of a plane figure is also a *measure*; that is, the area of a plane figure is a nonnegative real number such that if a plane figure P is contained in a plane figure Q, then the area of P, denoted $a(P)$, is less than or equal to the area of Q, $a(Q)$. If the plane figures P and Q are disjoint, then the area of the union of the two plane figures is equal to the sum of the area of P and the area of Q. The "length" of a line segment, as a measure, has these properties. The "volume" of an object in space, which will be discussed briefly in a later section, is also a measure and has these properties. Since we use these properties implicitly in what follows, we list them in the language of *area* and refer to them briefly.

Let P denote a plane figure and $a(P)$ its area; then

1. $a(P) \geq 0$. Nonnegative
2. If $P \subseteq Q$, then $a(P) \leq a(Q)$. Monotone
3. If $P \cap Q = \emptyset$, then $a(P \cup Q) = a(P) + a(Q)$. Finitely additive

Since we are interested in how to determine the area of a plane figure, we review what is meant by the area of a rectangle.

First choose a unit of length. Then a *square* one unit of length on a side is said to have *one square unit of area*. From Figure 17 we see that if we had three unit squares side by side we would want to count the area 3 square units. Similarly, if we have a rectangle with sides of length 2 and 4, by counting squares we would want to call its area 8 square units. A square $\frac{1}{2}$ unit length on each side is $\frac{1}{4}$ square unit in area, since four such squares precisely fill a unit square. A rectangle whose sides are 2 and $2\frac{1}{2}$ units in length has area 5 square units, as we see by counting.

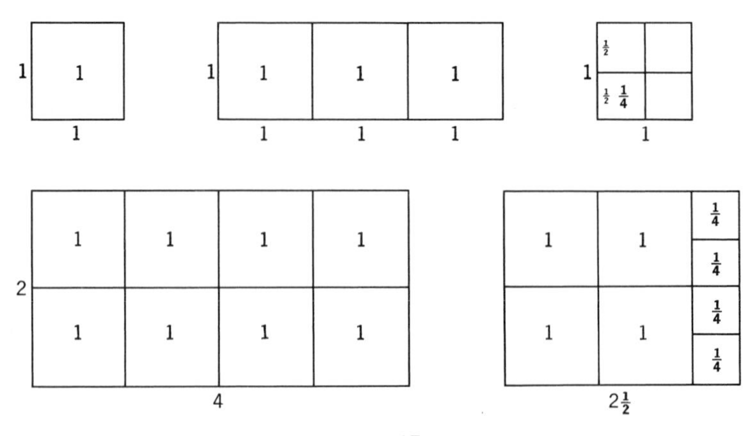

Figure 17

Using our imagination we can see how this idea of counting could be extended to any rectangle whose *length* and *width* were given in terms of rational numbers, and we find that the counting procedure gives us the same result as multiplying length by width (see problem 4, Exercise 6.12c).

We therefore *define* the area of a rectangle with length l and width w to be lw square units, where the real numbers l and w are given in the same units of length. If we use A to designate the area, then

$$A = lw.$$

For a square, which is just a special case of the rectangle with $l = w$, we usually use the symbol s to denote the length of a side. Then the area of a square is given by

$$A = s^2.$$

The area of plane figures, in general, can be defined in terms of the areas of rectangles. (See Section 8.13 for an example of the general procedure.) For our purposes we derive the formula for the area of a triangle and use this information to derive the formulas for the area of other plane figures.

8.9a Area of Triangles

The diagonal of a rectangle divides the rectangle into two equal parts. (See if you can recall enough of plane geometry to prove this. Try the theorem, "Two right triangles are congruent if the hypotenuse and a side of one are equal, respectively, to the hypotenuse and a side of the other.")

Each of these parts is a right triangle (see Figure 18). If the area of the rectangle is lw, then the area of the right triangle is $\frac{1}{2}lw$. We usually designate the length of the two legs of a right triangle by b and h, b for the base and h for the height. Then the area of a right triangle is the product of $\frac{1}{2}$, the base, and the height.

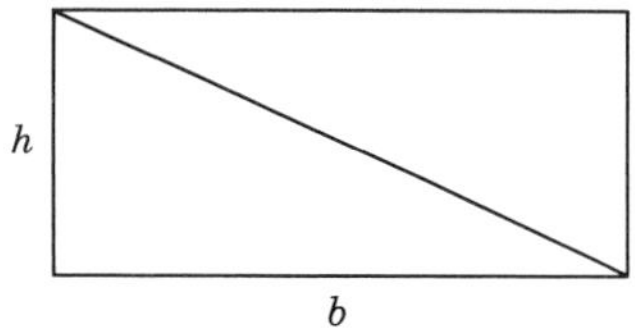

Figure 18

$$A = \tfrac{1}{2}(bh).$$

Notice that we are using the *additive property* of area, that is, the area of the rectangle is equal to the sum of the two equal numbers which represent the area of the right triangles.

To obtain the area of a general triangle, drop a perpendicular from a vertex to the opposite side (see Figure 19).

The length of the segment $\overline{CD}$ is designated as an altitude or height of the triangle and denoted h. The segment $\overline{CD}$ divides the triangle into two triangles, (1) and (2), each of which is a right triangle. Let us denote the

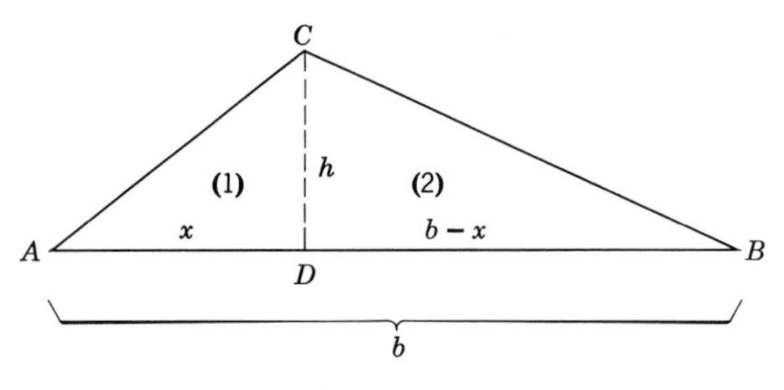

Figure 19

length of the base of (1) by x and the length of the base of (2) by $b-x$. They both have height h. Applying the known information about right triangles, we have

$$\text{Area of }(1) = \tfrac{1}{2}(xh),$$
$$\text{Area of }(2) = \tfrac{1}{2}(b-x)h = \tfrac{1}{2}(bh) - \tfrac{1}{2}(xh),$$

and

$$\text{Area }(1) + \text{Area }(2) = \tfrac{1}{2}(xh) + \tfrac{1}{2}(bh) - \tfrac{1}{2}(xh) = \tfrac{1}{2}(bh).$$

Hence the area of the general triangle is also given by the product of $\frac{1}{2}$, the base, and the height.

$$A = \tfrac{1}{2}(bh).$$

Notice that we rely on the additive property of area.

8.9b Area of Parallelograms

Let us turn to the parallelogram, a four-sided plane figure whose opposite sides are parallel and equal (see Figure 20).

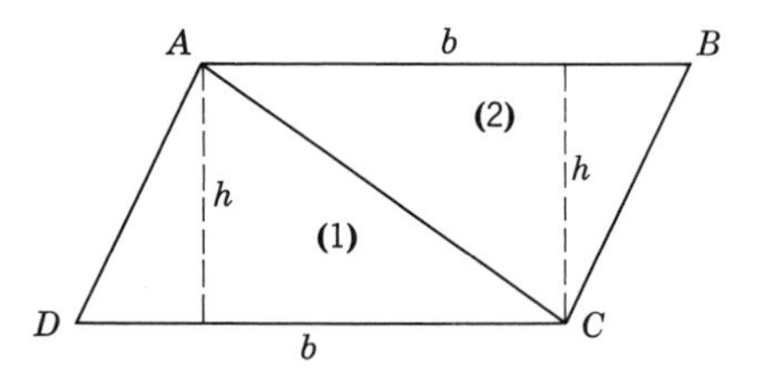

Figure 20

Here we designate the length of the perpendicular line segment between two parallel sides, called the "altitude" of the parallelogram, h, and the length of one of these parallel sides, called the base, b. If we draw the diagonal AC, we note that this divides the parallelogram into two triangles, each of which has base of length b and height of length h. Thus

$$\text{Area of triangle (1)} = \tfrac{1}{2}(bh),$$
$$\text{Area of triangle (2)} = \tfrac{1}{2}(bh),$$
and $$\text{Area (1)} + \text{Area (2)} = \tfrac{1}{2}(bh) + \tfrac{1}{2}(bh) = bh.$$

Hence the area of a parallelogram is the product of the lengths of the base and the height.

$$A = bh.$$

8.9c Area of Trapezoids

The trapezoid is a plane figure bounded by four straight-line segments, two of which are parallel (see Figure 21).

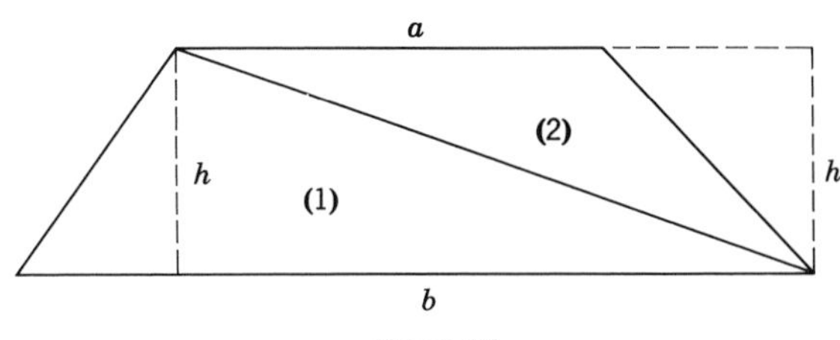

Figure 21

Again we designate the length of the perpendicular line segment between the two parallel sides by h, called the height. The length of one of the parallel sides is denoted b and the other a. By drawing a diagonal, as indicated in Figure 21, we see that the trapezoid is divided into two parts, each of which is a triangle. One triangle has base of length b and height of length h and the other has base of length a and height of length h. We have

$$\text{Area of triangle (1)} = \tfrac{1}{2}(bh),$$
and $$\text{Area of triangle (2)} = \tfrac{1}{2}(ah).$$

Then $$\begin{aligned}\text{Area of trapezoid} &= \text{Area (1)} + \text{Area (2)} = \tfrac{1}{2}(bh) + \tfrac{1}{2}(ah)\\ &= \tfrac{1}{2}(b+a)h \qquad \text{Distributive law}\end{aligned}$$

This is usually written

$$A = \tfrac{1}{2}h(a+b).$$

In words, the area of a trapezoid is found as the product of $\frac{1}{2}$, the height, and the sum of the two "bases." Notice that we again rely on the additive property of area.

8.9d The Area of Regular Polygons

Next we establish the formula for the area of a regular n-gon, a regular polygon of n sides. This is a plane figure that has equal sides and equal angles. The simplest plane figures of this type are the equilateral triangle and the square. In establishing a formula that yields the area of any regular n-gon we will illustrate the regular hexagon for references (see

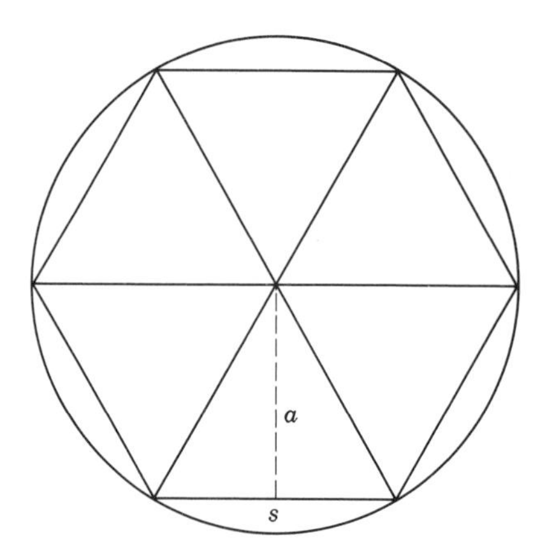

Figure 22

Figure 22). Associated with each regular polygon is a circumscribed circle. Let us designate the center of this circle as the center of the regular polygon. If we draw straight lines from the center to each of the vertices, we construct a triangle for each of the sides of the polygon. The distance from the center of a regular polygon to one of its sides (that is, perpendicular distance) is called the *apothem*. We shall designate its length by a. This is also the altitude of the triangle formed by one of the sides of the polygon and the lines from its endpoints to the center. Let us designate the length of one of the sides of the polygon by s; then the area of one of the triangles is $\frac{1}{2}(as)$. If the polygon has n sides, there are n such triangles in the polygon. Using the additive property of area, we have

$$A = \tfrac{1}{2}a(ns).$$

But ns is the "perimeter" (distance around) the regular polygon; hence

$$A = \tfrac{1}{2}(ap).$$

In words, the area of a regular polygon is the product of $\frac{1}{2}$, the apothem, and the perimeter.

Exercise 8.9

1. Find the areas of the triangles shown, using the dimensions given.

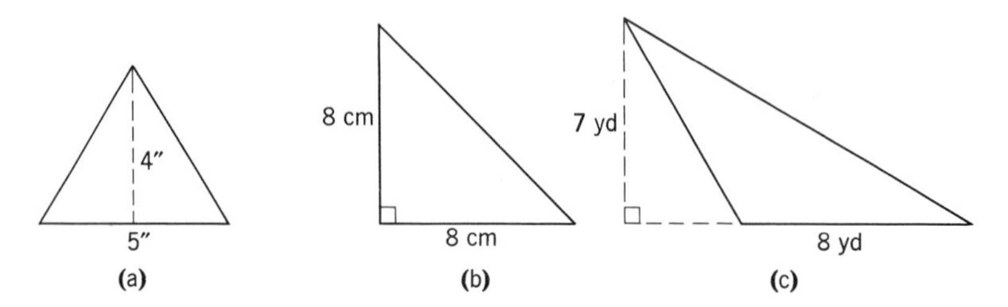

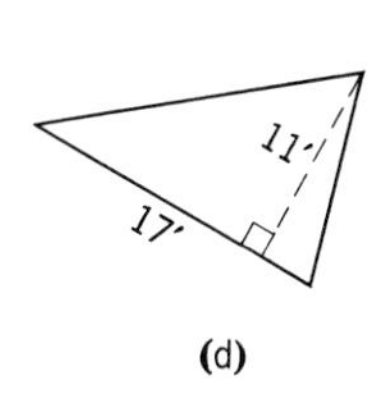

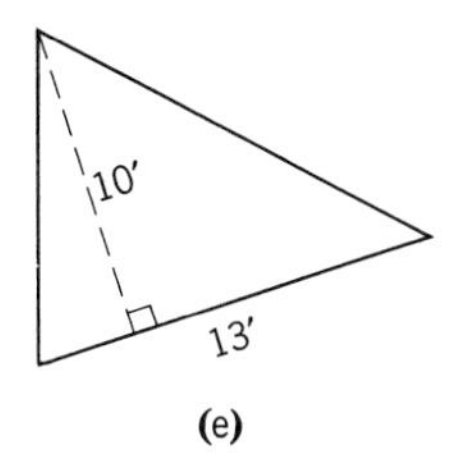

2. Find the areas of the parallelograms shown, using the dimensions given.

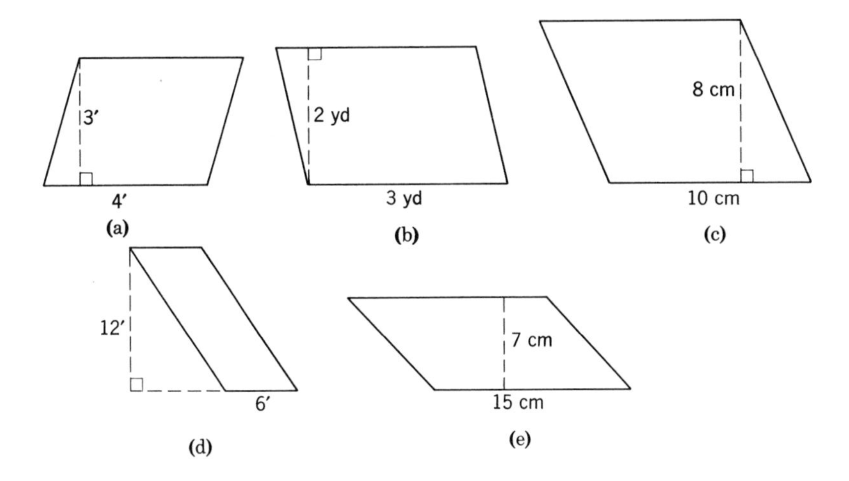

3. A man owned a rectangular lot 150 ft by 100 ft. From one corner A, a fence is placed to a point M in the center of the longer opposite side as shown.

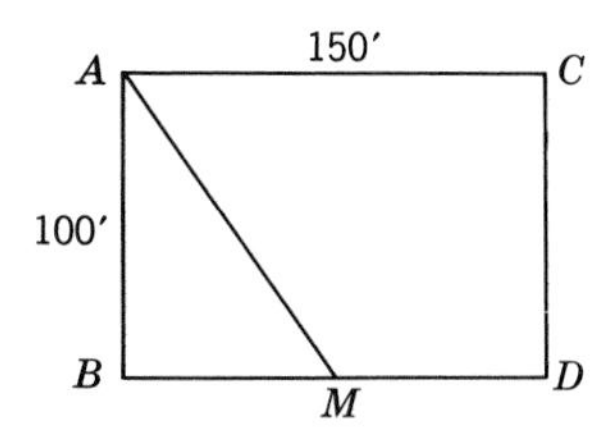

(a) Find the area of $ABCD$.
(b) Find the area of AMB.
(c) Find the area of $AMDC$.

4. (a) In the drawing below, measure $\overline{AB}$ and $\overline{DS}$. Using these measures, find the area of the parallelogram.

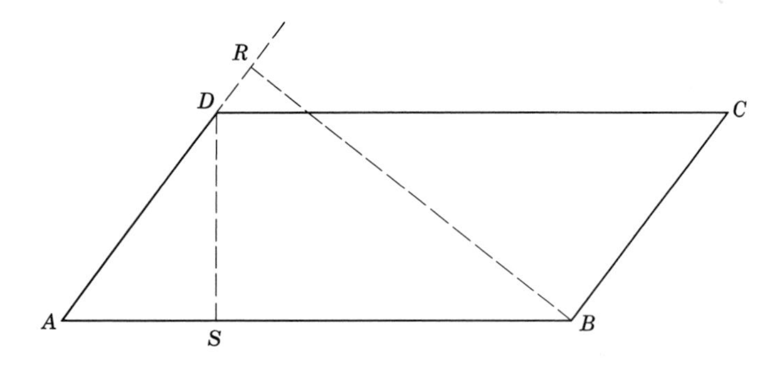

(b) Measure $\overleftrightarrow{AD}$ and $\overleftrightarrow{RB}$. Using these measures, find the area of the parallelogram.
(c) Do your results in (a) and (b) agree? Since measurement is approximate, they may not be exactly the same, but they should be close.

5. Find the area of the following trapezoids, using the dimensions given:

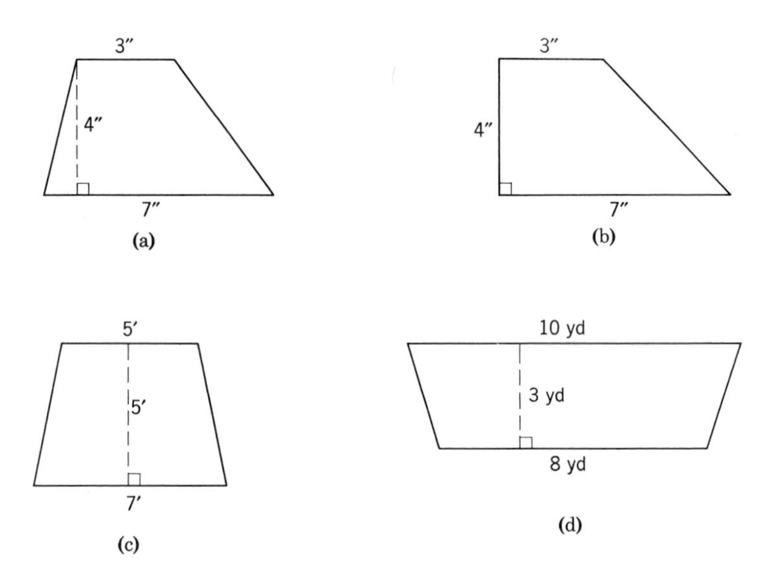

6. Using rational approximations for the irrational numbers, find an approximation to the area of a rectangle whose length is $\sqrt{2}$ ft and whose width is $\sqrt{3}$ ft.

8.9e Area of a Circular Region

The area of a circular region which is bounded by a circle of radius r is given by the formula

$$A = \pi r^2.$$

The formula is derived by using the tools of calculus but it can and should be made plausible to the students who have not had calculus. One of the first questions that students often ask is "Why is the area of a circular region given in *square* units?" There are several ways of responding to this question. Here is one which has worked successfully.

Let a circle have a radius of 5 units. From the formula for the area we get

$$A = \pi \cdot 5^2 = \pi \cdot 25 \cong (3.1416)(25)$$

The area is *approximately* 78.54 square units.

Draw a circle with a radius of 5 units. Construct squares on the two horizontal radii above and below the diameter as shown in Figure 22a. Each of the squares has an area of $r^2 = 25$ square units. The area of the large square is $4r^2$ square units, since it consists of the four smaller squares each with area $r^2 = 25$ *square* units. The units are *squares*.

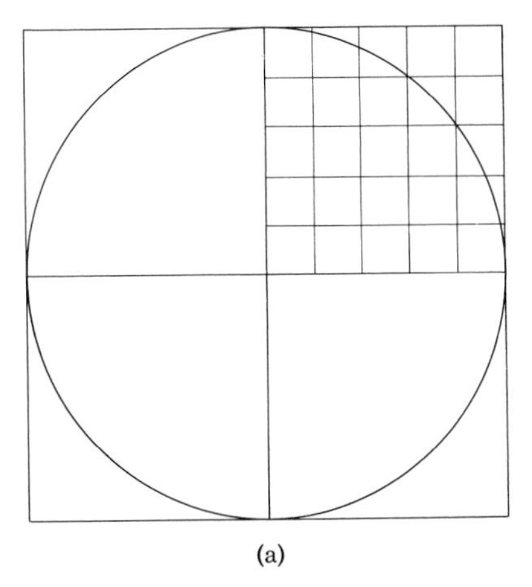

(a)

Figure 22a

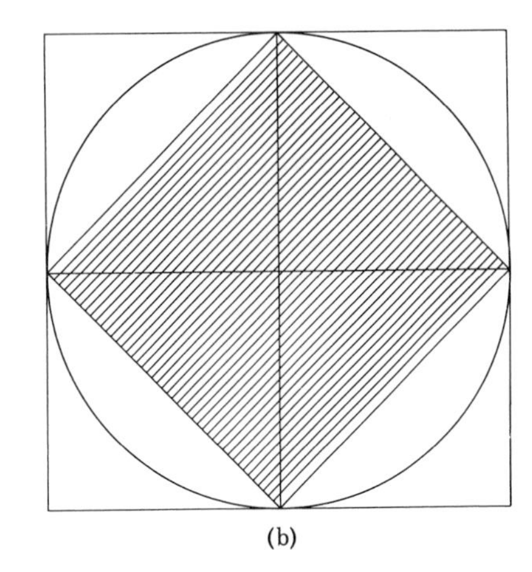

(b)

Figure 22b

The circular region is entirely contained in the large square whose area is $4r^2$ units. By property 2 (Section 8.9), the area of the circular region is less than $4r^2$. By constructing the diagonals (Figure 22b) and using *symmetry*, it is easy to see that the area of the circular region is greater than $2r^2$. The question is: How much less than $4r^2$ and how much greater than $2r^2$ is the area of the circular region? Just by comparing that portion of the circular region lying in the outer triangles it is plausible to estimate that the area is close to $3r^2$.

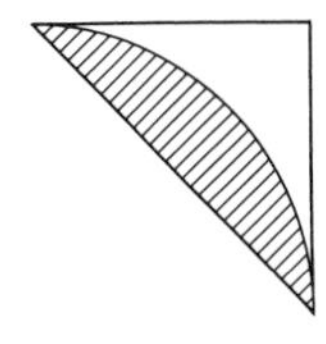

The validity of the formula may now be made plausible with the help of the following argument. Draw and cut out a disc, then cut it into two parts along a diameter (Figure 22c).

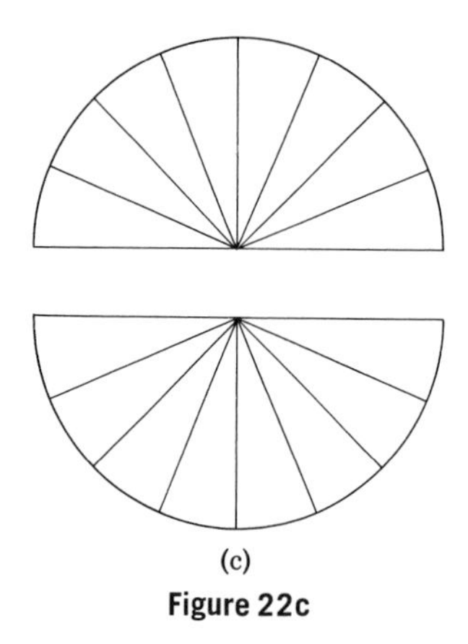

(c)

Figure 22c

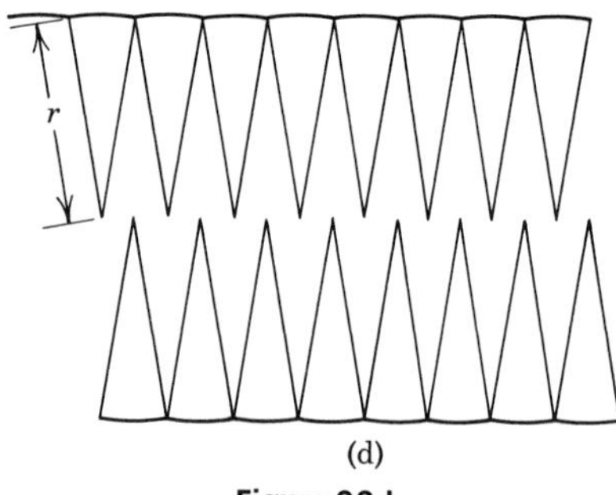

(d)

Figure 22d

Now cut matching slices and spread out each half disc as shown in Figure 22d.

Notice that the side of each pie-shaped slice has length r, the radius of the disc. Notice also that the length of the curved edges on one half of the disc add up to πr.

Now fit the two halves together as in Figure 22e. The resulting figure resembles a parallelogram with scalloped sides.

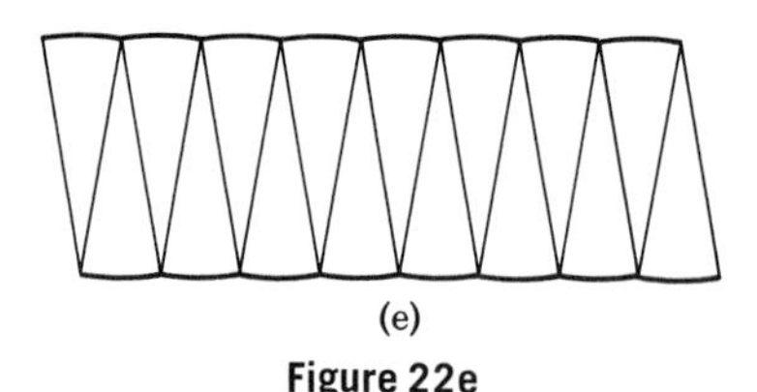

(e)

Figure 22e

If we had made narrower slices, the slanted sides would be more vertical and the serrated edges would be less bumpy. Try to imagine cutting the half discs into one million slices and fitting them together. The rectangular shaped figure would actually approximate a rectangle of dimension r by πr quite closely. That is, the area of the disc is the limiting value of this process which approaches $(r)(\pi r) = \pi r^2$ square units.

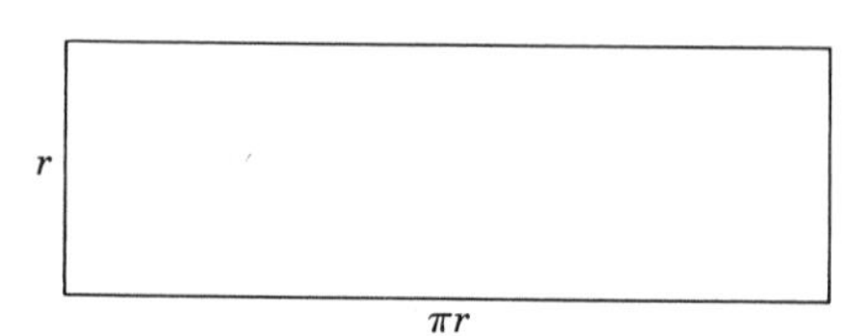

Exercise 8.9e

1. What is the area of a circle of radius 10 inches?

2. What is the area of a circle of radius 1000 inches?

3. What is the circumference of a circle of radius 10 inches?

4. What is the area of a circle of radius 20 inches?

5. What is the circumference of a circle of radius 20 inches?

6. How does the circumference of a circle of radius 20 inches compare with a circle of radius 10 inches? How do their areas compare?

7. How does the ratio of the circumference to the diameter of a circle of radius 10 inches compare with one of radius 20 inches?

8.10 VOLUME AND SURFACE AREA

In this section we present some of the standard formulas for volume. First we define a *unit of volume.* As with area, choose a unit of length. A *cube* one unit of length on each edge is said to have *one cubic unit of volume.* From Figure 23 we see that if we had 3 cubes side by side, we would want

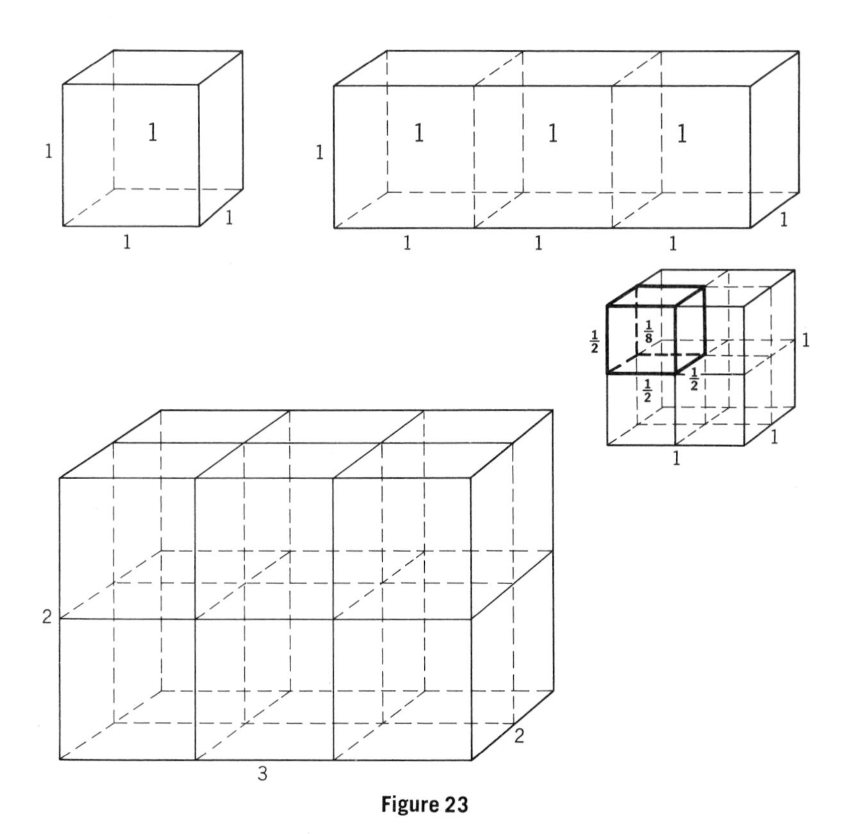

Figure 23

to count this as 3 cubic units. Similarly, by counting, we see that if we had a rectangular solid 2 units by 3 units by 2 units, we would want to call this 12 cubic units. A cube $\frac{1}{2}$ unit on an edge would be counted as $\frac{1}{8}$ cubic unit, for it takes precisely 8 of them to "fill" one cubic unit.

As with area, if the units of measure of the dimensions of the rectangular solid are given in terms of rational numbers, we can obtain the number of units of volume by "counting," and the result is the same as if we multiplied "length" by "width" by "height." If we designate the length, width, and height by l, w, and h, respectively, l, w, and h real numbers, then the volume V of the rectangular solid is *defined* by

$$V = lwh.$$

The derivation of the formulas for each of the solids given in the following paragraphs can be made precise with the use of calculus. We shall simply describe the type of figure under discussion and present the formula for that figure.

8.10a Prism

In the same sense that we referred to a two-dimensional figure bounded by straight lines as a polygon, we speak of a three-dimensional figure bounded by planes as a *polyhedron*.

A *prism* is a polyhedron two of whose faces (called bases) are congruent polygons (exactly the same size and shape) in parallel planes, and whose remaining faces (called lateral faces) are parallelograms (see Figure 25).

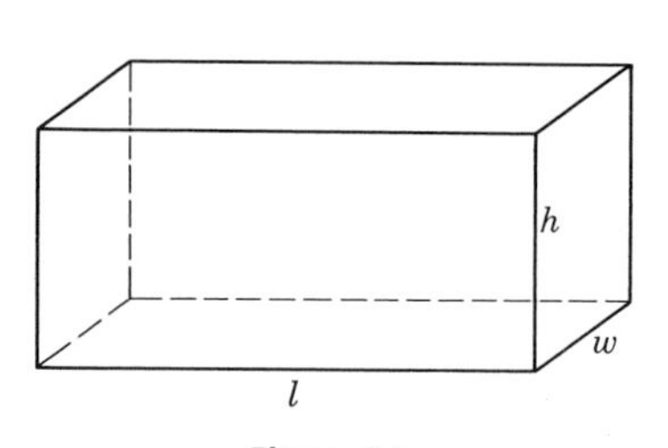

Figure 24

A prism, each of whose faces and bases is a rectangle, is called a *rectangular solid* or *a rectangular parallelepiped*. The volume of a parallelepiped (box-shaped figure) is given by the product of the lengths of three concurrent edges. (Concurrent means "meeting at one point")(see Figure 24).

$$V = lwh.$$

The *total area* of such a figure is the sum of the areas of the faces and bases. The *lateral area* is the total area less the area of the bases. Letting T denote the total area and S the lateral area, we have

$$T = 2(lw + lh + wh),$$

and $$S = 2(lw + wh).$$

To speak in general of prisms it is convenient to define what is meant by a "right section." A *right section* of a prism is the polygon formed by the intersection of the prism with a plane which is perpendicular to each of the lateral faces (see Figure 25).

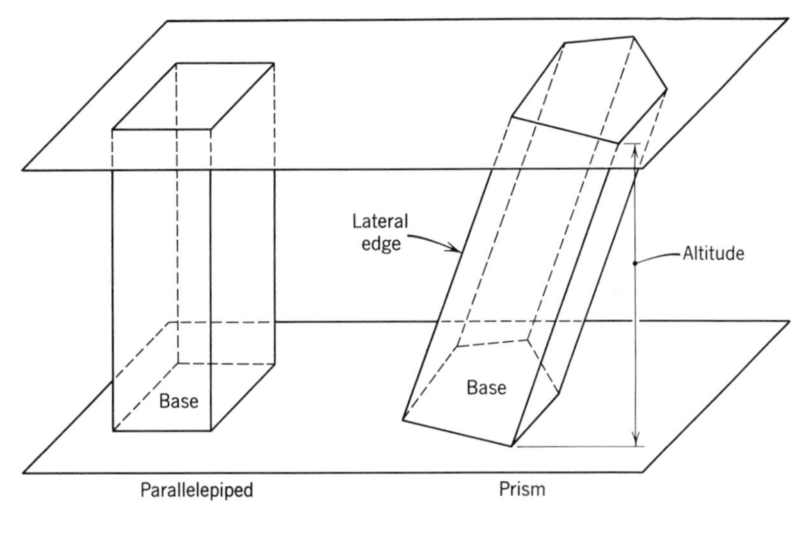

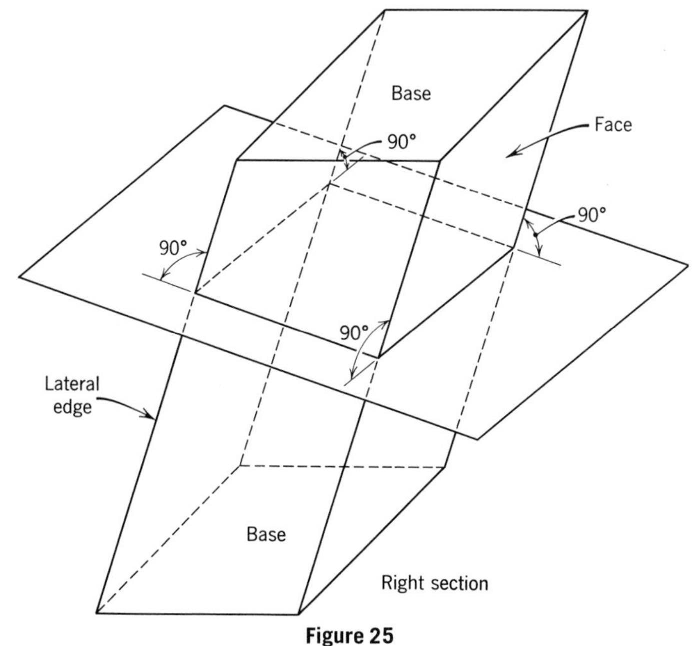

Figure 25

Then, in general, for a prism we have

$T =$ (perimeter of a right section) $\times$ (lateral edge) $+$ (area of the bases),

$S =$ (perimeter of a right section) $\times$ (lateral edge),

and

$V =$ (area of a right section) $\times$ (lateral edge),

$V =$ (area of the base) $\times$ (altitude).

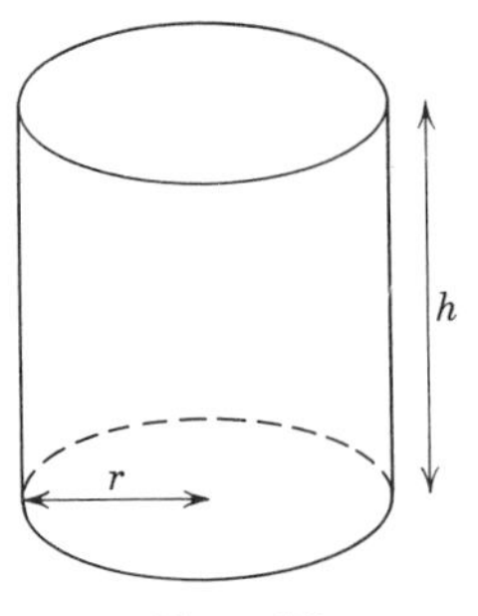

Figure 26

8.10b Circular Cylinder

A *right circular cylinder* resembles a prism whose bases are right sections except that the bases are circles rather than polygons. The formulas for the prism also hold true for the right circular cylinder, but they may be stated in language involving π (see Figure 26).

If r is the radius of the base and h the altitude, then (see Section 9.13a)

$$S = 2\pi rh,$$
$$T = 2\pi rh + 2\pi r^2,$$
and $$V = \pi r^2 h.$$

8.10c Pyramid

A *pyramid* is a three-dimensional figure whose base is a polygon and whose lateral faces are triangles. If the base is a regular polygon, and a line from the vertex of the pyramid to the center of the base is perpendicular to the base, then the pyramid is called a *regular pyramid* (see Figure 27).

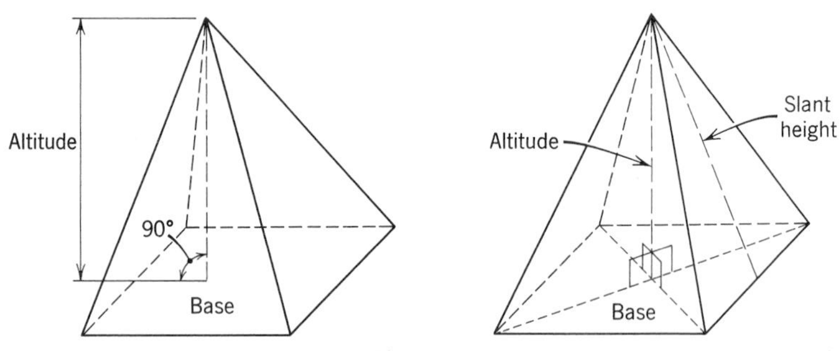

Figure 27

For a regular pyramid, Figure 27.

$$V = \tfrac{1}{3}\,(\text{area of base}) \times (\text{altitude}),$$
and $$S = \tfrac{1}{2}\,(\text{perimeter of base}) \times (\text{slant height}).$$

For the general pyramid this formula for the volume continues to hold, but the given formula for lateral surface area does not.

8.10d Cone

The *right circular cone* resembles a regular pyramid except that the base is a circle. Formulas for the regular pyramid will also serve for the cone. As with the cylinder, we can express these formulas in terms of π, namely (Figure 28).

$$S = \pi r \sqrt{r^2 + h^2},$$
and $$V = \tfrac{1}{3}\pi r^2 h.$$

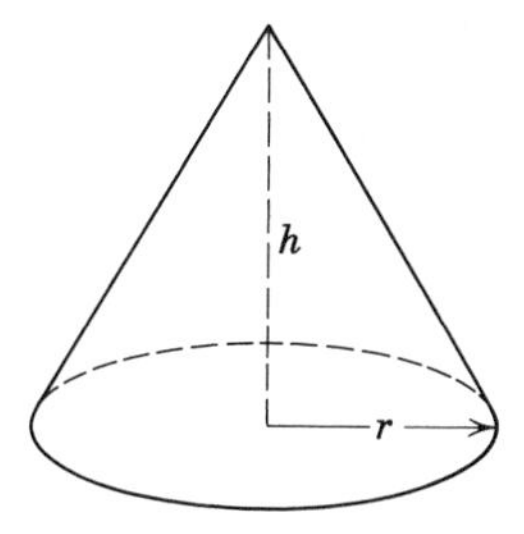

Figure 28

A *frustrum* of a regular pyramid or cone, or a truncated pyramid or cone, is illustrated in Figure 29. Let b, b' denote the areas of the bases and h the altitude, then

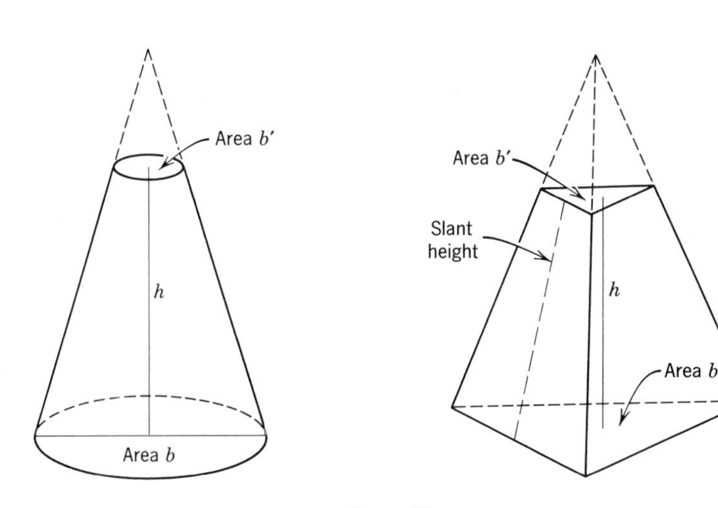

Figure 29

$$S = \tfrac{1}{2}(\text{sum of perimeters of bases}) \times (\text{slant height}),$$

and $$V = \tfrac{1}{3}h(b + b' + \sqrt{b \cdot b'}).$$

8.10e Regular Polyhedra

A *regular polyhedron* is a polyhedron whose faces are congruent regular polygons.

There are only five (except for size) *regular polyhedra*, as shown in Figure 30. (See Table 1.)

Although the argument required to show that there are five and only five regular polyhedra is not difficult, it will not be given here. (Consult *What is Mathematics?*, by Courant and Robbins.)

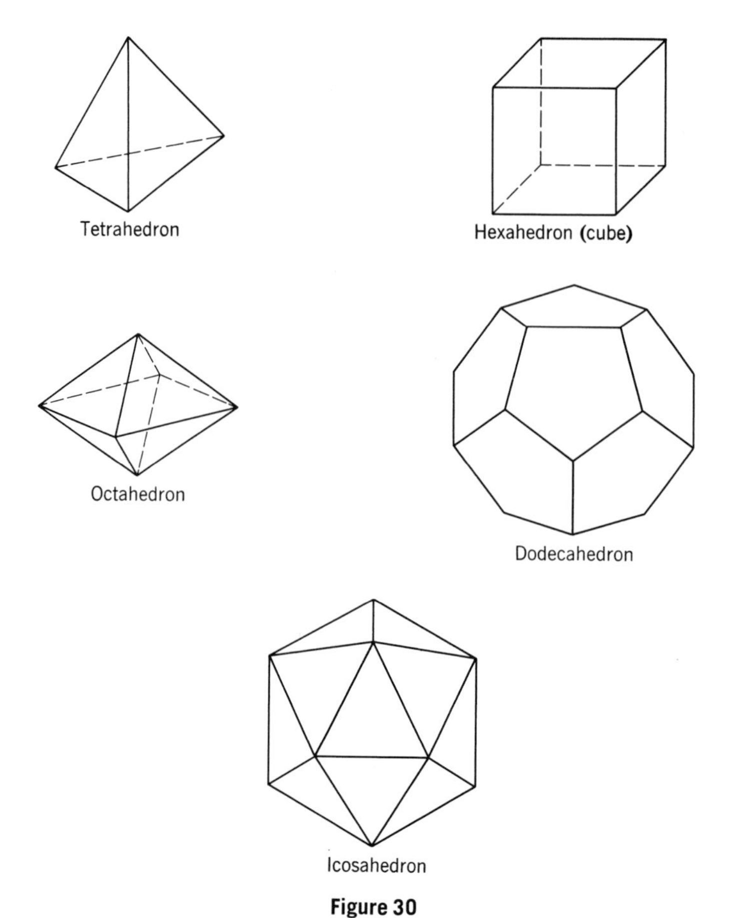

Figure 30

Table 1 Regular Polyhedra

Name	Nature of surface	Total Area	Volume
Tetrahedron	4 equilateral triangles	$1.73205a^2$*	$0.11785a^3$
Hexahedron	5 squares	$6.00000a^2$	$1.00000a^3$
Octahedron	8 equilateral triangles	$3.46410a^2$	$0.47140a^3$
Dodecahedron	12 pentagons	$20.64573a^2$	$7.66312a^3$
Icosahedron	20 equilateral triangles	$8.66025a^2$	$2.18170a^3$

*"a" denotes the length of an edge.

8.10f Sphere

For a *sphere* (see Figure 31) let r represent the radius, d the diameter; then

$$S = 4\pi r^2 = \pi d^2,$$

and $$V = \tfrac{4}{3}\pi r^3 = \tfrac{1}{6}\pi d^3.$$

8.10g Circular Torus

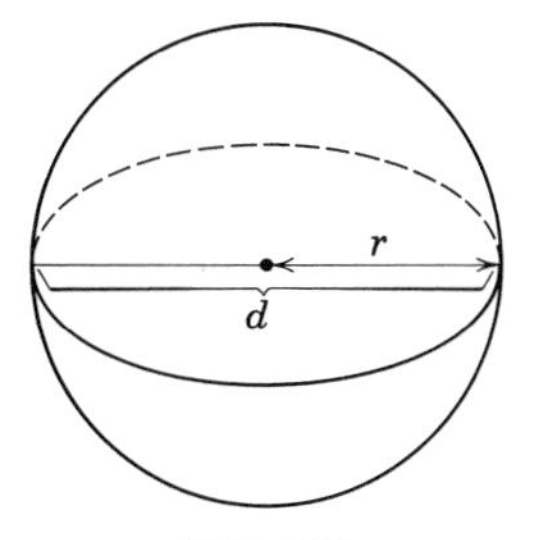

Figure 31

Consider a line segment, OA, of length a and a circle with center O and radius r, where $r < a$. Fix the point A and cause the line segment, with the circle, to rotate in a plane perpendicular to the plane of the circle. The circle will "sweep out" a doughnut-shaped figure called a *circular torus* (see Figure 32).

$$S = 4\pi^2 ar$$

and $$V = 2\pi^2 ar^2.$$

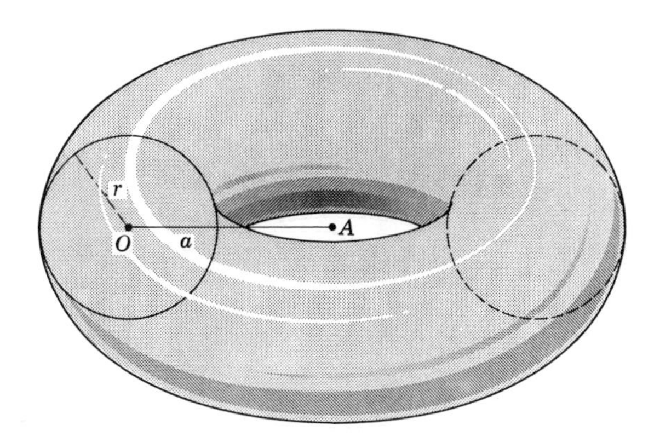

Figure 32

Exercise 8.10

1. Find the volume and total surface area of rectangular solids of the following dimensions:

(a) 4 in. by 5 in. by 12 in.
(b) 3.5 ft by 7.2 ft by 8.6 ft.
(c) 3 ft by 5 ft by 18 in.
(d) $4\frac{1}{2}$ yd by $2\frac{2}{3}$ yd by 8 yd.

2. Find the surface area and volume of a sphere whose radius is

(a) 3 in. (b) 4 ft
(c) 14 in. (d) $4\frac{1}{2}$ cm
(Use 3.14 or $\frac{22}{7}$ as an approximation to π.)

3. Find the total surface area, lateral surface area, and volume of the following:

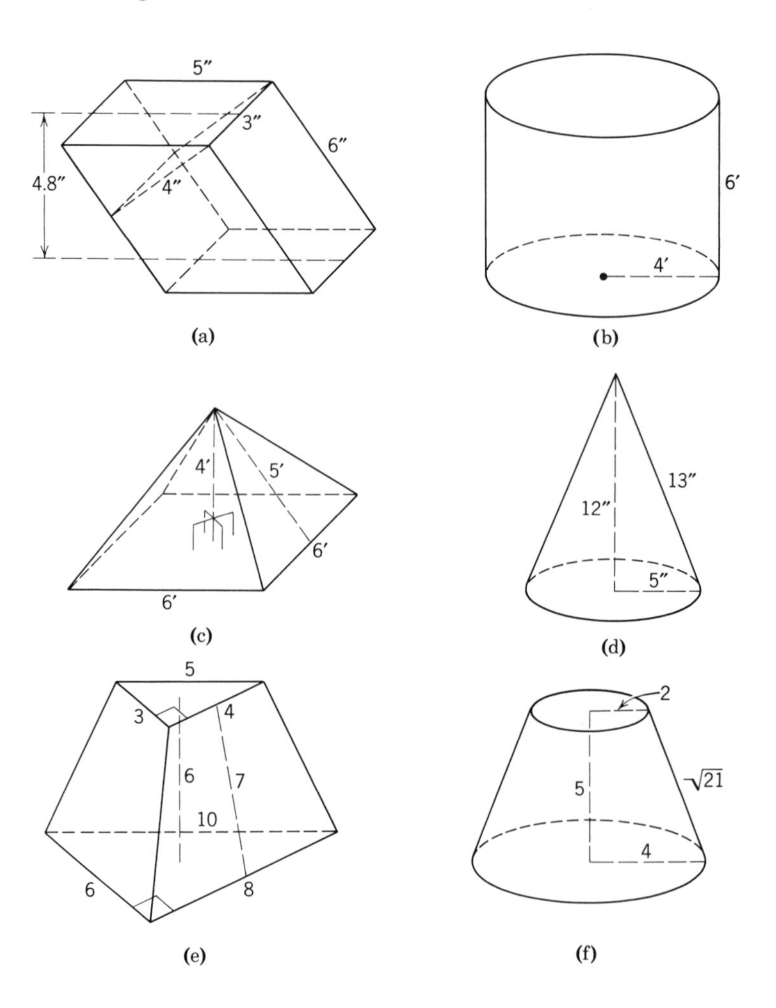

8.11 SIMILAR PLANE FIGURES

In this section we discuss what is meant by similar plane figures with the primary objective of using similar right triangles in indirect measurement.

Two triangles are said to be *similar* to each other if corresponding angles are equal in measure and corresponding sides are proportional in measure. This relation is symbolized $\sim$ (see Figure 33).

$$\triangle ABC \sim \triangle A'B'C'$$

$$m\angle A = m\angle A',\ m\angle B = m\angle B',\ m\angle C = m\angle C'.$$

$$\frac{m(\overset{\circ\!-\!\circ}{AB})}{m(\overset{\circ\!-\!\circ}{A'B'})} = \frac{m(\overset{\circ\!-\!\circ}{AC})}{m(\overset{\circ\!-\!\circ}{A'C'})} = \frac{m(\overset{\circ\!-\!\circ}{BC})}{m(\overset{\circ\!-\!\circ}{B'C'})} \quad \text{or} \quad \frac{m(\overset{\circ\!-\!\circ}{AB})}{m(\overset{\circ\!-\!\circ}{AC})} = \frac{m(\overset{\circ\!-\!\circ}{A'B'})}{m(\overset{\circ\!-\!\circ}{A'C'})}, \text{ etc.}$$

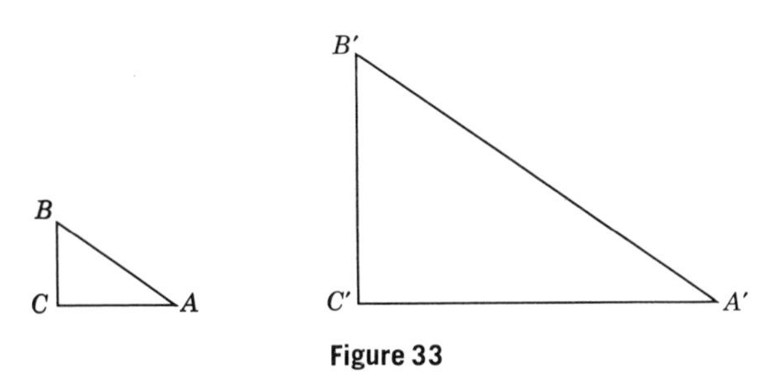

Figure 33

Two polygons are *similar* to each other if their corresponding angles are equal in measure and their corresponding sides are proportional in measure (see Figure 34).

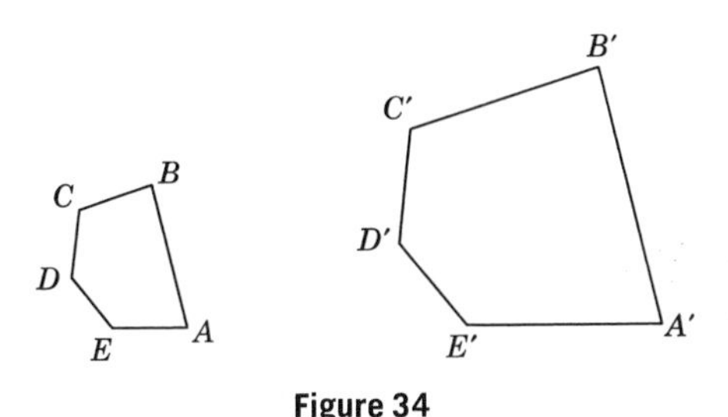

Figure 34

Polygons $ABCDE$ and $A'B'C'D'E'$ are similar.

$$m\angle A = m\angle A', \quad m\angle B = m\angle B', \quad m\angle C = m\angle C',$$
$$m\angle D = m\angle D', \quad m\angle E = m\angle E'.$$
$$\frac{m(\overset{\circ\!-\!\circ}{AB})}{m(\overset{\circ\!-\!\circ}{A'B'})} = \frac{m(\overset{\circ\!-\!\circ}{BC})}{m(\overset{\circ\!-\!\circ}{B'C'})} = \frac{m(\overset{\circ\!-\!\circ}{CD})}{m(\overset{\circ\!-\!\circ}{C'D'})} = \frac{m(\overset{\circ\!-\!\circ}{DE})}{m(\overset{\circ\!-\!\circ}{D'E'})} = \frac{m(\overset{\circ\!-\!\circ}{EA})}{m(\overset{\circ\!-\!\circ}{E'A'})}.$$

As a special case we can show that if an acute angle of one right triangle is equal to an acute angle of another right triangle, then the triangles are similar. This simple criterion for determining similarity of right triangles leads to their use in indirect measurement.

Example 1

A lake lies between points A and C (Figure 35). The problem is to determine the distance from A to C by using the properties of similar right triangles.

First we locate a point B such that $\overset{\circ\!-\!\circ}{AB}$ and $\overset{\circ\!-\!\circ}{BC}$ are at right angles to one another, and we pick a point D on $\overset{\circ\!-\!\circ}{AC}$ and a point E on $\overset{\circ\!-\!\circ}{BC}$ such that $\overset{\circ\!-\!\circ}{DE}$ is perpendicular to $\overset{\circ\!-\!\circ}{BC}$. Then triangles ABC and DEC are similar right triangles because $\angle C$ is common to both. We measure and find $m(\overset{\circ\!-\!\circ}{EC}) = 4$ *units of length*, $m(\overset{\circ\!-\!\circ}{BC}) = 52$ units of length, and $m(\overset{\circ\!-\!\circ}{DC}) = 5$ units of

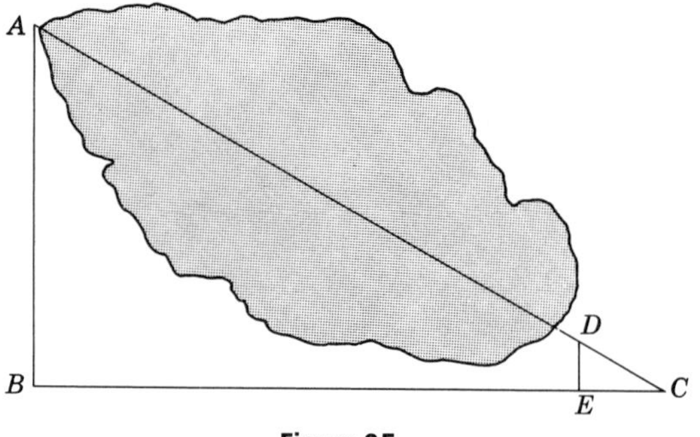

Figure 35

length. From the properties of similar triangles we have

$$\frac{m(\overleftrightarrow{AC})}{m(\overleftrightarrow{DC})} = \frac{m(\overleftrightarrow{BC})}{m(\overleftrightarrow{EC})}, \qquad \text{or} \qquad \frac{m(\overleftrightarrow{AC})}{5} = \frac{52}{4},$$

therefore

$$m(\overleftrightarrow{AC}) = \frac{260}{4} = 65 \text{ units of length.}$$

Exercise 8.11

1. Find the height of the tree in the accompanying diagram.

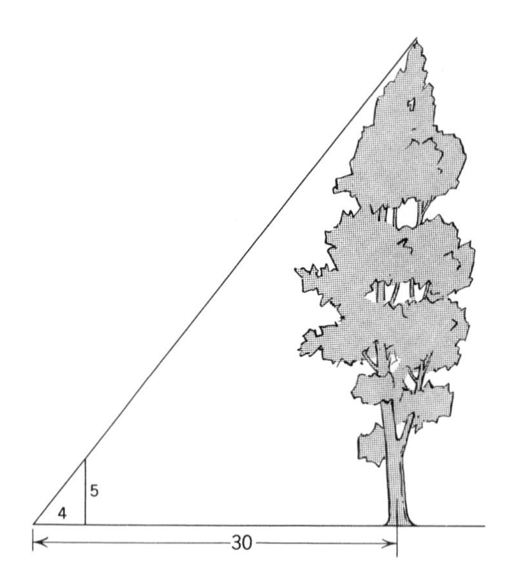

2. A man 6 ft tall casts a shadow of 9 ft when standing 24 ft away from a point directly under a street lamp. How high above the ground is the street lamp?

3. In the following diagram, if $\triangle ABC \sim \triangle DCE$ and $m(AC) = 12$ yd, $m(AB) = 20$ yd, $m(CE) = 100$ yd. Find DE. the length of the lake.

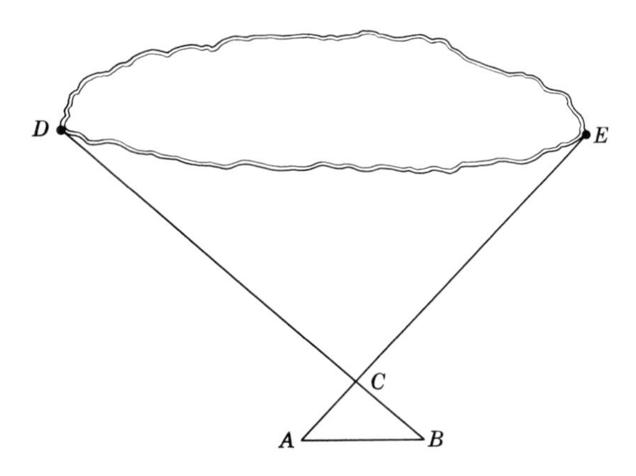

4. In the following diagram, $\triangle ABC \sim \triangle ADE$. Find x.

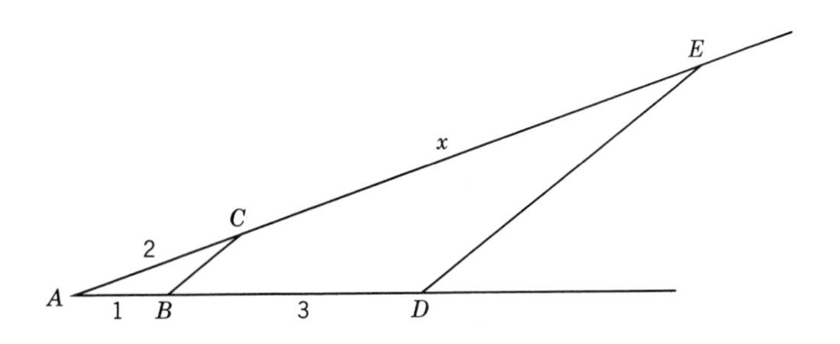

5. In the following diagram, $\triangle ABC \sim \triangle ADE$. Find x.

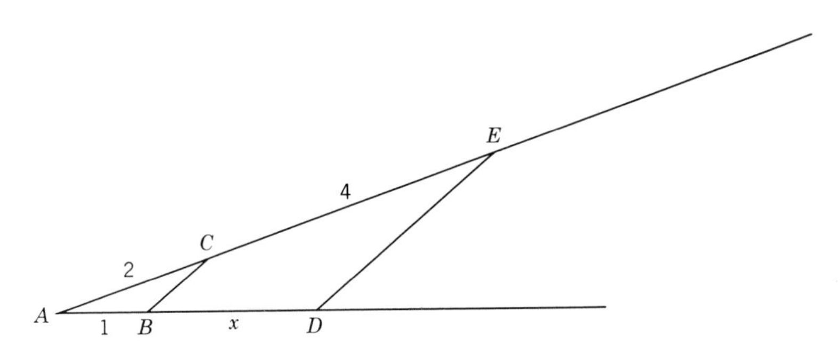

6. In the following diagram, $\triangle ABC \sim \triangle ADE$. Find x.

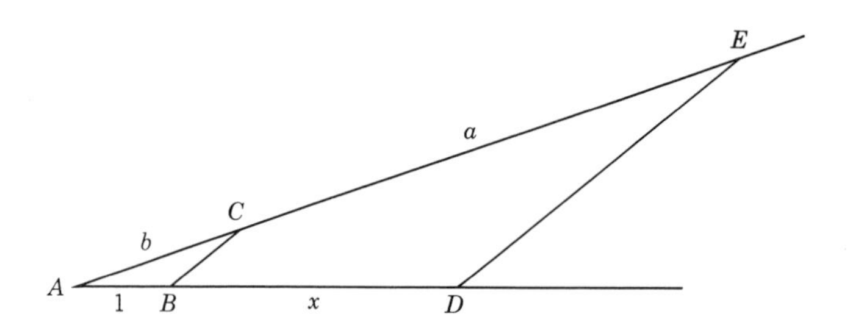

7. In the following diagram, $\triangle ABC \sim \triangle ADE$. Find x.

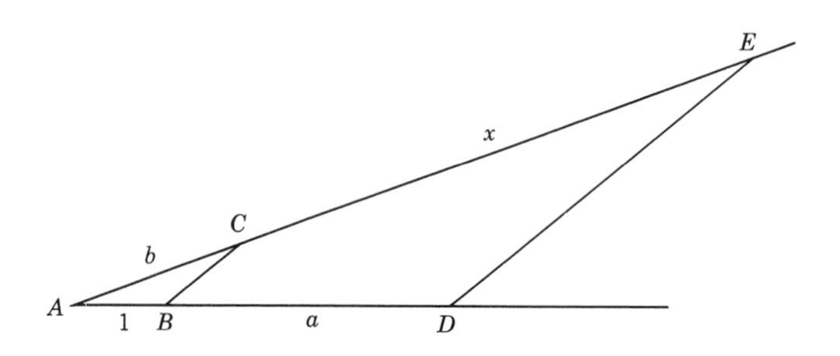

8. Examine your results of problems 4, 5, 6, and 7 and give a general geometric interpretation of multiplication and division.

9. Give a geometric interpretation of addition and subtraction.

8.12 THE PYTHAGOREAN THEOREM

Pythagorean Theorem. The square of the length of the hypotenuse of a right triangle is equal to the sum of the squares of the lengths of the legs.

Before we give any proofs of the Pythagorean Theorem we state a theorem about right triangles for which the reader can supply a proof.

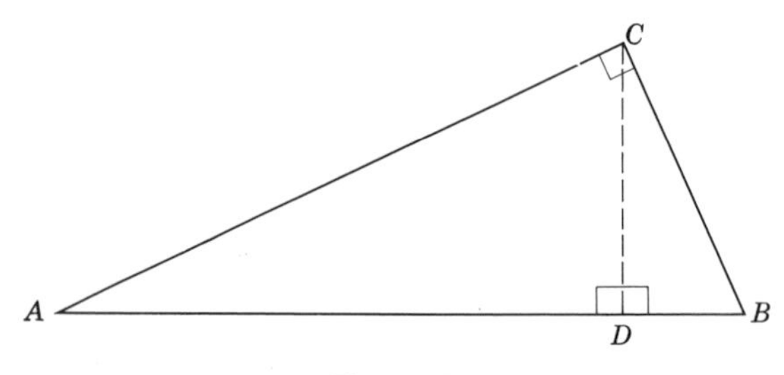

Figure 36

Theorem. The altitude to the hypotenuse of a right triangle forms two right triangles which are similar to the given triangle (see Figure 36):

$$\triangle ADC \sim \triangle ABC \sim \triangle BDC$$

From this and by using the properties of similar triangles, we obtain (see Figure 37):

1. The length of the altitude $\overline{CD}$ to the hypotenuse is a mean proportional between the lengths of the two segments $\overline{AD}$ and $\overline{DB}$ formed on the hypotenuse. (*Mean proportional*: x is a mean proportional between a and b if $a/x = x/b$, or $x^2 = ab$.)
2. The length of leg $\overline{AC}$ of a right triangle ABC is the mean proportional between the length of the hypotenuse $\overline{AB}$ and the length of an adjacent segment $\overline{AD}$ on the hypotenuse formed by the altitude $\overline{CD}$ from the right angle (see Figure 37).

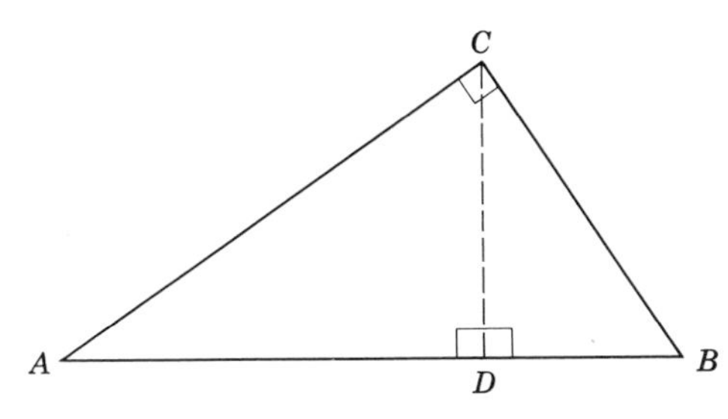

Figure 37

$$(1)\ \frac{m(\overline{AD})}{m(\overline{CD})} = \frac{m(\overline{CD})}{m(\overline{DB})} \qquad (2)\ \frac{m(\overline{AD})}{m(\overline{AC})} = \frac{m(\overline{AC})}{m(\overline{AB})} \text{ or}$$

$$m(\overline{AC})^2 = m(\overline{AD}) \cdot m(\overline{AB})$$

Let us now consider the proof of the theorem of Pythagoras.

Let $\overline{CD}$ be the altitude from the right angle to the hypotenuse $\overline{AB}$ (see Figure 38). Let d, e, a, b, and c be the lengths of $\overline{AD}$, $\overline{DB}$, $\overline{BC}$, $\overline{AC}$, and $\overline{AB}$, respectively. Then $d/b = b/c$, or $d = b^2/c$, and $e/a = a/c$, or $e = a^2/c$. But

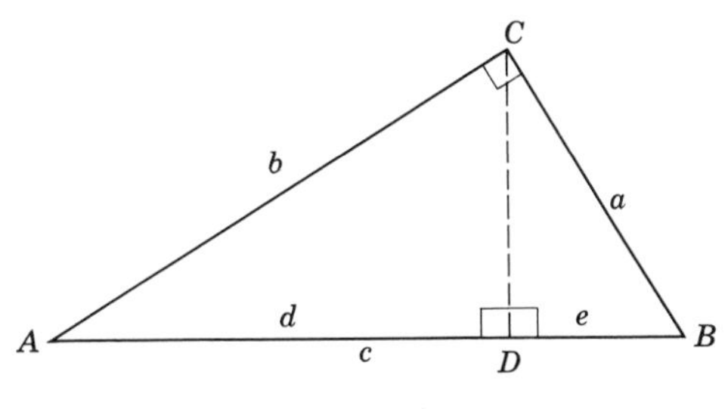

Figure 38

$d+e=c$. By substitution

$$d+e=c,$$

or
$$\frac{b^2}{c}+\frac{a^2}{c}=c,$$

therefore
$$b^2+a^2=c^2 \quad \text{or} \quad a^2+b^2=c^2.$$

Interpreting the Pythagorean Theorem in terms of "area" (Figure 39) we have:

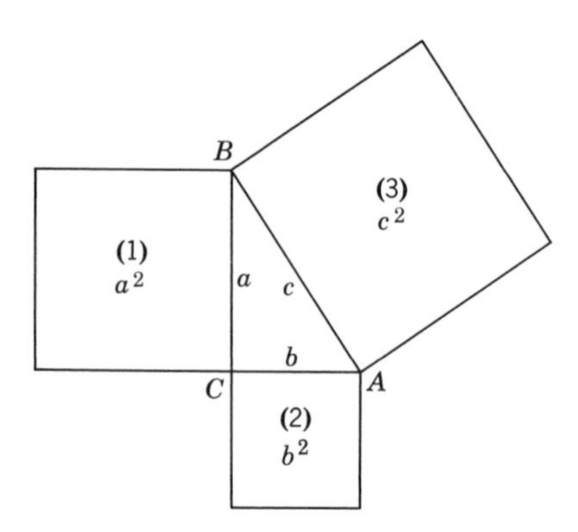

Figure 39. Area (1) + Area (2) = Area (3).

For every right triangle the square on the hypotenuse has an area equal to the sum of the areas of the squares on the other two sides (see Section 7.1).

For a proof of the theorem based on this interpretation, consider the diagram of Figure 40.

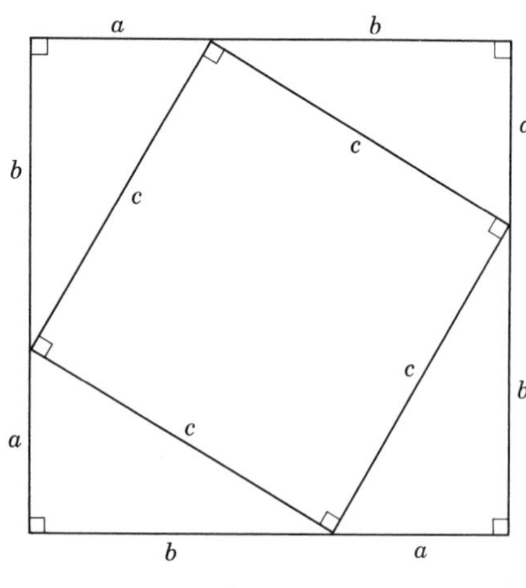

Figure 40

The area of the large square of side $a+b$ is equal to the area of the small square of side c plus the area of the four right triangles with legs a and b.

$$(a+b)^2 = c^2 + 4(\tfrac{1}{2}ab)$$
$$a^2 + 2ab + b^2 = c^2 + 2ab$$
$$a^2 + b^2 = c^2.$$

A demonstration involving areas which may be used at the elementary level involves cutting and matching. First construct the diagram shown in Figure 41 in relation to any right triangle ABC.

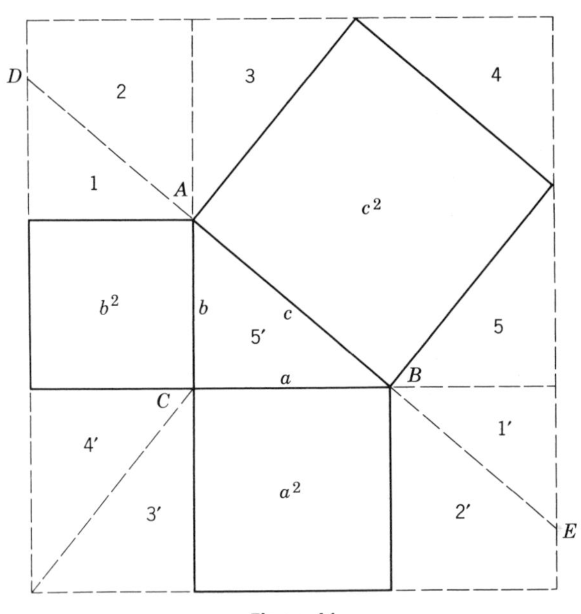

Figure 41

Cut out the figure on the line $DABE$ and compare the two sections by matching. Carefully done, this will illustrate that the areas are identical. By pairing like areas on opposite sides of this line and "subtracting" them from the original equal areas, we obtain

$$a^2 + b^2 = c^2.$$

The reader is to be cautioned that this does not form a "proof." It is merely a demonstration to add credibility to the statement of the theorem.

As an example of how "proofs" by cutting and fitting may be misleading, consider Figure 42.

The square is 8 by 8, hence has area 64. The rectangle is 5 by 13, hence has area 65. Where did the additional unit of area come from? Before reading further, see if you can develop an explanation.

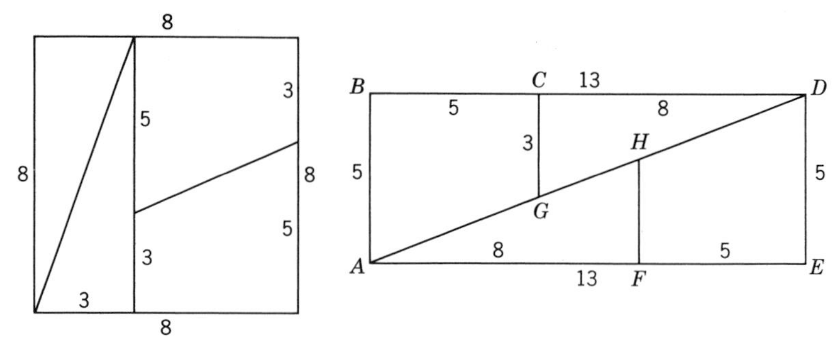

Figure 42

The angles at C and F are right angles, which means that the "fit" along the lines $\overleftrightarrow{HF}$ and $\overleftrightarrow{CG}$ is perfect. This leaves only the diagonal $\overleftrightarrow{AD}$ as a possible place for the extra square unit to hide. It may be that $\overleftrightarrow{AGHD}$ is not a straight line as it appears to be, or we might investigate angles BAF and CDE to make sure that they are right angles.

Let us use the properties of similar triangles. We know that $\overleftrightarrow{AFE}$ is a straight line. Why? Looking at the figure we see that $\overleftrightarrow{HF}$ is parallel to $\overleftrightarrow{DE}$ and *if* $\overleftrightarrow{AGHD}$ is a straight line, then $\triangle AFH \sim \triangle AED$. *If* these triangles are similar, then their corresponding sides are proportional. But $\frac{8}{3} \neq \frac{13}{5}$; hence these triangles are *not* similar. If the triangles are not similar, then $\overleftrightarrow{AGHD}$ is not a straight line. A slight bend in the line at G and H would account for the extra unit of area (see Figure 43).

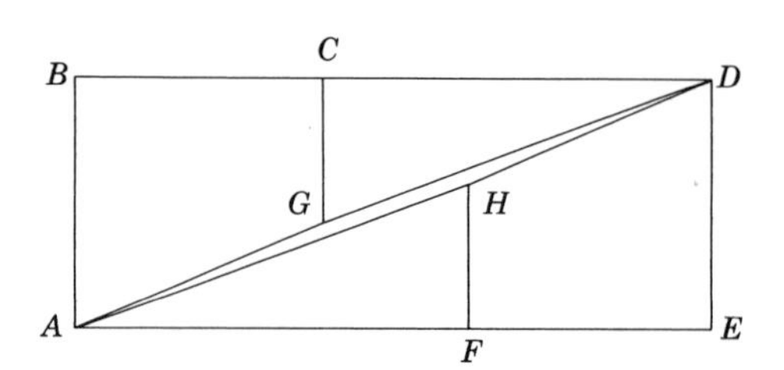

Figure 43. A square unit of area "hiding in a crack."

Exercise 8.12

1. Find the area of a square inscribed in a circle of radius

(a) 1 in. (b) 5 ft.

2. Find the area of a regular polygon of six sides (regular hexagon) inscribed in a circle of radius

(a) 10 in. (b) 8 ft.

3. Find the area of an equilateral triangle inscribed in a circle of radius

(a) 10 in. (b) 18 in.

4. In a 30, 60, 90° right triangle, the side opposite the 30° angle is equal to one-half the hypotenuse.

(a) What is the area of a parallelogram with sides 20 and 16 and included angle of 30°?
(b) What is the area of a parallelogram with sides 20 and 16 and included angle of 60°?

5. (a) The length of the diagonal of a square is 18. Find its area.
(b) The length of the diagonal of a square is 50. Find its area.

6. Prove that if two triangles are similar, the ratio of their areas is equal to the ratio of the squares of the lengths of two corresponding sides.

7. An alternate proof of the Pythagorean Theorem based on the area concept can be developed from the following figure:
Label the figure and complete the proof.

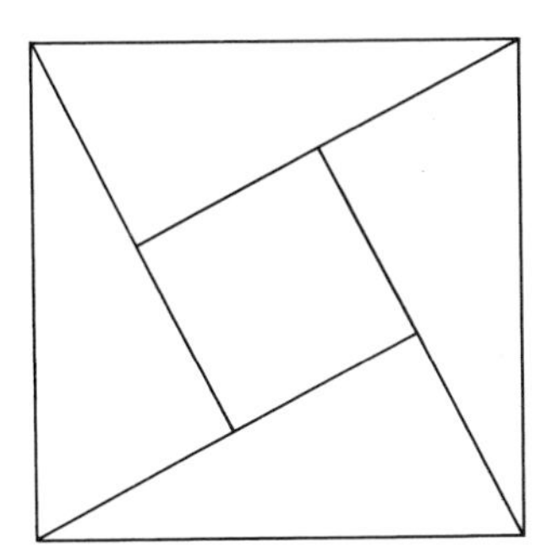

8.13 THE NUMBER π

The symbol π is much misunderstood. Many people think that it is something used in arithmetic to "stand for" $3\frac{1}{7}$, or 3.1416, and they are surprised to learn that π is a symbol for a particular number, just as 3 and $\sqrt{5}$ are symbols for numbers. π is an *irrational number* and as such would be an *infinite nonrepeating decimal.* The numbers symbolized by $3\frac{1}{7}$, 3.14, and 3.1416 are *rational approximations* to π.

The number symbolized by the Greek letter π, has a very interesting history. We will present a brief chronology of this number later in this section. First let us see if we can establish the *existence* of such a number.

8.13a The Existence of π and the Area of a Circular Region

Consider *any* two circles with centers O and O' of radii r and r', respectively, and $r < r'$ (see Figure 44). For simplicity let us reconstruct the circle with center O' so that it will be concentric with the circle with center O, that is, have the same center. From the common center O draw a line

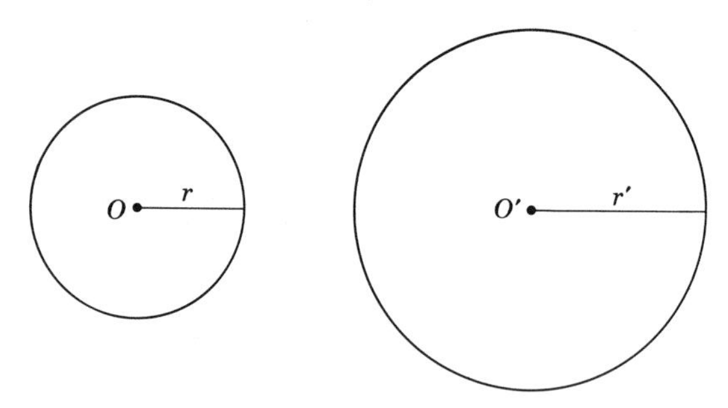

Figure 44

cutting the smaller circle at A and the larger circle at A'. From O draw a second line, different from $\overleftrightarrow{OA'}$, cutting the smaller circle at B and the larger circle at B'. Draw the segments $\overline{AB}$ and $\overline{A'B'}$. It is easy to show that $\triangle OAB \sim \triangle OA'B'$ (see Figure 45). Using the properties of similar triangles, we have

$$\frac{m(\overline{AB})}{n} = \frac{m(\overline{A'B'})}{n'}.$$

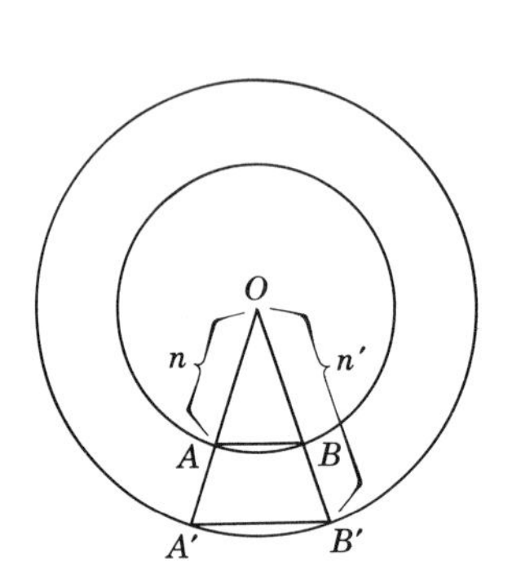

Figure 45

But this is an equality of ratios of real numbers; hence for any number n

$$\frac{(n)m(\overline{AB})}{n} = \frac{(n)m(\overline{A'B'})}{n'},$$

or

$$\frac{(n)m(AB)}{(2)n} = \frac{(n)m(A'B')}{(2)n'}.$$

If the circle were divided into n equal arcs and triangles formed by drawing in the chords and radii, then $(n)m(AB)$ would be the perimeter of the inscribed polygon and $(2)n$ would be the diameter of the smaller circle. Similarly, $(n)m(A'B')$ and $(2)n'$ would be the perimeter of the inscribed polygon and diameter of the larger circle. Then the equation

$$\frac{(n)m(\overline{AB})}{(2)n} = \frac{(n)m(\overline{A'B'})}{(2)n'}$$

means that the ratio of the perimeter of a regular inscribed polygon of n sides to the diameter of the circle is the same regardless of the size of the circle.

Now let us suppose n is allowed to become very, very large. The perimeters of the inscribed polygons come closer and closer to the circumference of the circle. In fact, the least upper bound of the sequence of real numbers denoting the perimeters is the length of the circumference of the circle (see Section 8.8). We would then have

$$\frac{\text{circumference of smaller circle}}{\text{diameter of smaller circle}} = \frac{\text{circumference of larger circle}}{\text{diameter of larger circle}}$$

This means that the ratio of the circumference of a circle to its diameter is the same regardless of the size of the circle, that is, this ratio is a *constant.* The name given to this constant is *pi* and it is symbolized π.

Definition 8.13a. The number π is the ratio of the circumference of any circle to its diameter.
Let c represent the circumference of any circle and d its diameter; then

$$\pi = \frac{c}{d}.$$

This approach to the existence of pi also leads to a reasonable explanation of the formulas $c = 2\pi r$ or $c = \pi d$ for the circumference and $A = \pi r^2$ for the area of a circle of radius r. Recall the formula developed for the area of a regular polygon with n sides of length s, namely,

$$A = (\tfrac{1}{2}ap) \qquad \text{or} \qquad A = \tfrac{1}{2}a(ns),$$

where a denotes the apothem and p the perimeter ns. As n becomes very large, the apothem a approaches the radius r of the circle and the perimeter ns approaches the circumference c of the circle. In fact, the least upper bounds of the sequences representing a and p are r and c, respectively. Then

$$A = \tfrac{1}{2}(rc).$$

But from the definition of π, $c = \pi d$ or $c = 2\pi r$, and

$$A = \tfrac{1}{2}r(2\pi r),$$

or

$$A = \pi r^2.$$

8.13b Calculation of π

The classical method of computing a numerical approximation to π makes use of inscribed and circumscribed regular polygons. Since $\pi = c/d$, the circumference of a circle of unit diameter is π. The computation may be begun by using an equilateral triangle or a square as the initial polygon; then, by doubling the number of sides, obtain polygons of 6, 12, 24, 36, . . . or 8, 16, 32, 64, . . . sides, respectively. Let us examine the procedure, using the square as the initial polygon.

As our first approximation we have (see Figure 46)

$$2\sqrt{2} < \pi < 4,$$

or

$$2.82 < \pi < 4.$$

For our second approximation let us double the number of sides of the inscribed polygon and the circumscribed polygon (see Figure 47). The length of a side of the inscribed octagon is

$$s=\sqrt{\left(\frac{\sqrt{2}}{4}\right)^2+\left(\frac{2-\sqrt{2}}{4}\right)^2}=\sqrt{\frac{2}{16}+\frac{4-4\sqrt{2}+2}{16}}$$

$$=\sqrt{\frac{2-\sqrt{2}}{4}}=\frac{1}{2}\sqrt{2-\sqrt{2}}$$

The perimeter is then $8(\frac{1}{2}\sqrt{2-\sqrt{2}})$ or approximately

$$4(0.7654)\cong 3.06.$$

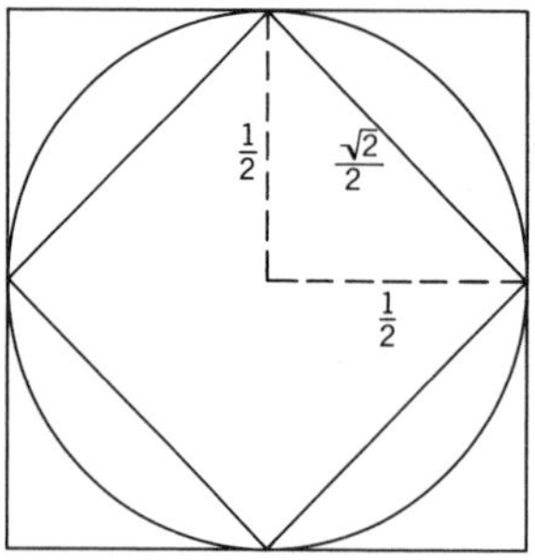

Figure 46

For the circumscribed polygon of side s',

$$\frac{s'/2}{\frac{(\sqrt{2-\sqrt{2}})}{4}}=\frac{1/2}{x},$$

where

$$x=\sqrt{\left(\frac{1}{2}\right)^2-\left(\frac{\sqrt{2-\sqrt{2}}}{4}\right)^2}.$$

Solving for s' and using approximate square roots yields

$$\frac{s'}{2}\cong 0.2071.$$

Then the perimeter of the circumscribed polygon is

$$(16)\left(\frac{s'}{2}\right)\cong 16(0.2071)\cong 3.31.$$

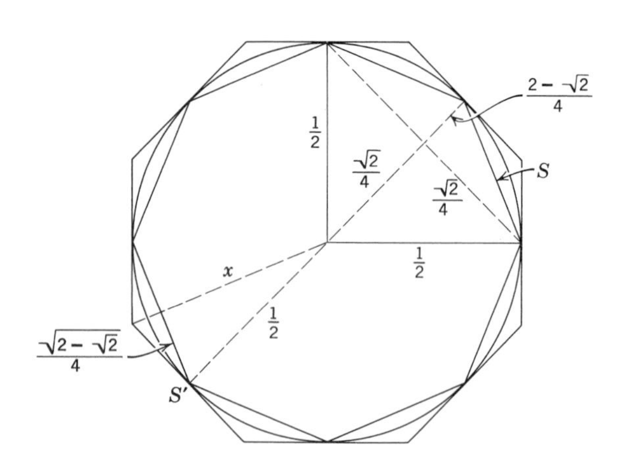

Figure 47

Hence, for our second approximation, we have

$$3.06 < \pi < 3.31.$$

This procedure can be continued indefinitely. Each successive computation yields a closer approximation to π.

The method of using inscribed polygons is called the "classical method" for computing π. It dates back to the time of Archimedes, about 240 B.C. (see Section 8.13c and Eves).

A method for computing π that involves area is based on the fact that a circle of radius 1 will have area π.

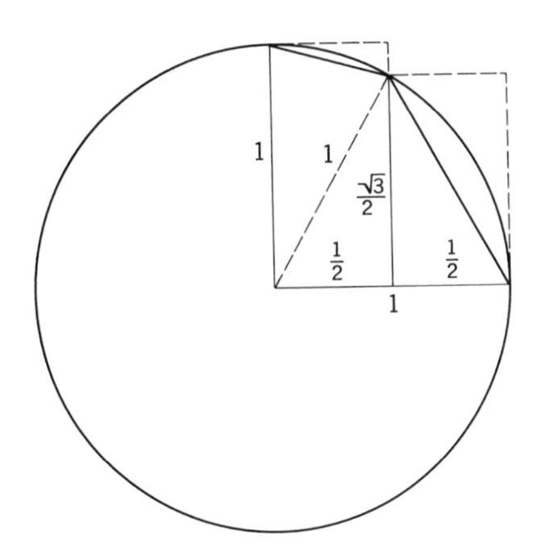

Figure 48

Considering just one quadrant to simplify calculations, we have, from the trapezoid and triangle, our first "lower" approximation to the area of the circle (see Figure 48).

4 (area of trapezoid + area of triangle)

$$= 4\left[\left(\frac{1}{2}\right)\left(\frac{1}{2}\right)\left(1+\frac{\sqrt{3}}{2}\right)+\left(\frac{1}{2}\right)\left(\frac{1}{2}\right)\left(\frac{\sqrt{3}}{2}\right)\right]$$

$$= 4\left(\frac{2+\sqrt{3}}{8}+\frac{\sqrt{3}}{8}\right) = 4\left(\frac{2+2\sqrt{3}}{8}\right),$$

which is approximately 2.732.

Using the rectangles, we have as our first "upper" approximation

$$4\left[1\left(\frac{1}{2}\right)+\frac{1}{2}\left(\frac{\sqrt{3}}{2}\right)\right] = 2+\sqrt{3},$$

which is approximately 3.732. Then

$$2.732 < \pi < 3.732.$$

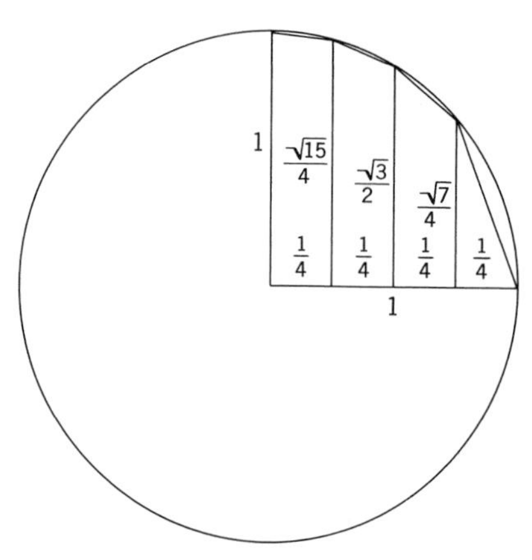

Figure 49

By dividing the radius into four equal parts we have three trapezoids and a triangle approximating the area (see Figure 49).

"Lower" approximation:

$$4\left[\frac{1}{8}\left(\frac{4+\sqrt{15}}{4}\right)+\frac{1}{8}\left(\frac{\sqrt{15}+2\sqrt{3}}{4}\right)+\frac{1}{8}\left(\frac{2\sqrt{3}+\sqrt{7}}{4}\right)+\frac{1}{2}\left(\frac{1}{4}\right)\left(\frac{\sqrt{7}}{4}\right)\right]$$

$$=\frac{1}{2}\left(\frac{4+\sqrt{15}}{4}+\frac{\sqrt{15}+2\sqrt{3}}{4}+\frac{2\sqrt{3}+\sqrt{7}}{4}+\frac{\sqrt{7}}{4}\right)$$

$$=\frac{1}{8}(4+2\sqrt{15}+4\sqrt{3}+2\sqrt{7}),$$

which is approximately 2.9957.

"Upper" approximation:

$$4\left[1\left(\frac{1}{4}\right)+\left(\frac{\sqrt{15}}{4}\right)\left(\frac{1}{4}\right)+\left(\frac{\sqrt{3}}{2}\right)\left(\frac{1}{4}\right)+\left(\frac{\sqrt{7}}{4}\right)\left(\frac{1}{4}\right)\right]$$

$$=1+\frac{\sqrt{15}}{4}+\frac{\sqrt{3}}{2}+\frac{\sqrt{7}}{4},$$

which is approximately 3.4957.

Our second approximation is

$$3.00 < \pi < 3.50.$$

As with the inscribed polygons, this procedure can be continued indefinitely, each successive computation yielding a closer approximation to π.

There are other geometric procedures that may be used at elementary levels which do not involve so much computation. Many experiments are suggested in standard texts to give the pupil information about π beyond the definition. One simple one is to construct a unit circle on squared paper (graph paper) and count squares to obtain an approximation to the area. This will be left as an exercise for the student.

8.13c Chronology of π

This will not be a true "chronology" of π in that it is not a date-by-date itemizing of the history of π. It is, rather, an informal discussion of selected items in the chronology of π.

We are led to believe that in ancient times the ratio of the circumference to the diameter of a circle was taken to be 3 (see the Biblical references: I Kings 7:23; II Chronicles 4:2). The Rhind papyrus gives us $\pi = (4/3)^4 = 3.1604....$ It is believed the first scientific attempt to compute π was made by Archimedes about 240 B.C. Using the classical method, he determined that π was between 223/71 and 22/7. This represents remarkable accuracy in approximating square roots with rational numbers.

About 400 years later Ptolemy of Alexandria in this famous *Syntaxis Mathamatica* developed a table of chords of a circle subtended by central angles of each degree and half degree. From this, using a regular inscribed polygon of 360 sides, he obtained a value, given in sexagesimal notation, as $3\ 8'30''$. Transcribed to decimal language this would be 377/120 or 3.1416 rounded to four places.

About 480 A.D. Tsu Ch'ung-chih, Chinese, gave the rational approximation 355/113. In decimal language this is 3.1415929 . . . , Which is accurate to six places. Around 1150 A.D. the Hindu mathematician, Bhaskara, gave 3927/1250 as an accurate value of π, 22/7 as an inaccurate value, and $\sqrt{10}$ for ordinary work.

In the late 1500's and early 1600's the following computations of π were carried out:

François Vieta, a French mathematician, found π correct to 9 decimal places by the classical method, using polygons having $6(2^{16}) = 393{,}216$ sides.

Adriaen van Roomen of the Netherlands found π correct to 15 decimal places by the classical method, using polygons having $2^{30} = 1{,}073{,}741{,}824$ sides.

Ludolph van Ceulen of Germany computed π to 35 decimal places by the classical method, using polygons having $2^{62} = 4{,}611{,}686{,}018{,}427{,}387{,}904$ sides. This was considered such an unusual accomplishment that, for a time in Germany, this 35-place approximation to π was called the Ludolphian Number.

One is led to wonder why so much time and effort were spent in the computation of π. One reason might be that these men were looking for a repeating sequence in the decimal approximation to π. If it could be found, they would then know that π was a rational number.

Later computations were based on infinite series:

> Abraham Sharp—71 correct places. 1699.
> De Lagny—112 correct places. 1719.

Much speculation came to an end when in 1767 Johann Heinrich Lambert proved that π is irrational. This did not stop the "π computers," however. William Shanks of England computed π to 707 places. For a long time this remained the most fabulous piece of calculation ever performed. It occupied Shanks for more than 15 years. In 1946 D. F. Ferguson of England found errors in Shanks' value of π. He published a corrected value to 710 places. In the same month J. W. Wrench, Jr., of the United States published an 808-place value of π, but Ferguson found an error in the 723rd place. In January 1948 they jointly published a corrected and checked value for π to 808 places.

To conclude this summary on the calculations of π, we turn to the electronic computer. The ENIAC (Electronic Numerical Integrator and Computer) at the Army Ballistic Research Laboratories in Aberdeen, Maryland, in about *70 hours* gave π to *2035 places*, checking the Ferguson-Wrench result of 808 places. (Compare this with the efforts of Shanks.) For a more recent discussion relating to π, see the article by R. K. Pathria cited in the references.

Also, recently computers were used to compute the decimal fraction for π to one million digits. The objective was to test the sequence of digits for randomness. That is, in an arbitrary place in the decimal approximation for π, does each of the ten digits, 0–9, have equal probability of occurring? Studies are continuing.

Among the curiosities connected with π are various mnemonics that have been devised for the purpose of remembering π to a large number of decimal places. In the following, by A. C. Orr, one has merely to replace each word by the number of letters it contains to obtain π correct to 30 decimal places.

> Now I, even I, would celebrate
> In rhymes unapt, the great
> Immortal Syracusan, rivaled nevermore
> Who in his wondrous lore,
> Passed on before,
> Left men his guidance
> How to circles mensurate.

Another similar mnemonic is

> See, I have a rhyme
> Assisting my feeble brain,
> Its tasks ofttimes resisting.

For a more complete chronology of π, see Eves and Schepler.

Exercise 8.13

1. Draw a circle of radius 10 units ($\frac{1}{4}$ in.) on graph paper of $\frac{1}{4}$ in. squares.

(a) Cross out all squares the circle passes through.
(b) Count all squares inside the circle not crossed out.
(c) Count the number of crossed-out squares.
(d) From the foregoing draw some conclusions about the value of π.

2. Calculate the third approximation for π, using the "area" method.

3. If a is the side of a regular polygon inscribed in a circle of radius r, then

$$b = \sqrt{2r^2 - r\sqrt{4r^2 - a^2}}$$

is the side of a regular inscribed polygon having twice the number of sides.

(a) Use this information to calculate the third approximation to π with an inscribed polygon of 16 sides.

4. In the Middle Ages, a common approximation for a square root was

$$\sqrt{n} = \sqrt{a^2 + b} \cong a + \frac{\text{b}}{2a+1}.$$

By taking $n = 10 = 3^2 + 1$, show why it may be that $\sqrt{10}$ was so frequently used for π.

5. Find the perimeter of a regular 12-sided polygon inscribed in a circle of radius 1, thus finding an approximation for π.

6. What does the irrationality of π imply about its decimal representation?

7. A *radian* is the measure of the central angle in a circle subtended by an arc equal in length to the radius of the circle.

(a) If θ is the measure of a central angle of a circle given in radian measure, show that the length of the arc cut off by the sides of this angle is given by $r\theta$.
(b) How many radians are equivalent to 360°?

8.14 INTRODUCTION TO COORDINATE GEOMETRY

We have used the concept of the number line to strengthen the concept of numbers themselves. There is also an interesting relation between the concept of the number line and geometry. When the terms "point" and "line" are used in plane geometry, they are undefined objects endowed with certain properties determined by the axiom. In order to have some intuitive idea of how these terms might be interpreted in the physical world, expressions such as "a point is like the tip of a pin" or "a straight-line segment is like the edge of a ruler" are used. Postulates such as "there is one and only one line through two distinct points" and "two distinct lines are parallel or meet in a point" sharpen our concept of "point" and "line." The "plane" can be discussed informally in the same way and made precise by a set of postulates. Since the points and lines are undefined, we might have a geometry in which the "line" is the number line for the integers or the number line for the rational numbers. We discussed the *number line* for the *integers*, the *rational numbers*, and the *real numbers*. The number "line" for the integers should be depicted as the set of points equally spaced with no points between consecutive pairs:

$$\begin{array}{ccccccccc} \cdot & \cdot & \cdot & \cdot & \cdot & \cdot & \cdot \\ \ldots -3 & -2 & -1 & 0 & 1 & 2 & 3\ldots \end{array}$$

$$\begin{array}{ccccccccccc} \cdot & \cdot & \cdot & \cdot & \cdot & \cdot & \cdot & \cdot & \cdot & \cdot & \cdot \\ \ldots -5 & -4 & -3 & -2 & -1 & 0 & 1 & 2 & 3 & 4 & 5\ldots \end{array}$$

The rational number "line" is difficult to depict because the rational numbers are *dense*, that is, between any two distinct rational numbers there is an infinite number of distinct rational numbers. Furthermore, the set of rational numbers is *not* complete. This would imply there are "holes" in the rational number "line."

The real number line is the "continuous" line one usually visualizes. This is the "line" referred to in plane geometry. We indicated in problem 7, Exercise 7.2a, that the *rational number "line"* has too many "holes" for the purposes of ordinary geometry. It is possible to have two rational number "lines" in a plane which are not parallel and which do not have a point in common (see Figure 50).

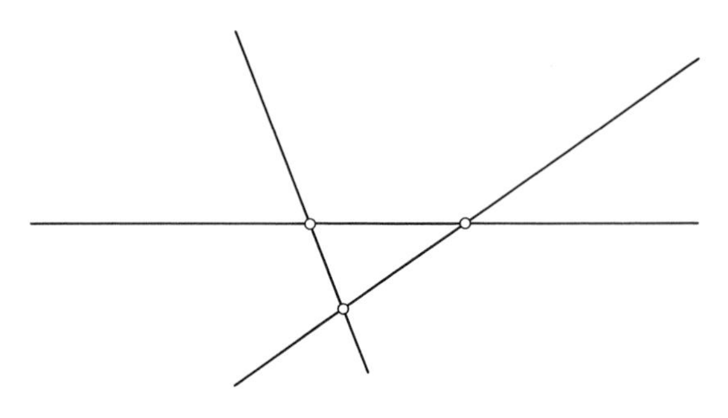

Figure 50

8.14a The Euclidean Plane

The pictorial representation of $R \times R$, where R is the set of real numbers, is called the *Euclidean Plane.* Just as we labeled the points on the real number line with the real numbers, so also we label the points in the plane with the ordered pairs of $R \times R$. We use the ordered pair (x, y) to denote an arbitrary ordered pair of real numbers and refer to it as the point (x, y) in the geometric representation. To establish this one-to-one correspondence between the set of ordered pairs of real numbers and the points in the plane, we choose a pair of perpendicular lines (usually horizontal and vertical) and call these the *axes* of the system. The horizontal line is called the x-axis and the vertical line the y-axis. Their point of intersection is called the origin and the ordered pair $(0, 0)$ is made to correspond to this point. The ordered pairs $(x, 0)$ are made to correspond to points on the x-axis in the same manner as the real numbers x were made to correspond to the points on the real number line. The ordered pairs $(0, y)$ are made to correspond to the points on the y-axis in a similar manner, with the upward direction taken as the positive direction.

We obtain the point corresponding to the ordered pair (x, y) by drawing a line parallel to the y-axis through the point $(x, 0)$ and a line parallel to the x-axis through the point $(0, y)$. The point of intersection of these lines corresponds to the ordered pair (x, y) (see Figure 51).

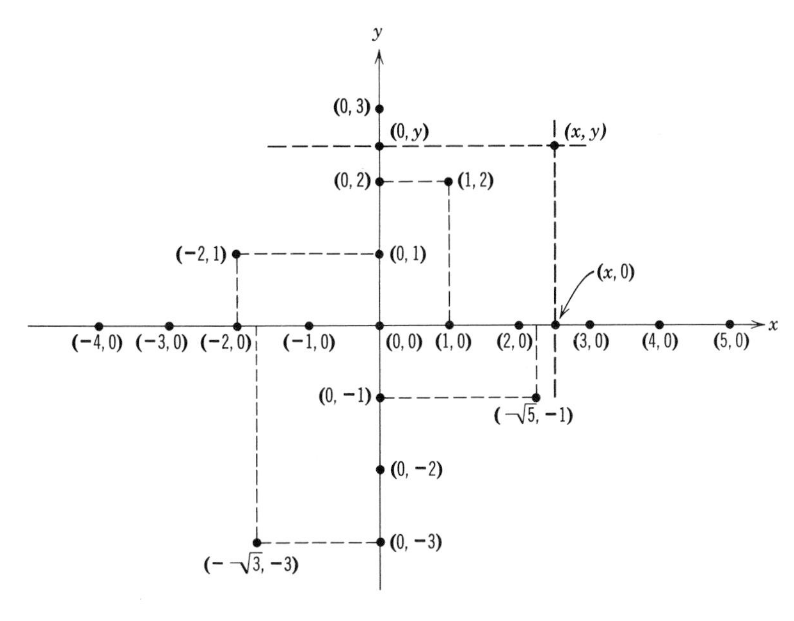

Figure 51

Exercise 8.14a

1. Describe the set $J \times J$, where J is the set of integers.

2. If the set $J \times J$ was represented pictorially, as in Section 2.6, indicate what this would look like by plotting a few points. Circle those points in the pictorial representation of $J \times J$ whose distance from the origin $(0, 0)$ is less than or equal to 5.

3. Plot several of the points in the pictorial representation of $R \times R$ whose coordinates are related as follows:

(a) The second coordinate is twice the first coordinate, that is, $(0, 0)$,

(1, 2), (2, 4), etc. This relation can be described by the equation $y=2x$. The set of points in the pictorial representation is called the *graph* of $y=2x$ and is a straight line.

(b) The second coordinate is the same as the first. This relation can be described by the equation $y=x$.

4. Plot the points (x, y) whose coordinates are given by the following equations:

(a) $y=2x+3$ (b) $y=x+4$
(c) $y=-x$ (d) $y=-3x+5$

5. Indicate, by shading, the set of points (x, y) satisfying the following inequalities:

(a) $x \leqq 3$ (b) $y \leqq 0$
(c) $-3 \leqq x \leqq 3$ (d) $|x| \leqq 3$
(e) $0 \leqq x \leqq 4$ *and* $0 \leqq y \leqq 3$ (f) $x+y \leqq 5$

8.14b Point Sets in the Plane

In problems 3, 4, and 5, Exercise 8.14a, we discussed particular *point sets* in the plane. The circle in plane geometry can also be described as a point set. It is usually defined as the set of points equidistant from a given point. What can we say about the coordinates of those points equidistant from the origin (0, 0)? We saw earlier that there are many ways of defining "distance." These were examples of distance, "following the streets" and "as the crow flies." The "distance" in ordinary geometry is the latter, the "straight-line distance." Since the coordinates of a point in the plane are given by the ordered pair of real numbers (x, y), we can compute the distance "as the crow flies" from (0, 0) to (x, y) (see Figure 52).
If we let d denote the distance from (0, 0) to (x, y), then

$$d=\sqrt{x^2+y^2}.$$

Let us consider the circle centered at (0, 0) with radius 5 (see Figure 53). What are the coordinates of some of the points on the circle? We list

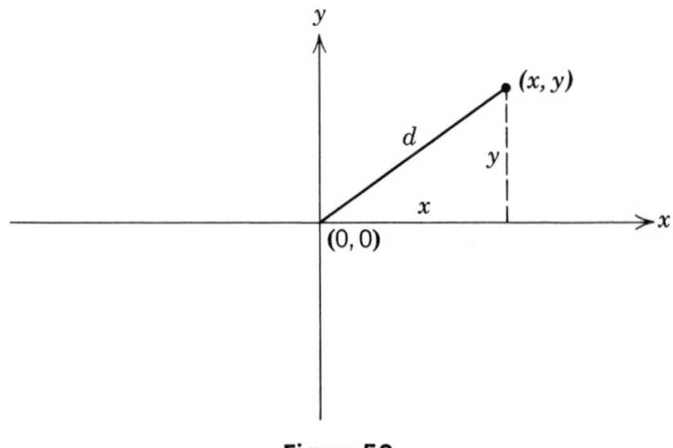

Figure 52

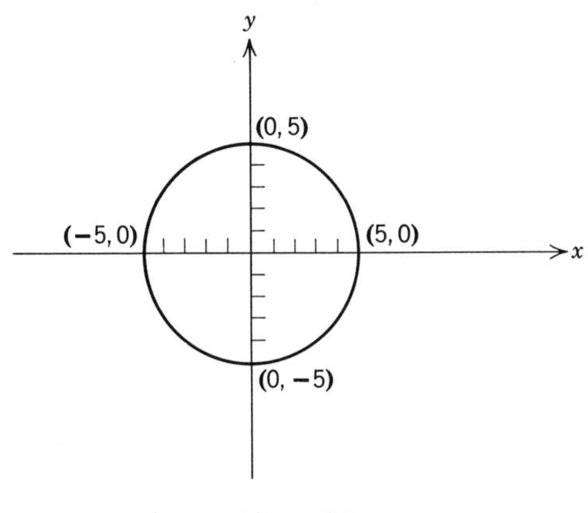

Figure 53

a few: (0, 5), (0, −5), (5, 0), and (−5, 0). Notice that each of these points is five units from the origin along one of the axes.

Any other point (x, y) on the circle must also be at a distance of 5 units from the origin, that is, its coordinates must satisfy the relation

$$\sqrt{x^2 + y^2} = 5.$$

We can write this equation without the radical sign by squaring both sides. We get the equation whose graph is a circle of radius 5, centered at the origin (Figure 54):

$$x^2 + y^2 = 25.$$

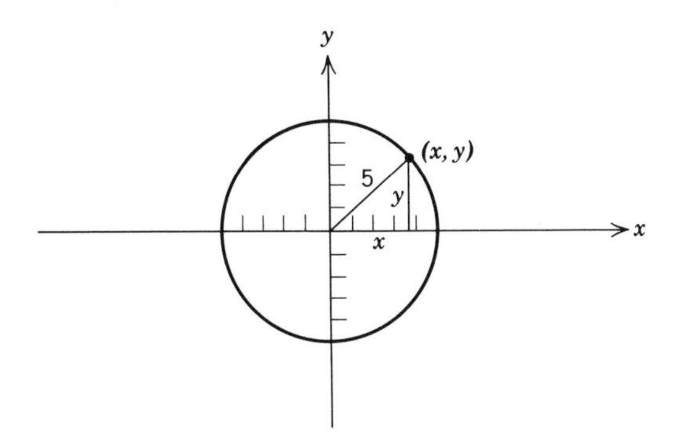

Figure 54

The relation of the algebra of real numbers to the study of point sets in the plane (plane geometry) leads to the investigation of geometry by analytic methods. This is formally called analytic geometry and further development of the concepts involved is beyond the scope of this book. (See any recent textbook in introductory college mathematics.)

Exercise 8.14b

1. (a) How far is the point (5, 12) from the point (0, 0) in straight-line distance?

(b) How far is the point (−12, 5) from the point (0, 0) in straight-line distance?

(c) Plot these points and two other points whose distance from (0, 0) is the same as the points in (a) and (b).

(d) Write the equation describing the set of *all* points whose distance from (0, 0) is the same as in (a), (b), and (c).

2. Write the equation of a circle of radius r centered at (a) the origin, (b) the point (a, b).

3. (a) What would be the distance from the point (0, 0) to the point (3, 4), using the "along the street" distance?

(b) What would be the distance from the point (0, 0) to the point (x, y), using the "along the street" distance?

(c) Using the "along the street" distance, plot several points whose distance from (0, 0) is equal to 7.

4. If we retain the definition of circle as the set of points equidistant from a fixed point, and use the "along the street" interpretation of distance as in problem 3(c), describe this new "circle."

REFERENCES

Brumfiel, Charles F., Robert E. Eicholz, and Merril E. Shanks, *Geometry*, Addison-Wesley Publishing Co., Reading, Mass., 1960.

Courant, R. and H. Robbins, *What is Mathematics*, Oxford University Press, New York, 1941.

Eves, Howard, *An Introduction to the History of Mathematics*, revised-edition, Holt, Rinehart and Winston, New York, 1964.

Newman, James R., *The World of Mathematics*, Simon and Schuster, New York, 1956.

Pathria, R. K., "A Statistical Analysis of Randomness among the First 10,000 Digits of π, *Math. Comp.*, **16**, 1962, pp. 188–197.

Schepler, H. C., "The Chronology of Pi," *Mathematics Magazine*, January–February, March–April, and May–June issues, 1950.

School Mathematics Study Group, *Mathematics for Junior High School*, vols. I, II, and III.

Appendix

Table 1 Powers and roots

N	N^2	$\sqrt{N}$	N	N^2	$\sqrt{N}$
1	1	1	26	676	5.099
2	4	1.414	27	729	5.196
3	9	1.732	28	784	5.292
4	16	2	29	841	5.385
5	25	2.236	30	900	5.477
6	36	2.449	31	961	5.568
7	49	2.646	32	1,024	5.657
8	64	2.828	33	1,089	5.745
9	81	3	34	1,156	5.831
10	100	3.162	35	1,225	5.916
11	121	3.317	36	1,296	6
12	144	3.464	37	1,369	6.083
13	169	3.606	38	1,444	6.164
14	196	3.742	39	1,521	6.245
15	225	3.873	40	1,600	6.325
16	256	4	41	1,681	6.403
17	289	4.123	42	1,764	6.481
18	324	4.243	43	1,849	6.557
19	361	4.359	44	1,936	6.633
20	400	4.472	45	2,025	6.708
21	441	4.583	46	2,116	6.782
22	484	4.690	47	2,209	6.856
23	529	4.796	48	2,304	6.928
24	576	4.899	49	2,401	7
25	625	5	50	2,500	7.071

Table 1 *(continued)*

N	N^2	$\sqrt{N}$	N	N^2	$\sqrt{N}$
51	2,601	7.141	76	5,776	8.718
52	2,704	7.211	77	5,929	8.775
53	2,809	7.280	78	6,084	8.832
54	2,916	7.348	79	6,241	8.888
55	3,025	7.416	80	6,400	8.944
56	3,136	7.483	81	6,561	9
57	3,249	7.550	82	6,724	9.055
58	3,364	7.616	83	6,889	9.110
59	3,481	7.681	84	7,056	9.165
60	3,600	7.746	85	7,225	9.220
61	3,721	7.810	86	7,396	9.274
62	3,844	7.874	87	7.569	9.327
63	3.969	7.937	88	7,744	9.381
64	4.096	8	89	7,921	9.434
65	4.225	8.062	90	8,100	9.487
66	4,356	8.124	91	8,281	9.539
67	4,489	8.185	92	8,464	9.592
68	4,624	8.246	93	8,649	9.644
69	4,761	8.307	94	8,836	9.695
70	4,900	8.367	95	9,025	9.747
71	5,041	8.426	96	9,216	9.798
72	5,184	8.485	97	9,409	9.849
73	5,329	8.544	98	9,604	9.899
74	5,476	8.602	99	9,801	9.950
75	5.625	8.660	100	10,000	10

Table 2 Prime Numbers Between 1 and 1000

2	109	269	439	617	811
3	113	271	443	619	821
5	127	277	449	631	823
7	131	281	457	641	827
11	137	283	461	643	829
13	139	293	463	647	839
17	149	307	467	653	853
19	151	311	479	659	857
23	157	313	487	661	859
29	163	317	491	673	863
31	167	331	499	677	877
37	173	337	503	683	881
41	179	347	509	691	883
43	181	349	521	701	887
47	191	353	523	709	907
53	193	359	541	719	911
59	197	367	547	727	919
61	199	373	557	733	929
67	211	379	563	739	937
71	223	383	569	743	941
73	227	389	571	751	947
79	229	397	577	757	953
83	233	401	587	761	967
89	239	409	593	769	971
97	241	419	599	773	977
101	251	421	601	787	983
103	257	431	607	797	991
107	263	433	613	809	997

Answers to Selected Exercises

Exercise 1.3a

1. (a) 𓎆𓎆𓎆𓎆𓎆𓎆𓎆 𓏺𓏺𓏺𓏺𓏺𓏺𓏺 (c) 𓂭𓂭𓂭𓂭𓂭𓂭𓂭𓂭𓂭 𓍢𓍢𓍢𓍢𓍢𓍢𓍢𓍢𓍢 𓏺𓏺𓏺𓏺𓏺𓏺𓏺𓏺𓏺 (e) 𓆼𓆼𓍢 𓎆𓎆 𓏺𓏺𓏺𓏺 (g) 𓍢𓍢𓍢𓍢𓍢𓍢𓍢𓍢 𓏺𓏺𓏺𓏺𓏺𓏺𓏺𓏺

2. (a) 1,220,453 (c) 1,033,306 **3.** (a) 𓁨𓁨𓆐𓆐𓆐𓆐𓆐𓂭𓂭 𓍢𓍢𓍢𓍢𓍢𓍢 𓏺𓏺𓏺𓏺𓏺𓏺

4. (b) is largest. (d) is smallest.

Exercise 1.3b

1. (a) XXVI (c) XLIX (e) CDXXXI (g) MDLI (i) MMCDIX
2. (a) 37 (c) 94 (e) 457 (g) 1151 (i) 2999
4. (a) MMVI (c) DXLIII **5.** (a) MCML **6.** MMCMXXIX
7. (a) LXXIV (c) CLXXXVIII (e) CMXIV (g) MMCCCII
(i) MMMMMCMXCVIII

Exercise 1.3c

1. (a) λd (c) $\mu\theta$ (e) $\nu\lambda\alpha$ (g) $\alpha'\phi\nu\alpha$ (i) $\beta'\ \theta$
2. (a) 44 (c) 653 (e) 172 (g) 6435 (i) 54,567
4. (a) $\omega\pi\delta$ **5.** $\chi\pi\eta$ is largest. $\chi\nu\delta$ is smallest. **6.** (a) $\pi\eta$ (g) $\alpha M\beta' o$

Exercise 1.4a

1. (a) 四十二 (c) 三十六 (e) 二百四十六 (g) 二千百四十六 (i) 五千四百六十九

2. (a) 36 (c) 208 (e) 2535

Exercise 1.5b

1. (a) 10^7 (c) 2^5 **2.** 100,000 (c) 1/3 (e) 1
3. (a) 1/8 (c) $3^{27} = (19{,}683)^3 = 7{,}625{,}597{,}484{,}987$ (e) $2^8 = 256$

Exercise 1.5c

1. $1 \cdot 10^6 + 0 \cdot 10^5 + 2 \cdot 10^4 + 0 \cdot 10^3 + 3 \cdot 10^2 + 0 \cdot 10^1 + 4 \cdot 10^0$
3. $1 \cdot 10^1 + 0 \cdot 10^0$ **5.** $3 \cdot 10^2$; $3 \cdot 10^1$; $3 \cdot 10^0$
6. (a) 2^9 (c) a^8 (e) m^2 (g) $1/x^4 = x^{-4}$
7. (a) 100,000 (c) $8 - 1 = 7$ (e) 625 (g) 1/8

Exercise 1.5d

1. (a) $1 \cdot 10^1 + 2 \cdot 10^0$ (c) $3 \cdot 10^2 + 0 \cdot 10^1 + 2 \cdot 10^0$
(e) $1 \cdot 10^4 + 0 \cdot 10^3 + 0 \cdot 10^2 + 0 \cdot 10^1 + 0 \cdot 10^0$ (g) $1 \cdot 10^1 + 1 \cdot 10^0$
2. (a) 10^5; 100,000 (c) 10^{10}; 10,000,000,000 (e) 2^7; 128 (g) 2^1; 2
3. (12) (a) XII; (b) ∩ ||; (c) $\iota\beta$. (302) (a) CCCII; (b) ∩ ||; (c) $\tau\beta$
(10,000) (a) X; (b) ∩ |; (c) *M*. (11) (a) XI; (b) ∩ |; (c) $\iota\alpha$
5. 13 **6.** (a) $6 \cdot 10^{12}$ (c) $18 \cdot 10^5 = 1.8 \cdot 10^6$
7. (a) $6.5 \cdot 10^{-6}$ (c) $8 \cdot 10^{-9}$ (e) $6 \cdot 10^{12}$
8. (a) 8,700,000,000 (c) 0.000000087 (e) 0.000000005
9. (a) 10^2; 100 (c) 10^0; 1
(e) $(7.25)(2.16)(10^{-3}) = 15.66(10^{-3}) = 1.566(10^{-2}) = 0.01566$

Exercise 1.7

1. 385 (a) 999 ∩∩∩∩∩∩∩∩ ||||| (c) $\tau\pi\epsilon$

(b) 三百八十五

2. (a) 27
4. (a) 1, 10, 11, 100, 101, 110, 111, 1000, 1001, 1010, 1011, 1100, 1101, 1110, 1111, 10000, 10001, 10010, 10011, 10100, 10101, 10110, 10111, 11000, 11001,
(b) $10101_{\text{two}} = 21_{\text{ten}}$; $1110011_{\text{two}} = 115_{\text{ten}}$
5. *A*, *C*, *D*
6. See answer to problem 5.
7. (a) $2^8 \cdot 3^6$ **8.** (a) 10,000 (c) $2{,}000{,}000 = 2 \cdot 10^6$
9. (a) 2^9 **11.** A symbol for one; the additive property.
12. (a) 25 = △▭▭*I*; 100 = ⊡△△▭; 197 = ⊡⊡⊡▭*I* (b) 25 = *I*△*L*▭*I*;
100 = *I*□*L*△▭; 197 = *F*□*I*▭*I* (c) 25 = *ILI*; 100 = *ILIO*; 197 = *FOII*
13. (a) 360 = *EEFFFFLL* (b) 360 = *LELO*

Exercise 2.2

1. 32 is a numeral and {32} is a singleton set.
2. (a) Easy; {April, June, September, November} (c) Easy; {44, 46, 48, 50,...}
(e) Easy; {0} (g) Difficult (i) Difficult
3. (a) 0, 2, 4, 6, 8, 10, 12, 14, 16, 18 (c) The even whole numbers
(e) $m = 2k$ and k a whole number
4. (a) 1, 3, 5, 7, 9, 11, 13, 15, 17, 19 (c) The odd whole numbers

Exercise 2.3

1. (a) P, S, Q, and $\emptyset$ (c) P
3. Yes, every element of S is an element of P.
5. Yes, every element of S is an element of S
6. (a) For example, $S_{(\text{even})} = \{0, 2, 4, 6, 8\}$; $S_{(\text{odd})} = \{1, 3, 5, 7\}$ (c) $\{2\}$; ϕ; $\{0, 1, 2, 3, 4, 5\}$; S itself; $\{5, 6, 7, 8\}$ (e) $\{0, 1\}$
7. $S = \{x|x \text{ is an even number}\}$; $T = \{x|x \text{ is an odd number}\}$
10. (a) No (c) Yes
11. (a) {The first president of the United States}, {The "father of our country"}, {The husband of Martha Washington}.

Exercise 2.4

1.

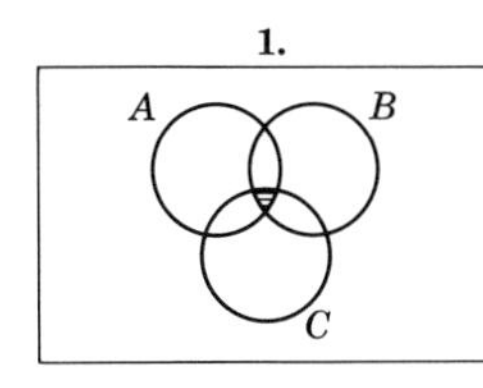

3.

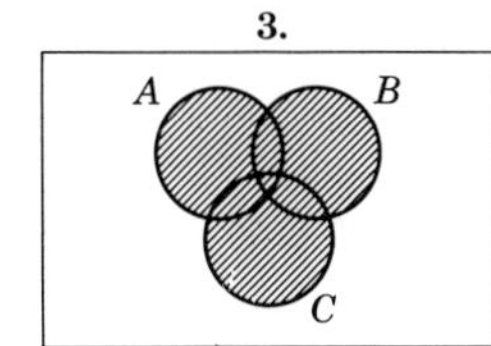

5.

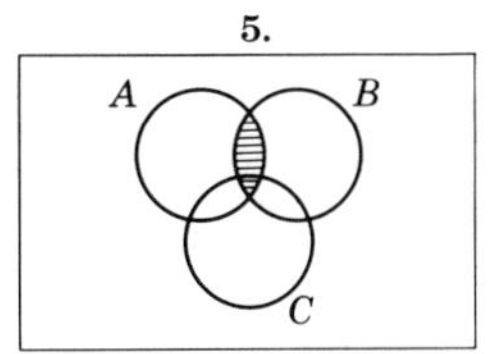

7.

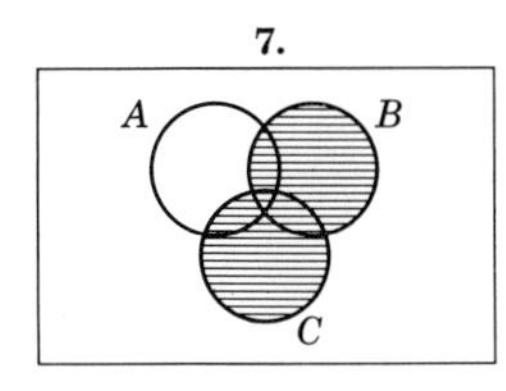

9.

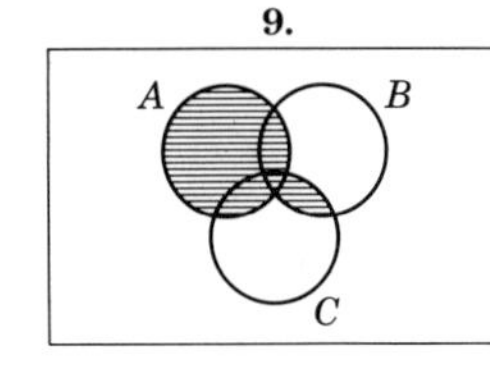

11.

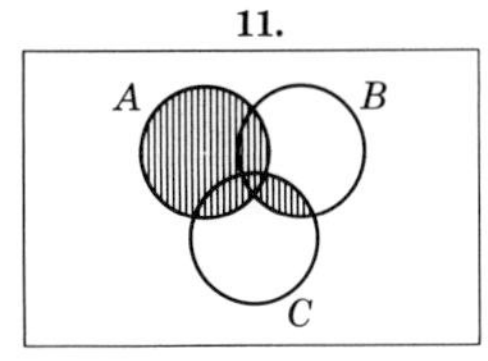

13. (a) $B \cap (A \cup C) = (B \cap A) \cup (B \cap C)$
    (c) $A \cup (B \cap C) = (A \cup B) \cap (A \cup C)$
14. (a) Ted, Tim (c) Ted, Tim, John, Jill

Exercise 2.6

1. (a) The union of a set with itself is the set itself. The union of a set with the empty set is the set itself. The union of a set with the universal set is the universal set. (c) If a set is a proper subset of each of two other sets then it is a proper subset of their union. (e) The intersection of a set with itself is the set itself. The intersection of a set with the universal set is the set itself. The intersection of a set with the empty set is the empty set. (g) If a set is a proper subset of each of two others, then it is a subset of their intersection.
2. (a) $A \cup B = \{1, 2, 3, 4, 5\}$; $A \cap B = \phi$
   (c) $A \cup B = \{1, 2, 3, 5\}$; $A \cap B = \{1, 3\}$ (e) $A \cup B = \{1, 2, 3\}$; $A \cap B = \{2, 3\}$
3. (a) $E \cup U = U$ (c) $A \cup B$ is the set of odd natural numbers and 2, 4, and 6
   (e) $E \cap U = E$ (g) $A \cap B$ is the first three odd natural numbers; $\{1, 3, 5\}$
   (i) $(A \cup B) \cap E$ is the first three even natural numbers; $\{2, 4, 6\}$
   (k) $(E \cap A) \cup B = B$; $\{1, 2, 3, 4, 5, 6\}$
4. (a) ϕ (c) A (e) $A \cap B$
5. (a) $A \cup B$ (c) B (e) $A \cup B$
6. (a) $A = \{2, 3, 4, 5\}$ (c) $A = \{3, 4\}$; $B = \{2, 3\}$

7. (a)

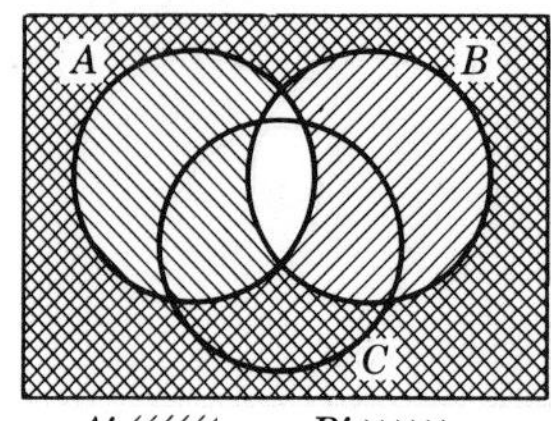

A' ////// B' \\\\\\

$A' \cap B'$ ✕✕✕✕

A B C

$(A \cup B)'$ |||||||

9. $2^5 = 32$ subsets; 31 proper subsets

10. (a) The set of specimens that react to both the A-test and B-test.
(b) The set of specimens that react to the B-test.
(c) O-Neg, A-Neg, B-Neg, and AB-Neg.
(d) O-Neg and B-Neg.
(e) A' B' Rh'
(f) $0'$
(g) A B Rh
(h) All of them.

11. (a) $A, B, A \cap B$ (c) $A \cup B \cup Rh$

Exercise 2.7

1. {(blue, 31), (blue, 43), (blue, 47), (blue, 59), (green, 31), (green, 43), (green, 47), (green, 59), (gray, 31), (gray, 43), (gray, 47), (gray, 59)}

2. (a) $\{0, 1\}, \{0\}, \{1\}, \emptyset$ (c) $\{(0, 2), (1, 2)\}$

3. (a) $X = \{n | n \text{ is an even whole number}\}$
(c) $Z = \{k | k \text{ is an odd whole number}\}$

(a)

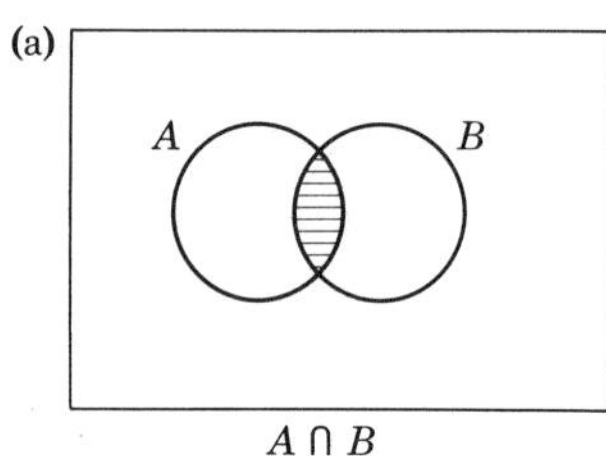

$A \cap B$

(c)

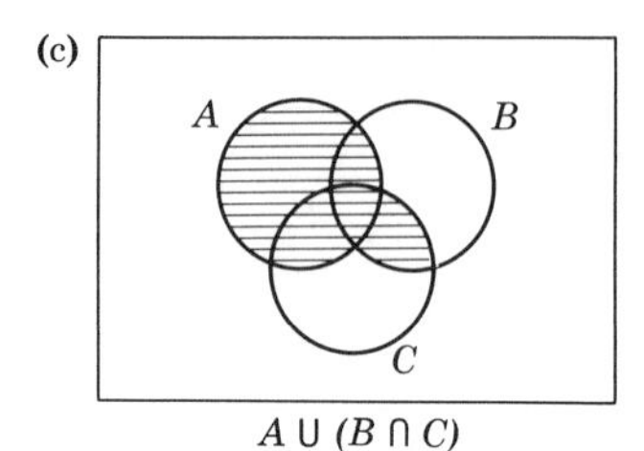

$A \cup (B \cap C)$

5. $2^{16} = 65{,}536$ subsets

6. (a) $\{2, 3\}$ (c) $\{3\}$ (e) ϕ (g) $\{(2, 2), (2, 3), (3, 2), (3, 3)\}$
(i) $\{(4, 4), (4, 5), (5, 4), (5, 5)\}$

7. (a) $A \cap C$ (c) A (e) $A \cap B$

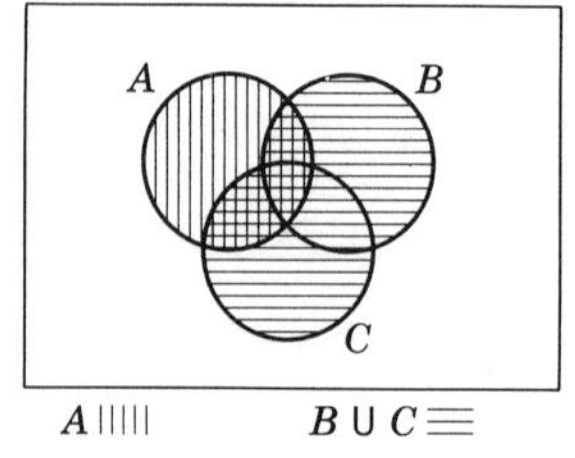

A ||||| $B \cup C$ ☰

$A \cap (B \cup C)$ ⊞

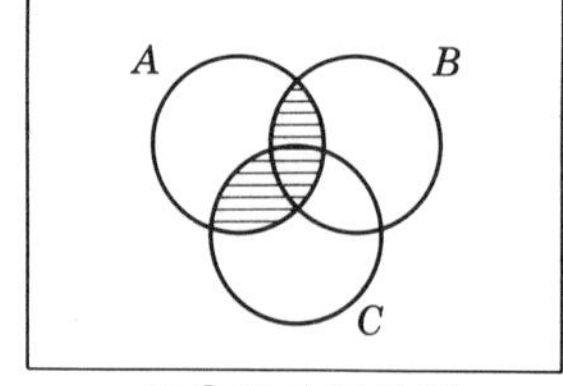

$(A \cap B) \cup (A \cap C)$

10. (a) Ted, Tim, Jack, June (c) Ted, Tim (e) Ted, Tim (g) Tom, Tim, Ted, Tobe, Jim, Joan, Jack, June, John, Jill, Jane, Jan, Sam, Sono, Sue, Sara; 16 attended party

11. Total 932.

Exercise 3.4

1. Ⓣ is not reflexive, not symmetric, but is transitive
3. Ⓒ is not reflexive, not symmetric, and not transitive
5. They are all multiples of 7, or exactly divisible by 7

Exercise 3.5a

2. 120 ways
3. (a) $7|35$ because $35 = 7 \cdot 5$ (c) $3|51$ because $51 = 3 \cdot 17$
5. Let $n \leftrightarrow 2n$ for $n = 1, 2, \ldots$
7. 5040 ways
8. 1,307,674,368,000 possible permutations. At the rate of 6 per min (360/hr) the time required is 3,632,428,800 hr, or 454,053,600 8-hr days, or 1,816,214.4 250-working-day years.

Exercise 3.5b

1. "$1-1$" is reflexive since a set may be placed in one-to-one correspondence with itself. "$1-1$" is symmetric by the definition of the relation, that is, "$1-1$" means "both ways." A "$1-1$" B implies there is a $1-1$ correspondence of A to B and B to A; hence B "$1-1$" A. "$1-1$" is transitive. Consider sets A, B, and C. For $a \in A$, $b \in B$, and $c \in C$, let $a \leftrightarrow b$, $b \leftrightarrow c$; then from $a \leftrightarrow b \leftrightarrow c$ we can obtain $a \leftrightarrow c$. Hence $A\ 1-1 B$ and $B\ 1-1\ C$ implies $A\ 1-1\ C$.
3. Reflexive since any student is "same sex as" himself. Symmetric since if student A is the same sex as student B, then student B is the same sex as student A. Transitive since if student A is the same sex as student B, for example, both boys, and student B is the same sex as student C, then C must be a boy. Hence A "is the same sex as" C. The equivalence classes are the *boys* and the *girls*.
5. Call the relation Ⓡ. For a, b, and c integers: a Ⓡ a because $a-a=0=0 \cdot 4$; hence Ⓡ is reflexive. a Ⓡ b means $a-b=k \cdot 4$ for some integer k; then $b-a=(-k)4$; hence b Ⓡ a. Ⓡ is symmetric. a Ⓡ b means $a-b=k \cdot 4$ and b Ⓡ c means $b-c=m \cdot 4$; then $(a-b)+(b-c)=k \cdot 4+m \cdot 4$ or $a-c=(k+m)4$; but $k+m$ is an integer; hence a Ⓡ c. Ⓡ is transitive. It is an equivalence relation. The equivalence classes are those integers related to 0, 1, 2, and 3, respectively.

Exercise 3.6

1. 10 **3.** 12 **5.** 20 **7.** 0 **9.** $16 = 10 + 12 - 6$
11. 100 **13.** 56, or more **15.** 24

Exercise 3.7

1. (a) If A and B have exactly the same elements (c) If $a = c$ and $b = d$
4. One cut and the other choose first

Exercise 3.8

2. When the ordered pairs of f have no two second components the same

4. (b) is a function of x and has domain $\{2, 3, 4\}$ and range $\{2, 3, 4\}$; (d) is a function of x and has domain $\{1, 2, 3, 4\}$ and range $\{1, 3\}$; (a) is a function of y and has domain $\{1, 2, 3, 4\}$ and range $\{4\}$; (b) is a function of y and has domain $\{2, 3, 4\}$ and range $\{2, 3, 4\}$

6. (a) True since $1-13=-12$ and -12 is divisible by 3
(c) True since $1-4=-3$ and -3 is divisible by 3
(e) True since $a \circledR b$ implies $a-b=k \cdot 3$, $k \in W$. Then $b-a=(-k) \cdot 3$, or $b \circledR a$

Exercise 4.4

1. (a) 11 (c) 19 (e) Those with complete uniforms
3. $132+97-43=186$ **4.** (a) 0 (c) Yes **5.** (a) 1 (c) $R=Q$
6. (a) 8 (c) 25 **7.** (a) $n(A \cup \emptyset)=n(A)=3$
8. (a) The set can be placed in $1-1$ correspondence with the set $\{1, 2, 3, 4, 5, \ldots, n\}$
9. Cardinal: There are 3 boys in the group
Ordinal: He is 3rd in line
Cardinal: I have read 10 pages.
Ordinal: I stopped reading on page 10.
10. (b) Orange
11. Equivalence classes:

0	12	24	36	48	...
1	13	25	37	49	...
2	14	26	38	50	...
3	15	27	39	51	...
4	16	28	40	52	...
5	17	29	41	53	...
6	18	30	42	54	...
7	19	31	43	55	...
8	20	32	44	56	...
9	21	33	45	57	...
10	22	34	46	58	...
11	23	35	47	59	...

Exercise 4.5

1. See Section 1.3
3. Letters for symbols instead of special characters; use of "intermediate" symbols; use of the multiplicative principle applied in large numbers.
5. 21_{three} (c) 20_{three} (e) 12_{three}
6. (a) Same (c) $10^{(10^{10})}$
7. (a) For example, both outdoor games, both use balls, both involve kicking the ball, and both have goals.
(b) For example, balls are different shape, different rules for moving the ball, and different ways of scoring.

Exercise 4.6

1. (a) If $A \subset B$ and $B \subset A$
2. (a) (0, 1), (0, 2), (0, 3), (1, 1), (1, 2), (1, 3)
3. $(2, 3) \equiv (3, 4)$ since $2+4=3+3$
   Reflexive: $(2, 3) \equiv (2, 3)$ since $2+3=3+2$
   Symmetric: $(2, 3) \equiv (3, 4)$ implies $2+4=3+3$ or $3+3=4+2$ which implies $(3, 4) \equiv (2, 3)$
   Transitive: If $(2, 3) \equiv (3, 4)$ and $(3, 4) \equiv (4, 5)$, is $(2, 3) \equiv (4, 5)$? Yes, since $2+5=3+4$.

Exercise 4.8

1. (a) Yes, 0 (c) No, there is no whole number x such that $6+x=0$.
2. (a) No (c) No
3. (a) No identity element for addition. 1 is the identity for multiplication.
   (c) No inverses for addition or multiplication.
4. (a) 0 is the additive identity. There is no multiplicative identity.
   (c) No inverses for addition or multiplication.
5. (a) $5^3=125$ (c) 64 (e) 3^{27} (g) A 1 followed by 100 zeros.
   (i) No, since $2 \oplus 3 = 2^3 = 8$, $3 \oplus 2 = 3^2 = 9$.
6. (a) 3 (c) No. $3 \odot 7 = 3$ but $7 \odot 3 = 7$
7. (a) Yes (c) No
8. (a) Yes. $\{a, b\} \cup \{b, c\} = \{a, b, c\} \in S$
   (c) Yes. $(\{a, b\} \cup \{b, c\}) \cup \{d\} = \{a, b, c\} \cup \{d\} = U$
   $\{a, b\} \cup (\{b, c\} \cup \{d\}) = \{a, b\} \cup \{b, c, d\} = U$
9. (a) Yes. $\{a, b\} \cap \{b, c\} = \{b\} \in S$
   (c) Yes. $(\{a, c\} \cap \{b, c\}) \cap \{c, d\} = \{c\} \cap \{c, d\} = \{c\}$
   $\{a, c\} \cap (\{b, c\} \cap \{c, d\}) = \{a, c\} \cap \{c\} = \{c\}$

Exercise 4.10d

1. Commutative law.
3. Commutative law.
5. Commutative law.
7. Associative law.
9. Zero is used as the cardinal number of the empty set. As an element of a number system zero is the additive identity.
10. Assume there is a $\overline{0} \in W$, $\overline{0} \neq 0$, such that for any $n \in W$, $\overline{0}+n=n+\overline{0}=n$, that is, assume there is more than one additive identity in W. Then $\overline{0}+0=\overline{0}$ since 0 is an additive identity; and $\overline{0}+0=0$ by the foregoing assumption. Then $\overline{0}=0$ by the transitive property of equals. This contradicts $\overline{0} \neq 0$, so the assumption must be false. In other words, 0 is the only additive identity in W.
11. (a) $A \cup B = \{u, v, w, a, b, c\}$. $B \cup A = \{a, b, c, u, v, w\} = A \cup B$
    (c) $A \cup \emptyset = \{u, v, w\} \cup \emptyset = \{u, v, w\} = A$
    $\emptyset \cup A = \emptyset \cup \{u, v, w\} = \{u, v, w\} = A$

Exercise 4.11b

1. For any $(a, b) \in (A \times B)$, $(b, a) \in (B \times A)$. Let $(a, b) \leftrightarrow (b, a)$
3. For any $w \in A$ let $w \leftrightarrow (e, w) \in \{e\} \times A$
5. $\emptyset \times A = \{(x, y) | x \in \emptyset \text{ and } y \in A\}$ Since there is no x such that $x \in \emptyset$, $\emptyset \times A = \emptyset$

Exercise 4.11e

1. (a) Commutative law of multiplication
(c) Commutative law of multiplication
(e) Associative law of multiplication

3. (a) Commutative law of addition
(c) Commutative law of multiplication
(e) Commutative law of multiplication
(g) Associative law of addition

Exercise 4.11h

1. $ab+a\cdot 2$ or $ab+2a$
3. $23\cdot 2+23\cdot 1=69$
5. $(30+2)10$
7. $ac+bc+ad+bd$
9. $xy+x\cdot 2$
11. $(2a+3)x$
13. $(3\cdot 10+2)10$
15. $(2xb+1)a$
17. $a\triangle b=b\triangle a$
19. $(a\triangle b)\triangle c=a\triangle(b\triangle c)$
21. $a\triangle(b\oplus c)=(a\triangle b)\oplus(a\triangle c)$

Exercise 4.12

1. Let 1 represent the multiplicative identity and suppose there is another, call it $1'$. Then $1\cdot 1'=1'$ because 1 is the multiplicative identity; also $1\cdot 1'=1$ since we are supposing $1'$ is another multiplicative identity. Then $1'=1$ by the transitive property of equals. Hence the multiplicative identity is unique.

4. (a) No

5. (a) Yes, 5 "divides" 0 because $0=5\cdot 0$.

8.
$(a+b)^2=(a+b)(a+b)$ by defn. of exponents
$=a(a+b)+b(a+b)$ by the distributive law
$=a^2+ab+ba+b^2$ by the distributive law
$=a^2+ab+ab+b^2$ by the com. law of mult.
$=a^2+(ab+ab)+b^2$ by the assoc. law of add.
$=a^2+(1\cdot ab+1\cdot ab)+b^2$ 1 is the mult. identity
$=a^2+(1+1)ab+b^2$ by the dist. law
$=a^2+2ab+b^2$ by the tables of elem. facts

9. (a) $2^{(3^0)}$

Exercise 4.13

1. $A'=\{n\in W|n\leq 7\}$ $\quad A'=\{0,1,2,3,4,5,6,7\}$

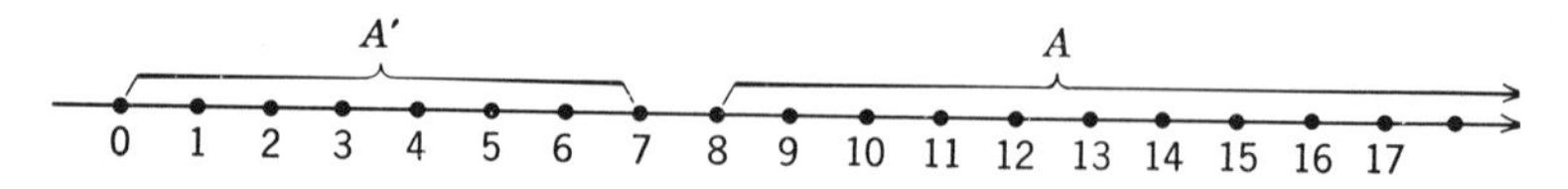

3. $Y\cap Z=\{7,8,9\}$ $\quad Y\cap Z'=\{4,5,6\}$

5. $3\cdot n<7\cdot n$ if $n\neq 0$, $\quad 3\cdot n=7\cdot n$ if $n=0$

Exercise 4.13a

1. (a) $A = \{4, 5, 6, 7, 8, 9, 10, 11\}$; l.u.b. 11; g.l.b. 4
(b) $B = \{1, 2, 3\}$; for example, 3 and 4

3. $A \cap B = \phi$

5. $C = \{6, 7, 8, 9, 10, 11, 12\}$; l.u.b. is 12; $\{1, 2, 3, 6, 7, 8, 9, 10, 11, 12\}$

7. (a) $\{n | 0 \leq n < 7\}$ (b) $\{n | 0 \leq n < 4\}$ (c) $\{n | 0 \leq n < 4\}$

12.

$(x+3)+(y+2) = x+(3+y)+2$	Assoc. law of add.
$= x+(y+3)+2$	Com. law of add.
$= (x+y)+(3+2)$	Assoc. law of add.
$= (3+2)+(x+y)$	Com. law of add.
$= (2+3)+(x+y)$	Com. law of add.
$(x+3)+(y+2) = (2+3)+(x+y)$	Trans. prop of equals

14.

$(3b)^2 = (3b)(3b)$	Defn. of exponent
$= 3(b3)b$	Assoc. law of mult.
$= 3(3b)b$	Com. law of mult.
$= (3 \cdot 3)(bb)$	Assoc. law of mult.
$= 9b^2$	Defn. of exponent and tables
$(3b)^2 = 9b^2$	Trans. prop. of equals

16.

$3 \cdot 10^0 + 4 \cdot 10^0 = (3+4)10^0$	Dist. law
$= 7 \cdot 10^0$	Tables

18.

$13 \cdot 10^2 = (1 \cdot 10^1 + 3 \cdot 10^0)10^2$	System of numeration
$= (1 \cdot 10^1)10^2 + (3 \cdot 10^0)10^2$	Dist. law
$= 1(10^1 \cdot 10^2) + 3(10^0 \cdot 10^2)$	Assoc. law of mult.
$= 1 \cdot 10^3 + 3 \cdot 10^2$	Law of exponents
$13 \cdot 10^2 = 1 \cdot 10^3 + 3 \cdot 10^2$	Trans. prop. of equals

20.

$(3+x)(2+y) = 3(2+y)+x(2+y)$	Dist. law
$= 3 \cdot 2 + 3y + x \cdot 2 + xy$	Dist. law
$= 6 + 3y + x \cdot 2 + xy$	Tables
$= 6 + 3y + 2x + xy$	Com. law of mult.

Exercise 4.13b

1. (a) $n > 2$ and ($n < 12$ or $n = 12$) (c) $n > 75$ and $n < 77$

2. (a) 12 (c) 76

3. (a) 0 1 2 3 4 5 6 7 8 9 10 11 12 13 14 15 16

(c) 69 70 71 72 73 74 75 76 77 78 79 80 81 82 83

4. (a) simple (c) simple (e) simple

Exercise 4.15

1. (a) 2342_{five} 347_{ten}

3. (a) $3 \cdot 5^2 + 2 \cdot 5^1 + 0 \cdot 5^0 = 85_{ten}$
(c) $2 \cdot 5^2 + 0 \cdot 5^1 + 3 \cdot 5^0 = 53_{ten}$
(e) $1 \cdot 5^0 + 1 \cdot 5^{-1} + 1 \cdot 5^{-2} = 1.24_{ten}$

4. (a) 24_{five} (c) 2302_{five}

5. (a) 0 1 2 3 10 11 12 13

6. 233.1332_{four} ounces.

7. (a) Two dens, four qens, one fen, and one sen. (c) Four mens, three dens, two qens, one fen, and two sens.
9. TE.
11. Odd, since it is not a multiple of 2.
13. (a) $T9$; $T9-9T=E$ (c) $E1$; $E1-9E=12$
15. 10

Exercise 4.17b

2. (a) 1144 (c) 22,010
3. (a) 13,131 (c) 1,130,011
4. (a) 132 (c) 302
5. (a) 232 (c) 20,432
7. (a) 1202; 1110 (c) 20,211; 122,120,021
8. 4 weights: 1 oz, 3 oz, 9 oz, and 27 oz. Use weights in both pans as needed.

Exercise 4.18

2. (a) 1100 (c) 11,101 (e) 1,001,110
3. (a) 1,001,101 (c) 101,011,111 (e) 1001

Exercise 4.19

1. Fewer symbols; fewer elementary facts; and adaptable to yes-no or on-off type problems.
3. (a) 123 (c) $935E8$ (e) 526 (g) 13,096 (i) 4646 (k) $13T0874$ (m) 49 (o) $239-15$ rem
4. (a)

$$\begin{aligned}
27+9 &= (2\cdot 10^1+7\cdot 10^0)+9\cdot 10^0 && \text{System of numeration}\\
&= 2\cdot 10^1+(7\cdot 10^0+9\cdot 10^0) && \text{Assoc. law of add.}\\
&= 2\cdot 10^1+(7+9)10^0 && \text{Dist. law}\\
&= 2\cdot 10^1+16\cdot 10^0 && \text{Tables}\\
&= 2\cdot 10^1+(1\cdot 10^1+6\cdot 10^0)10^0 && \text{System of numeration}\\
&= 2\cdot 10^1+(1\cdot 10^1)10^0+(6\cdot 10^0)10^0 && \text{Dist. law}\\
&= 2\cdot 10^1+1(10^1\cdot 10^0)+6(10^0\cdot 10^0) && \text{Assoc. law of mult.}\\
&= 2\cdot 10^1+1\cdot 10^1+6\cdot 10^0 && \text{Law of exponents}\\
&= (2\cdot 10^1+1\cdot 10^1)+6\cdot 10^0 && \text{Assoc. law of add.}\\
&= (2+1)10^1+6\cdot 10^0 && \text{Dist. law}\\
&= 3\cdot 10^1+6\cdot 10^0 && \text{Tables}\\
&= 36 && \text{System of numeration}\\
27+9 &= 36 && \text{Trans. prop. of equals}
\end{aligned}$$

(c)

$$\begin{aligned}
379+96 &= (3\cdot 10^2+7\cdot 10^1+9\cdot 10^0)+(9\cdot 10^1+6\cdot 10^0) && \text{System of numeration}\\
&= 3\cdot 10^2+7\cdot 10^1+(9\cdot 10^1+9\cdot 10^0)+6\cdot 10^0 && \text{Assoc. law of add.}\\
&= 3\cdot 10^2+7\cdot 10^1+(9\cdot 10^1+9\cdot 10^0)+6\cdot 10^0 && \text{Com. law of add.}\\
&= 3\cdot 10^2+(7\cdot 10^1+9\cdot 10^1)+(9\cdot 10^0+6\cdot 10^0) && \text{Assoc. law of add.}\\
&= 3\cdot 10^2+(7+9)\cdot 10^1+(9+6)\cdot 10^0 && \text{Dist. law}\\
&= 3\cdot 10^2+16\cdot 10^1+15\cdot 10^0 && \text{Elem. facts}
\end{aligned}$$

$= 3 \cdot 10^2 + (1 \cdot 10^1 + 6 \cdot 10^0)10^1 + (1 \cdot 10^1 + 5 \cdot 10^0) \cdot 10^0$ — System of numeration

$= 3 \cdot 10^2 + (1 \cdot 10^1)10^1 + (6 \cdot 10^0)10^1 + (1 \cdot 10^1)10^0 + (5 \cdot 10^0)10^0$ — Dist. law

$= 3 \cdot 10^2 + 1(10^1 \cdot 10^1) + 6 \cdot (10^0 \cdot 10^1) + 1(10^1 \cdot 10^0) + 5(10^0 \cdot 10^0)$ — Assoc. law of mult.

$= 3 \cdot 10^2 + 1 \cdot 10^2 + 6 \cdot 10^1 + 5 \cdot 10^0$ — Law of exponents

$= (3 \cdot 10^2 + 1 \cdot 10^2) + (6 \cdot 10^1 + 1 \cdot 10^1) + 5 \cdot 10^0$ — Assoc. law of add.

$= (3+1)10^2 + (6+1)10^1 + 5 \cdot 10^0$ — Dist. law

$= 4 \cdot 10^2 + 7 \cdot 10^1 + 5 \cdot 10^0$ — Elem. facts

$= 475$ — System of numeration

$379 + 96 = 475$ — Trans. prop. of equals

5. (a) $(36)(9) = (3 \cdot 10^1 + 6 \cdot 10^0)(9 \cdot 10^0)$ — System of numeration

$= (3 \cdot 10^1)(9 \cdot 10^0) + (6 \cdot 10^0)(9 \cdot 10^0)$ — Dist. law

$= 3(10^1 \cdot 9)10^0 + 6(10^0 \cdot 9)10^0$ — Assoc. law of mult.

$= 3(9 \cdot 10^1)10^0 + 6(9 \cdot 10^0)10^0$ — Com. law of mult.

$= (3 \cdot 9)(10^1 \cdot 10^0) + (6 \cdot 9)(10^0 \cdot 10^0)$ — Assoc. law of mult.

$= (3 \cdot 9)(10^1) + (6 \cdot 9)(10^0)$ — Law of exponents

$= 27 \cdot 10^1 + 54 \cdot 10^0$ — Elem. facts

$= (2 \cdot 10^1 + 7 \cdot 10^0)10^1 + (5 \cdot 10^1 + 4 \cdot 10^0)10^0$ — System of numeration

$= (2 \cdot 10^1)(10^1) + (7 \cdot 10^0)(10^1) + (5 \cdot 10^1)(10^0) + (4 \cdot 10^0)(10^0)$ — Dist. law

$= 2(10^1 \cdot 10^1) + 7(10^0 \cdot 10^1) + 5(10^1 \cdot 10^0) + 4(10^0 \cdot 10^0)$ — Assoc. law of mult.

$= 2 \cdot 10^2 + 7 \cdot 10^1 + 5 \cdot 10^1 + 4 \cdot 10^0$ — Law of exponents

$= 2 \cdot 10^2 + (7 \cdot 10^1 + 5 \cdot 10^1) + 4 \cdot 10^0$ — Assoc. law of add.

$= 2 \cdot 10^2 + (7+5)10^1 + 4 \cdot 10^0$ — Dist. law

$= 2 \cdot 10^2 + 12 \cdot 10^1 + 4 \cdot 10^0$ — Elem. facts

$= 2 \cdot 10^2 + (1 \cdot 10^1 + 2 \cdot 10^0)(10^1) + 4 \cdot 10^0$ — System of numeration

$= 2 \cdot 10^2 + (1 \cdot 10^1)10^1 + (2 \cdot 10^0)10^1 + 4 \cdot 10^0$ — Dist. law

$= 2 \cdot 10^2 + 1(10^1 \cdot 10^1) + 2(10^0 \cdot 10^1) + 4 \cdot 10^0$ — Assoc. law of mult.

$= 2 \cdot 10^2 + 1 \cdot 10^2 + 2 \cdot 10^1 + 4 \cdot 10^0$ — Law of exponents

$= (2 \cdot 10^2 + 1 \cdot 10^2) + 2 \cdot 10^1 + 4 \cdot 10^0$ — Assoc. law of mult.

$= (2+1)10^2 + 2 \cdot 10^1 + 4 \cdot 10^0$ — Dist. law

$= 3 \cdot 10^2 + 2 \cdot 10^1 + 4 \cdot 10^0$ — Elem. facts

$= 324$ — System of numeration

$(36)(9) = 324$ — Trans. prop. of equals

(c) $(36)(45) = (3 \cdot 10^1 + 6 \cdot 10^0)(4 \cdot 10^1 + 5 \cdot 10^0)$ — System of numeration

$= (3 \cdot 10^1)(4 \cdot 10^1 + 5 \cdot 10^0) + (6 \cdot 10^0) \times (4 \cdot 10^1 + 5 \cdot 10^0)$ — Dist. law

$= (3 \cdot 10^1)(4 + 10^1) + (3 \cdot 10^1)(5 \cdot 10^0) + (6 \cdot 10^0)(4 \cdot 10^1) + (6 \cdot 10^0)(5 \cdot 10^0)$ — Dist. law

$= 3(10^1 \cdot 4)10^1 + 3(10^1 \cdot 5)10^0 + 6(10^0 \cdot 4) \times 10^1 + 6(10^0 \cdot 5)10^0$ — Assoc. law of mult.

$= 3(4 \cdot 10^1)10^1 + 3(5 \cdot 10^1)10^0 + 6(4 \cdot 10^0) \times 10^1 + 6(5 \cdot 10^0)10^0$ — Com. law of mult.

$= (3 \cdot 4)(10^1 \cdot 10^1) + (3 \cdot 5)(10^1 \cdot 10^0)$
$+ (6 \cdot 4)(10^0 \cdot 10^1) + (6 \cdot 5)(10^0 \cdot 10^0)$ Assoc. law of mult.
$= (3 \cdot 4)10^2 + (3 \cdot 5)10^1 + (6 \cdot 4)10^1$
$+ (6 \cdot 5)10^0$ Law of exponents
$= 12 \cdot 10^2 + 15 \cdot 10^1 + 24 \cdot 10^1 + 30 \cdot 10^0$ Elem. facts
$= 12 \cdot 10^2 + (15 \cdot 10^1 + 24 \cdot 10^1) + 30 \cdot 10^0$ Assoc. law of add.
$= 12 \cdot 10^2 + (15 + 24)10^1 + 30 \cdot 10^0$ Dist. law
$= 12 \cdot 10^2 + 39 \cdot 10^1 + 30 \cdot 10^0$ Add. algorithm
$= (1 \cdot 10^1 + 2 \cdot 10^0)10^2 + (3 \cdot 10^1 + 9 \cdot 10^0)10^1$
$+ (3 \cdot 10^1 + 0 \cdot 10^0)10^0$ System of numeration
$= (1 \cdot 10^1)10^2 + (2 \cdot 10^0)10^2 + (3 \cdot 10^1)10^1$
$+ (9 \cdot 10^0)10^1 + (3 \cdot 10^1)10^0 + (0 \cdot 10^0)10^0$ Dist. law
$= 1(10^1 \cdot 10^2) + 2(10^0 \cdot 10^2) + 3(10^1 \cdot 10^1)$
$+ 9(10^0 \cdot 10^1) + 3(10^1 \cdot 10^0) + 0(10^0 \cdot 10^0)$ Assoc. law of mult.
$= 1 \cdot 10^3 + 2 \cdot 10^2 + 3 \cdot 10^2 + 9 \cdot 10^1 + 3 \cdot 10^1$
$+ 0 \cdot 10^0$ Law of exponents
$= 1 \cdot 10^3 + (2 \cdot 10^2 + 3 \cdot 10^2) + (9 \cdot 10^1 + 3 \cdot 10^1)$
$+ 0 \cdot 10^0$ Assoc. law of add.
$= 1 \cdot 10^3 + (2 + 3)10^2 + (9 + 3)10^1 + 0 \cdot 10^0$ Dist. law
$= 1 \cdot 10^3 + 5 \cdot 10^2 + 12 \cdot 10^1 + 0 \cdot 10^0$ Elem. facts
$= 1 \cdot 10^3 + 5 \cdot 10^2 + (1 \cdot 10^1 + 2 \cdot 10^0)10^1$
$+ 0 \cdot 10^0$ System of numeration
$= 1 \cdot 10^3 + 5 \cdot 10^2 + (1 \cdot 10^1)10^1 + (2 \cdot 10^0)10^1$
$+ 0 \cdot 10^0$ Dist. law
$= 1 \cdot 10^3 + 5 \cdot 10^2 + 1(10^1 \cdot 10^1) + 2(10^0 \cdot 10^1)$
$+ 0 \cdot 10^0$ Assoc. law of mult.
$= 1 \cdot 10^3 + 5 \cdot 10^2 + 1 \cdot 10^2 + 2 \cdot 10^1 + 0 \cdot 10^0$ Law of exponents
$= 1 \cdot 10^3 + (5 \cdot 10^2 + 1 \cdot 10^2) + 2 \cdot 10^1$
$+ 0 \cdot 10^0$ Assoc. law of add.
$= 1 \cdot 10^3 + (5 + 1)10^2 + 2 \cdot 10^1 + 0 \cdot 10^0$ Dist. law
$= 1 \cdot 10^3 + 6 \cdot 10^2 + 2 \cdot 10^1 + 0 \cdot 10^0$ Elem. facts
$= 1620$ System of numeration
$(36)(45) = 1620$ Trans. prop. of equals

7. (a) $9E7 + T = (9 \cdot 10^2 + E \cdot 10^1 + 7 \cdot 10^0) + T \cdot 10^0$ System of numeration
Note that the base has the same symbol but $10)_{12} = 12)_{10}$.
$= 9 \cdot 10^2 + E \cdot 10^1 + (7 \cdot 10^0 + T \cdot 10^0)$ Assoc. law of add.
$= 9 \cdot 10^2 + E \cdot 10^1 + (7 + T)10^0$ Dist. law
$= 9 \cdot 10^2 + E \cdot 10^1 + 15 \cdot 10^0$ Tables
$= 9 \cdot 10^2 + E \cdot 10^1 + (1 \cdot 10^1 + 5 \cdot 10^0)10^0$ System of numeration
$= 9 \cdot 10^2 + E \cdot 10^1 + (1 \cdot 10^1)10^0$
$+ (5 \cdot 10^0)10^0$ Dist. law
$= 9 \cdot 10^2 + E \cdot 10^1 + 1(10^1 \cdot 10^0)$
$+ 5(10^0 \cdot 10^0)$ Assoc. law of mult.
$= 9 \cdot 10^2 + E \cdot 10^1 + 1 \cdot 10^1 + 5 \cdot 10^0$ Law of exponents
$= 9 \cdot 10^2 + (E \cdot 10^1 + 1 \cdot 10^1) + 5 \cdot 10^0$ Assoc. law of add.
$= 9 \cdot 10^2 + (E + 1)10^1 + 5 \cdot 10^0$ Dist. law
$= 9 \cdot 10^2 + 10 \cdot 10^1 + 5 \cdot 10^0$ Tables
$= 9 \cdot 10^2 + (1 \cdot 10^1 + 0 \cdot 10^0)10^1 + 5 \cdot 10^0$ System of numeration
$= 9 \cdot 10^2 + (1 \cdot 10^1)10^1 + (0 \cdot 10^0)10^1$
$+ 5 \cdot 10^0$ Dist. law

$= 9 \cdot 10^2 + 1(10^1 \cdot 10^1) + 0(10^0 \cdot 10^1) + 5 \cdot 10^0$ Assoc. law of mult.
$= 9 \cdot 10^2 + 1 \cdot 10^2 + 0 \cdot 10^1 + 5 \cdot 10^0$ Law of exponents
$= (9 \cdot 10^2 + 1 \cdot 10^2) + 0 \cdot 10^1 + 5 \cdot 10^0$ Assoc. law of add.
$= (9+1)10^2 + 0 \cdot 10^1 + 5 \cdot 10^0$ Dist. law
$= T \cdot 10^2 + 0 \cdot 10^1 + 5 \cdot 10^0$ Tables
$= T05$ System of numeration
$9E7 + T = T05$ Trans. prop. of equals

(c) $EE + E = (E \cdot 10^1 + E \cdot 10^0) + E \cdot 10^0$ System of numeration
$= E \cdot 10^1 + (E \cdot 10^0 + E \cdot 10^0)$ Assoc. law of add.
$= E \cdot 10^1 + (E + E)10^0$ Dist. law
$= E \cdot 10^1 + 1T \cdot 10^0$ Elem. facts
$= E \cdot 10^1 + (1 \cdot 10^1 + T \cdot 10^0)10^0$ System of numeration
$= E \cdot 10^1 + (1 \cdot 10^1)10^0 + (T \cdot 10^0)10^0$ Dist. law
$= E \cdot 10^1 + 1(10^1 \cdot 10^0) + T(10^0 \cdot 10^0)$ Assoc. law of mult.
$= E \cdot 10^1 + 1 \cdot 10^1 + T \cdot 10^0$ Law of exponents
$= (E \cdot 10^1 + 1 \cdot 10^1) + T \cdot 10^0$ Assoc. law of add.
$= (E+1)10^1 + T \cdot 10^0$ Dist. law
$= 10 \cdot 10^1 + T \cdot 10^0$ Elem. facts
$= (1 \cdot 10^1 + 0 \cdot 10^1 + T \cdot 10^0$ System of numeration
$= (1 \cdot 10^1)10^1 + (0 \cdot 10^0)10^1 + T \cdot 10^0$ Dist. law
$= 1(10^1 \cdot 10^1) + 0(10^0 \cdot 10^1) + T \cdot 10^0$ Assoc. law of mult.
$= 1 \cdot 10^2 + 0 \cdot 10^1 + T \cdot 10^0$ Law of exponents
$= 10T$ System of numeration
$EE + E = 10T$ Trans. prop. of equals

8. (a) $(EE)(9) = (E \cdot 10^1 + E \cdot 10^0)(9 \cdot 10^0)$ System of numeration
$= (E \cdot 10^1)(9 \cdot 10^0) + (E \cdot 10^0)(9 \cdot 10^0)$ Dist. law
$= E(10^1 \cdot 9)10^0 + E(10^0 \cdot 9)10^0$ Assoc. law of mult.
$= E(9 \cdot 10^1)10^0 + E(9 \cdot 10^0)10^0$ Com. law of mult.
$= (E \cdot 9)(10^1 \cdot 10^0) + (E \cdot 9)(10^0 \cdot 10^0)$ Assoc. law of mult.
$= (E \cdot 9)10^1 + (E \cdot 9)10^0$ Law of exponents
$= 83 \cdot 10^1 + 83 \cdot 10^0$ Tables
$= (8 \cdot 10^1 + 3 \cdot 10^0)10^1 + (8 \cdot 10^1 + 3 \cdot 10^0)10^0$ System of numeration
$= (8 \cdot 10^1)10^1 + (3 \cdot 10^0)10^1 + (8 \cdot 10^1)10^0 + (3 \cdot 10^0)10^0$ Dist. law
$= 8(10^1 \cdot 10^1) + 3(10^0 \cdot 10^1) + 8(10^1 \cdot 10^0) + 3(10^0 \cdot 10^0)$ Assoc. law of mult.
$= 8 \cdot 10^2 + 3 \cdot 10^1 + 8 \cdot 10^1 + 3 \cdot 10^0$ Law of exponents
$= 8 \cdot 10^2 + (3 \cdot 10^1 + 8 \cdot 10^1) + 3 \cdot 10^0$ Assoc. law of add.
$= 8 \cdot 10^2 + (3+8)10^1 + 3 \cdot 10^0$ Dist. law
$= 8 \cdot 10^2 + E \cdot 10^1 + 3 \cdot 10^0$ Tables
$= 8E3$ System of numeration
$(EE)(9) = 8E3$ Trans. prop. of equals

(c) $(7E)(T5) = (7 \cdot 10^1 + E \cdot 10^0)(T \cdot 10^1 + 5 \cdot 10^0)$ System of numeration
$= (7 \cdot 10^1 + E \cdot 10^0)(T \cdot 10^1) + (7 \cdot 10^1 + E \cdot 10^0)(5 \cdot 10^0)$ Dist. law
$= (7 \cdot 10^1)(T \cdot 10^1) + (E \cdot 10^0)(T \cdot 10^1) + (7 \cdot 10^1)(5 \cdot 10^0) + (E \cdot 10^0)(5 \cdot 10^0)$ Dist. law

$$\begin{aligned}
&= 7(10^1 \cdot T)10^1 + E(10^0 \cdot T)10^1 \\
&\qquad + 7(10^1 \cdot 5)10^0 + E(10^0 \cdot 5)10^0 && \text{Assoc. law of mult.} \\
&= 7(T \cdot 10^1)10^1 + E(T \cdot 10^0)10^1 \\
&\qquad + 7(5 \cdot 10^1)10^0 + E(5 \cdot 10^0)10^0 && \text{Com. law of mult.} \\
&= (7 \cdot T)(10^1 \cdot 10^1) + (E \cdot T)(10^0 \cdot 10^1) \\
&\qquad + (7 \cdot 5)(10^1 \cdot 10^0) + (E \cdot 5)(10^0 \cdot 10^0) && \text{Assoc. law of mult.} \\
&= (7 \cdot T)10^2 + (E \cdot T)10^1 + (7 \cdot 5)10^1 \\
&\qquad + (E \cdot 5)10^0 && \text{Law of exponents} \\
&= 5T \cdot 10^2 + 92 \cdot 10^1 + 2E \cdot 10^1 + 47 \cdot 10^0 && \text{Elem. facts} \\
&= 5T \cdot 10^2 + (92 \cdot 10^1 + 2E \cdot 10^1) + 47 \cdot 10^0 && \text{Assoc. law of add.} \\
&= 5T \cdot 10^2 + (92 + 2E)10^1 + 47 \cdot 10^0 && \text{Dist. law} \\
&= 5T \cdot 10^2 + 101 \cdot 10^1 + 47 \cdot 10^0 && \text{Add. algorithm} \\
&= (5 \cdot 10^1 + T \cdot 10^0)10^2 + (1 \cdot 10^2 + 0 \cdot 10^1 \\
&\qquad + 1 \cdot 10^0)10^1 + (4 \cdot 10^1 + 7 \cdot 10^0)10^0 && \text{System of numeration} \\
&= (5 \cdot 10^1)10^2 + (T \cdot 10^0)10^2 + (1 \cdot 10^2)10^1 \\
&\qquad + (0 \cdot 10^1)10^1 + (1 \cdot 10^0)10^1 + (4 \cdot 10^1)10^0 \\
&\qquad + (7 \cdot 10^0)10^0 && \text{Dist. law} \\
&= 5(10^1 \cdot 10^2) + T(10^0 \cdot 10^2) + 1(10^2 \cdot 10^1) \\
&\qquad + 0(10^1 \cdot 10^1) + 1(10^0 \cdot 10^1) + 4(10^1 \cdot 10^0) \\
&\qquad + 7(10^0 \cdot 10^0) && \text{Assoc. law of mult.} \\
&= 5 \cdot 10^3 + T \cdot 10^2 + 1 \cdot 10^3 + 0 \cdot 10^2 + 1 \cdot 10^1 \\
&\qquad + 4 \cdot 10^1 + 7 \cdot 10^0 && \text{Law of exponents} \\
&= 5 \cdot 10^3 + (T \cdot 10^2 + 1 \cdot 10^3) + 0 \cdot 10^2 \\
&\qquad + 1 \cdot 10^1 + 4 \cdot 10^1 + 7 \cdot 10^0 && \text{Assoc. law of add.} \\
&= 5 \cdot 10^3 + (1 \cdot 10^3 + T \cdot 10^2) + 0 \cdot 10^2 \\
&\qquad + 1 \cdot 10^1 + 4 \cdot 10^1 + 7 \cdot 10^0 && \text{Com. law of add.} \\
&= (5 \cdot 10^3 + 1 \cdot 10^3) + (T \cdot 10^2 + 0 \cdot 10^2) \\
&\qquad + (1 \cdot 10^1 + 4 \cdot 10^1) + 7 \cdot 10^0 && \text{Assoc. law of add.} \\
&= (5+1)10^3 + (T+0)10^2 + (1+4)10^1 \\
&\qquad + 7 \cdot 10^0 && \text{Dist. law} \\
&= 6 \cdot 10^3 + T \cdot 10^2 + 5 \cdot 10^1 + 7 \cdot 10^0 && \text{Elem. facts} \\
&= 6T57 && \text{System of numeration} \\
(7E)(T5) &= 6T57 && \text{Trans. prop. of equals}
\end{aligned}$$

Exercise 5.3

1. No. Zero is neither positive nor negative.

3. Yes. $0 + {}^{-}0 = 0$ by the property of the additive inverse. $0 + {}^{-}0 = {}^{-}0$ by the property of the additive identity. Hence $0 = {}^{-}0$ by the transitive property of equals.

5. For any two integers m and n, either $m = n$, or $m < n$, or $n < m$.

7. ${}^{-}(m+n)$ is the additive inverse of $(m+n)$; also ${}^{-}(m+n) = {}^{-}m + {}^{-}n$.

9. ${}^{-}({}^{-}m) = m$.

Exercise 5.4a

1.
$$\begin{aligned}
9 + {}^{-}3 &= (6+3) + {}^{-}3 \\
&= 6 + (3 + {}^{-}3) && \text{Assoc. law of add.} \\
&= 6 + 0 && \text{Additive inverse} \\
&= 6 && \text{Additive identity}
\end{aligned}$$

2.
$$\begin{aligned}
{}^{-}7 + 4 &= {}^{-}(3+4) + 4 && \text{Tables} \\
&= ({}^{-}3 + {}^{-}4) + 4 && \textbf{A-1}
\end{aligned}$$

$= {}^{-}3 + ({}^{-}4 + 4)$ Assoc. law of add.
$= {}^{-}3 + 0$ Additive inverse
$= {}^{-}3$ additive identity

4. $(m+n) + {}^{-}(m+n) = 0$ Additive inverse
$(m+n) + ({}^{-}m + {}^{-}n) = m + (n + {}^{-}m) + {}^{-}n$ Assoc. law of add.
$= m + ({}^{-}m + n) + {}^{-}n$ Com. law of add.
$= (m + {}^{-}m) + (n + {}^{-}n)$ Assoc. law of add.
$= 0 + 0$ Additive inverse
$= 0$ Additive identity

Hence ${}^{-}(m+n) = ({}^{-}m + {}^{-}n)$ since the additive inverse is unique.

5. (a) ${}^{-}12$ (c) 3 (e) 0 (g) a (i) ${}^{-}({}^{-}2 + a) = 2 + {}^{-}a$
(k) ${}^{-}({}^{-}3 + 3) = 3 + {}^{-}3 = 0$ (m) $a + b + {}^{-}2$ (o) $a + {}^{-}b + {}^{-}3$

6. (a) ${}^{-}8$ (c) $a + {}^{-}4$ (e) ${}^{-}22$ (g) 8

7. (a) $3 + n = 10$ Given
${}^{-}3 + 3 + n = {}^{-}3 + 10$ Uniqueness of sums
$({}^{-}3 + 3) + n = {}^{-}3 + 10$ Assoc. law of add.
$0 + n = {}^{-}3 + 10$ Additive inverse
$n = {}^{-}3 + 10$ Additive identity
$n = 7$ Addition of integers

(c) $a + x = b$ Given
${}^{-}a + a + x = {}^{-}a + b$ Uniqueness of sums
$({}^{-}a + a) + x = {}^{-}a + b$ Assoc. law of add.
$0 + x = {}^{-}a + b$ Additive inverse
$x = {}^{-}a + b$ Additive identity
$x = b + {}^{-}a$ Com. law of add.
$x = b - a$ **A-2**

9. $372 - 176 = 372 + {}^{-}176$ **A-2**
$= (196 + 176) + {}^{-}176$ Substitution prin.
$= 196 + (176 + {}^{-}176)$ Assoc. law of add.
$= 196 + 0$ Additive inverse
$= 196$ Additive identity
$372 + 176 = 196$ Trans. prop. of equals

11. At 9 in. or 27 in.

13. No. $(12-5) - 2 = 7 - 2 = 5$; $12 - (5-2) = 12 - 3 = 9$

14. $0 + 0 = 0$ Additive identity
$m(0+0) = m \cdot 0$ Uniqueness of products
$m \cdot 0 + m \cdot 0 = m \cdot 0$ Dist. law
$m \cdot 0 + m \cdot 0 + {}^{-}m \cdot 0 = m \cdot 0 + {}^{-}m \cdot 0$ Uniqueness of sums
$m \cdot 0 + (m \cdot 0 + {}^{-}m \cdot 0) = (m \cdot 0 + {}^{-}m \cdot 0)$ Assoc. law of add.
$m \cdot 0 + 0 = 0$ Additive inverse
$m \cdot 0 = 0$ Additive identity

Exercise 5.4b

3. $(-2)(-3) = (2)(3)$ by M-2

5. $(-2)(-3)(-4) = [(-2)(-3)](-4)$ Assoc. law of mult.
$= [(2)(3)](-4)$ **M-2**
$= (6)(-4)$ Tables
$= -(6)(4)$ **M-1**
$= -24$ Tables

7. $(^-3)(^-4+^-5) = (^-3)(^-4)+(^-3)(^-5)$ Dist. law
$= (3)(4)+(3)(5)$ **M-2**
$= 12+15$ Tables
$= 27$ Addition algorithm

or

$(^-3)(^-4+^-5) = (^-3)[^-(4+5)]$ **A-1**
$= (^-3)(^-9)$ Tables
$= (3)(9)$ **M-2**
$= 27$ Tables

8. $8(7-3) = (8)(7+^-3)$ **A-2**
$= (8)(7)+(8)(^-3)$ Dist. Law
$= (8)(7)+^-(8)(3)$ **M-1**
$= 56+^-24$ Tables
$= (32+24)+^-24$ Substitution
$= 32+(24+^-24)$ Assoc. law of add.
$= 32+0$ Additive inverse
$= 32$ Additive identity

or

$8(7-3) = 8(7+^-3)$ **A-2**
$= 8[(4+3)+^-3]$ Tables and substitution
$= 8[4+(3+^-3)]$ Assoc. law of add.
$= 8(4+0)$ Additive inverse
$= 8(4)$ Additive identity
$= 32$ Tables

11. $^-(m-n) = ^-(m+^-n)$ **A-2**
$= ^-m+^-(^-n)$ **A-1**
$= ^-m+n$ **I-1**
$= n+^-m$ Com. law of add.
$= n-m$ **A-2**

13. **M-1**

15. See problem 14,5.4a

Exercise 5.5

1. Either $x=3$ or $x=7$. The product $(x-3)(x-7)=0$ if and only if $(x-3)=0$, in which case $x=3$, or $(x-7)=0$, in which case $x=7$.
3. $x \neq 1$. If $x=1$ is substituted into the expression, we would have $y=\frac{6}{0}$. If the expression $\frac{6}{0}$ is interpreted as division, it would be undefined.
5. Additive identity; Additive inverse and Substitution; $m-1$; Associative law of addition; Distributive law; Additive inverse; $0 \cdot m = 0$ for any m; and Additive identity.
7. $^-(^-a)$ is the additive inverse of ^-a; $^-(^-a)=a$
9. Yes, if $m \neq 0$
11. Uniqueness of sums; Associative law of addition; Additive inverse; and Additive identity.
13. If $a \cdot x = a \cdot y$ and $a \neq 0$, then $x=y$

Exercise 5.6

1. 2, 3, 5, 7, 11, 13, 17, 19, 23, 29, 31, 37, 41, 43, 47
3. 1, 2, 4, 13, 26, 52; primes 2, 13

5. 1, 3, 13, 39; primes 3, 13
7. See Section 5.7
9. (a) $2^3 \cdot 3^2$ (b) $2^2 \cdot 89$ (c) 2^9 (d) $2^3 \cdot 5^3$

Exercise 5.7

1. 1, 2, 3, 4, 6, 8, 9, 12, 18, 24, 36, 72
3. (a) $2^3 \cdot 3 \cdot 7$ (c) $3 \cdot 5^2 \cdot 13$ (e) $3^4 \cdot 5 \cdot 11$
5. $1, 3, -1, -3$
7. $2^3 \cdot 19 \cdot 29$
9. (a) None (c) 2 and 4

Exercise 5.9a

1. 6; 18
3. 12; 48
5. 18
7. 3
9. If $n = 0$ then g.c.d. $(p, n) = |p|$. If n is a multiple of p then g.c.d. $(p, n) = |p|$. If $n \neq 0$ and not a multiple of p then g.c.d. $(p, n) = 1$.

Exercise 5.9b

1. (a) 6 (c) 18
2. (a) 2 (b) 42
3. (a) Yes (b) Yes (c) Yes (d) Yes, 0
5. (a) 9 (c) 9
7. 1

Exercise 5.10

1. (a) 160 (c) 504
2. (a) 252 (c) 672
3. (a) 1 (c) 441 (e) 6 (g) 10^{10} (i) 62 (k) 8
4. (a) 144 (c) $2^3 \cdot 3^5 \cdot 5^3 \cdot 7^4$ (e) 630 (g) 10^{19} (i) 14,508 (k) $2^6 \cdot 3^2 \cdot 5^6$
5. (a) Yes (b) Yes (c) Yes (d) Yes, 1 since l.c.m. $(m, 1) = m$.

Exercise 5.11

1. (a) The set A consists of the integers $-2, -1, 0, 1, 2, 3, 4, 5$ (c) The set C is the set of negative integers (e) The set O is the singleton set $\{0\}$.
2. (a) $n > -3$ and $n \leqslant 5$
   (c) simple
   (e) simple
3. (a) The complement of A consists of the set $\{\ldots, -5, -4, -3, 6, 7, 8, 9, \ldots\}$
   (b) The complement of C consists of the nonnegative integers
   (e) The complement of O consists of all the integers except zero

4. (a) $A' = \{n|n \text{ is an integer and } n \leqslant -3 \text{ or } n > 5\}$
   (c) $C' = \{n|n \text{ is an integer and } n \geqslant 0\}$
   (e) $O' = \{n|n \text{ is an integer and } n \neq 0\}$
5. $N \times N = \{(m, n)|m > 0 \text{ and } n > 0\}$, $N \times N$ is the set of all ordered pairs of positive integers.
7. No; yes, 1
9. Yes, -1
11. $a < b$ if and only if $b-a > 0$. We must use this fact to show that $(b+c)-(a+c) > 0$ or that $a+c < b+c$. But $(b+c)-(a+c) = b+c+{}^{-}a+{}^{-}c = b+{}^{-}a+c+{}^{-}c = b+{}^{-}a+0 = b+{}^{-}a = b-a > 0$. That is, $(b+c)-(a+c) > 0$, so $a+c < b+c$.
13. If $a < b$ then $b-a > 0$. If $c < 0$ then $0-c = -c > 0$. $-c(b-a) > 0$ since the product of two positive integers is positive. But $-c(b-a) = {}^{-}c(b+{}^{-}a) = -cb+ca = ca-cb > 0$, which means $ca > cb$.
15. 15
17. $\dfrac{10 \cdot 11}{2} = 55$
19. $\dfrac{50 \cdot 51}{2} = 1275$
21. 9
23. 25
25. Each is the square of the number of terms in the indicated sum, and each differs from the next by the $(n+1)st$ term of the sum.
27. $11^2 = 121$
28. $1+3+5+7+\cdots+(2n-1) = n^2$, where n is the number of terms in the sum.

Exercise 5.12

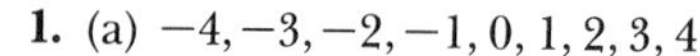

1. (a) $-4, -3, -2, -1, 0, 1, 2, 3, 4$

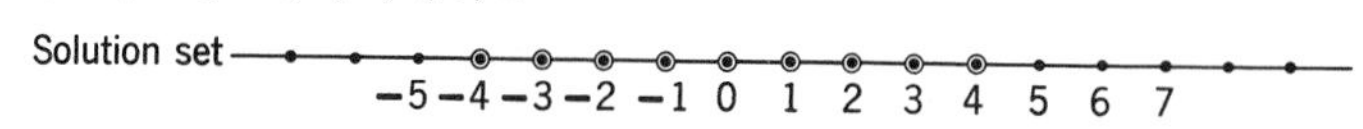

Complement
−5 −4 −3 −2 −1 0 1 2 3 4 5 6 7

   (c) $-1, 0, 1, 2, 3, 4$

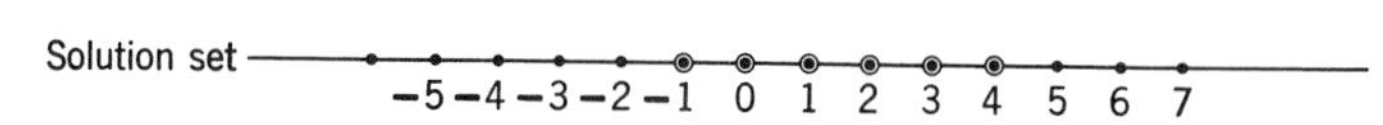

Complement
−5 −4 −3 −2 −1 0 1 2 3 4 5 6

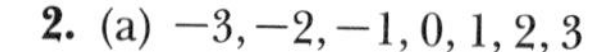

2. (a) $-3, -2, -1, 0, 1, 2, 3$

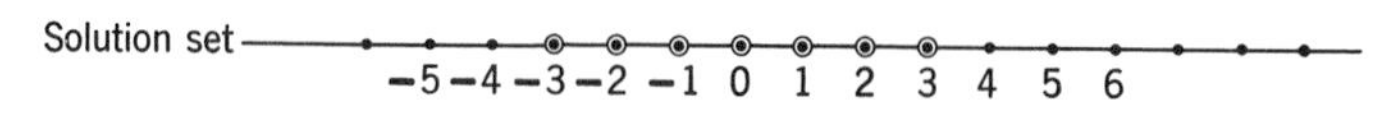

Complement
−5 −4 −3 −2 −1 0 1 2 3 4 5 6 7

   (c) $-1, 0, 1$

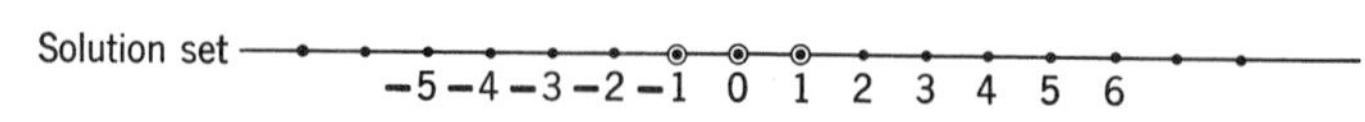

Complement
−6 −5 −4 −3 −2 −1 0 1 2 3 4 5 6 7

(e) 4, 8

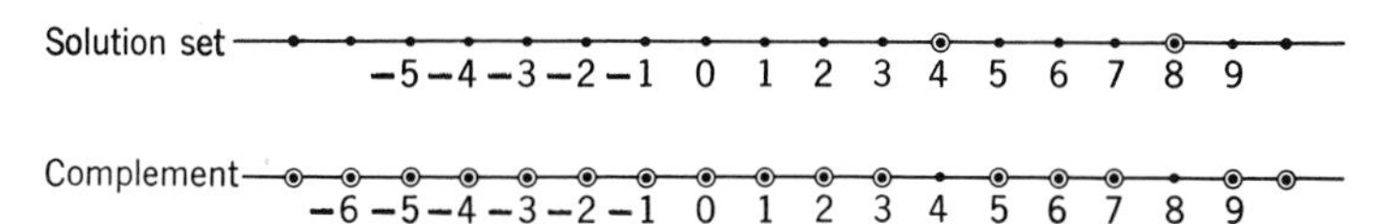

3. $a < b$ then $ac < bc$ if $c > 0$. $a < b$ then $ac > bc$ if $c < 0$.
5. (a) $|3-9|$
6. (a) $|4-n|$
7. 4; 10
9. Not necessarily

Exercise 5.13

1. (a) 6:00 A.M. (c) 8:00 A.M.
2. (a) 2 (c) 11
3. (a) 7:00 A.M. (c) 11:00 A.M.

Exercise 5.14a

2. (a) [4] (c) [4] (e) [6]

Exercise 5.14c

3. Not in general; $[4] \cdot [x] = [3]$ has no solution.
4. (a) $[x] = [0]$ or $[x] = [6]$ (c) $[x] = [1]$ or [5] or [9]
5. No, for example, $[3] \cdot [4] = [0]$

7.

+	[0]	[1]	[2]
[0]	[0]	[1]	[2]
[1]	[1]	[2]	[0]
[2]	[2]	[0]	[1]

·	[1]	[2]
[1]	[1]	[2]
[2]	[2]	[1]

9. [1]
11. 210; 87,780
13. $a-b=0$; cannot divide by 0

Exercise 6.2

1. (a) ordinal (b) cardinal (c) ordinal
3. 0 is the additive identity, that is, $0+a=a$
1 is the multiplicative identity, that is, $1 \cdot a = a$.

Exercise 6.4

1. (a) $\frac{-6}{-9}, \frac{-4}{-6}, \frac{-2}{-3}, \frac{2}{3}, \frac{4}{6}, \frac{6}{9}$, etc. (c) $\frac{-2}{-2}, \frac{-1}{-1}, \frac{1}{1}, \frac{2}{2}, \frac{3}{3}$, etc.

(e) $\frac{-34}{38}, \frac{-17}{19}, \frac{34}{-38}, \frac{51}{-57}, \frac{68}{-76}$, etc. (g) $\frac{-18}{-2}, \frac{-9}{-1}, \frac{18}{2}, \frac{27}{3}, \frac{36}{4}$, etc.

(i) $\frac{-9}{-12}, \frac{-6}{-8}, \frac{-3}{-4}, \frac{6}{8}, \frac{9}{12}$, etc. (k) $\frac{-8}{12}, \frac{-4}{6}, \frac{-2}{3}, \frac{2}{-3}, \frac{4}{-6}, \frac{8}{-12}, \frac{12}{-18}, \frac{16}{-24}$, etc.

2. First and third; fourth and fifth.

3. $\frac{33}{29} \doteq \frac{2 \cdot 33}{2 \cdot 29}$

4. (a) $(-2)(-3) = (2)(3)$ (c) $(-5)(-6) = +(5)(6) = +30$, $(3)(10) = 30$
(e) $(0)(6) = 0$; $(7)(0) = 0$ (g) $(0)(1) = 0$; $(-1)(0) = 0$

5. (a) $\frac{1982}{43{,}629}$ (c) $\frac{2}{7}$

6. (a) What number multiplied by 3 and added to 1 yields a sum of 7?
(c) What is the number which when multiplied by 3 and added to 1 gives 10?

7.
$\ldots, -39, -26, -13, 0, 13, 26, \ldots$
$\ldots, -38, -25, -12, 1, 14, 27, \ldots$
$\ldots, -37, -24, -11, 2, 15, 28, \ldots$
$\ldots, -36, -23, -10, 3, 16, 29, \ldots$
$\ldots, -35, -22, -9, 4, 17, 30, \ldots$
$\ldots, -34, -21, -8, 5, 18, 31, \ldots$
$\ldots, -33, -20, -7, 6, 19, 32, \ldots$
$\ldots, -32, -19, -6, 7, 20, 33, \ldots$
$\ldots, -31, -18, -5, 8, 21, 34, \ldots$
$\ldots, -30, -17, -4, 9, 22, 35, \ldots$
$\ldots, -29, -16, -3, 10, 23, 36, \ldots$
$\ldots, -28, -15, -2, 11, 24, 37, \ldots$
$\ldots, -27, -14, -1, 12, 25, 38, \ldots$

Exercise 6.5

1. If $a/b \doteq c/d$, then $ad = bc$; but if $ad = bc$, then $bc = ad$ by the symmetric property of equals and $cd = da$ by the commutative law of multiplication; then $cb = da$ implies $c/d \doteq a/b$ by the definition of $\doteq$.

3. (a) $\left[\frac{2}{3}\right]$ (c) $\left[\frac{19}{2}\right]$ (e) $\left[\frac{0}{1}\right]$

4. (a) $\left[\frac{-1}{2}\right]$ (c) $\left[\frac{-2}{3}\right]$ (e) $\left[\frac{0}{1}\right]$ (g) $\left[\frac{-2}{3}\right]$ (i) $\left[\frac{4}{5}\right]$ (k) $\left[\frac{-4}{1}\right]$

5. (a) $\frac{2}{3} \doteq \frac{2 \cdot 2}{2 \cdot 3}$ since $2(2 \cdot 3) = 2 \cdot 6 = 12$ and $3(2 \cdot 2) = 3 \cdot 4 = 12$

(c) $\frac{m}{n} \doteq \frac{2m}{2n}$ since $m(2n) = (m \cdot 2)n = (2m)n = 2(mn)$ and

$n(2m) = (n \cdot 2)m = (2n)m = 2(nm) = 2(mn)$

7. The two classes are equal (set equality)

9. The class to which 6/6 belongs is the same class as the one to which $-3/-3$ belongs.

Exercise 6.7

1. a, b, c, and d are integers; $ad + bc$ is an integer since the system of integers is closed under addition and multiplication; bd is also an integer; $bd \neq 0$ since $b \neq 0$ and $d \neq 0$; hence $(ad + bc)/bd$ is an ordered pair of integers.

3. $\frac{-3}{-4} + \frac{1}{6} = \frac{-18 - 4}{-24} = \frac{-22}{-24} = \frac{(-2)(11)}{(-2)(12)} \doteq \frac{11}{12}$

5. The class of $\frac{3}{4}+\frac{1}{6}$ is the same as the class of $\frac{3\cdot 6+4\cdot 1}{4\cdot 6}$

6. (a) $\frac{2}{3}$ (c) $\frac{0}{1}$

7. (a) $\frac{131}{72}$

9. $\left[\frac{m}{1}\right]+\left[\frac{n}{1}\right]=\left[\frac{m+n}{1}\right]$; $\frac{m+n}{1}$ corresponds to the integer $m+n$

Exercise 6.8

1. $\frac{14}{27}$

3. $\frac{1}{1}$

5. (a) $\left(\frac{7}{7}\right)\left(\frac{5}{7}\right)$ (b) $\left(\frac{3}{3}\right)\left(\frac{11}{13}\right)$

7. $\left[\frac{m}{1}\right]\cdot\left[\frac{n}{1}\right]=\left[\frac{m\cdot n}{1\cdot 1}\right]=\left[\frac{m\cdot n}{1}\right]$ which corresponds to $m\cdot n$

8. (a) $\frac{80}{63}$ (c) $\frac{4}{3}$ (e) $\frac{7}{1}$ (g) $\frac{0}{1}$ (i) $\frac{4}{9}$ (k) $\frac{355{,}630{,}706{,}103}{4{,}031{,}419{,}203{,}605}$

9. (a) $\frac{4}{5}$ (c) $\frac{1}{1}$ (e) $\frac{7}{6}$ (g) $\frac{8}{1}$ (i) $\frac{7}{1}$ (k) $\frac{1{,}167{,}543{,}234}{4{,}031{,}419{,}203{,}605}$

10. (a) Any member of the class containing $\frac{6}{8}$ added to any member of the class containing $\frac{4}{8}$ gives a sum which is in the class $\frac{5}{4}$.

11. (a) Any member from the class containing $\frac{9}{18}$ multiplied by any member from the class containing $\frac{6}{3}$ gives a product which is a member of the class containing $\frac{1}{1}$.

12. (a) for example, $\frac{2}{3}+\frac{2}{5}=\frac{2\cdot 5+3\cdot 2}{3\cdot 5}=\frac{10+6}{15}=\frac{16}{15}$;

$$\frac{2}{5}+\frac{2}{3}=\frac{2\cdot 3+5\cdot 2}{5\cdot 3}=\frac{6+10}{15}=\frac{16}{15};\quad \text{hence}\quad \frac{2}{3}+\frac{2}{5}=\frac{2}{5}+\frac{2}{3}.$$

13. (a) for example, $\frac{2}{3}\cdot\frac{3}{5}=\frac{2\cdot 3}{3\cdot 5}=\frac{6}{15}$; $\frac{3}{5}\cdot\frac{2}{3}=\frac{3\cdot 2}{5\cdot 3}=\frac{6}{15}$; hence $\frac{2}{3}\cdot\frac{3}{5}=\frac{3}{5}\cdot\frac{2}{3}$.

14. (a) $\left[\frac{395}{84}\right]$

15. (a) $\left[\frac{64}{15}\right]$

16. (a) $\begin{bmatrix} \frac{1}{1} \end{bmatrix}$

17. (a) $\dfrac{30{,}150{,}837{,}407{,}444}{5{,}954{,}621{,}431{,}472}$

Exercise 6.9

1. $\dfrac{1}{2}$

3. $\dfrac{40{,}816}{688{,}527}$

5. $\dfrac{81}{104}$

7. $\dfrac{13}{18}$

9. $-\dfrac{2}{3}$

11. $\dfrac{0}{1}$

Exercise 6.10

3. (a) $\dfrac{29}{10}$ (c) $\dfrac{275}{68}$ (e) $\dfrac{8}{1}$

6. (a) $\dfrac{0}{1}$ (c) $\dfrac{1}{5}$

Exercise 6.10a

3. Yes

4. (a) $\dfrac{a}{b}+\dfrac{c}{d}=\dfrac{a\cdot d+b\cdot c}{b\cdot d}$ Defn. of add. for ratl. nos.

$=\dfrac{d\cdot a+c\cdot b}{d\cdot b}$ Com. law for mult. of integers

$=\dfrac{c\cdot b+d\cdot a}{d\cdot b}$ Com. law for add. of integers

$=\dfrac{c}{d}+\dfrac{a}{b}$ Defn. of add. for ratl. nos.

$\dfrac{a}{b}+\dfrac{c}{d}=\dfrac{c}{d}+\dfrac{a}{b}$ Trans. prop. of equals

(b) $\frac{a}{b} \cdot \frac{c}{d} = \frac{a \cdot c}{b \cdot d}$ Defn. of mult. for ratl. nos.

$= \frac{c \cdot a}{d \cdot b}$ Com. law for mult. of integers

$= \frac{c}{d} \cdot \frac{a}{b}$ Defn. of mult. for ratl. nos.

$\frac{a}{b} \cdot \frac{c}{d} = \frac{c}{d} \cdot \frac{a}{b}$ Trans. prop. of equals

(c) $\frac{a}{b} \cdot \left(\frac{c}{d} + \frac{e}{f}\right) = \frac{a}{b} \cdot \left(\frac{cf + de}{df}\right)$ Defn. of add.

$= \frac{a(cf + de)}{b(df)}$ Defn. of mult.

$= \frac{a(cf) + a(de)}{b(df)}$ Dist. law for integers

$= \frac{a(cf) + a(de)}{b(df)} \cdot \frac{b}{b}$ Mult. by the identity

$= \frac{[a(cf) + a(de)] \cdot b}{[b(df)] \cdot b}$ Defn. of mult.

$= \frac{[a(cf)] \cdot b + [a(de)] \cdot b}{[b(df)] \cdot b}$ Dist. law for integers

$= \frac{(ac)(bf) + (bd)(ae)}{(bd)(bf)}$ Assoc. and com. laws for mult. of integers

$= \frac{ac}{bd} + \frac{ae}{bf}$ Defn. of add.

$= \frac{a}{b} \cdot \frac{c}{d} + \frac{a}{b} \cdot \frac{e}{f}$ Defn. of mult.

$\frac{a}{b} \cdot \left(\frac{c}{d} + \frac{e}{f}\right) = \frac{a}{b} \cdot \frac{c}{d} + \frac{a}{b} \cdot \frac{e}{f}$ Trans. prop. of equals

6. Cancellation law for multiplication of rational numbers:

If $\frac{a}{b} \cdot \frac{c}{d} = \frac{a}{b} \cdot \frac{e}{f}$, then $\frac{c}{d} = \frac{e}{f}$.

Proof: $\frac{a}{b} \cdot \frac{c}{d} = \frac{a}{b} \cdot \frac{e}{f}$ Hypothesis

$\frac{b}{a} \cdot \left(\frac{a}{b} \cdot \frac{c}{d}\right) = \frac{b}{a} \cdot \left(\frac{a}{b} \cdot \frac{e}{f}\right)$ Uniqueness of products

$$\left(\frac{b}{a}\cdot\frac{a}{b}\right)\cdot\frac{c}{d}=\left(\frac{b}{a}\cdot\frac{a}{b}\right)\cdot\frac{e}{f} \qquad \text{Assoc. law of mult.}$$

$$\frac{1}{1}\cdot\frac{c}{d}=\frac{1}{1}\cdot\frac{e}{f} \qquad \text{Multiplicative inverse}$$

$$\frac{c}{d}=\frac{e}{f} \qquad \text{Multiplicative identity}$$

Exercise 6.11a

1. (a) $\frac{32}{51}-\frac{3}{5}=\frac{32\cdot 5-51\cdot 3}{5\cdot 51}=\frac{160-153}{255}=\frac{7}{255}>0;$ hence $\frac{3}{5}<\frac{32}{51}$

2. (a) $\frac{2}{3}-\frac{-2}{3}=\frac{2\cdot 3-3(-2)}{3\cdot 3}=\frac{6+6}{9}=\frac{12}{9}>0;$ hence $\frac{-2}{3}<\frac{2}{3}$

3. $\frac{1}{a}>\frac{1}{b}$

5. (a) $-2,-1,0,1,2,3,4,5$ (c) $-4<x\leqq 3$

Exercise 6.11b

1. (a) $-3,-2,-1,0,1,2,3$ (c) $-8,-7,-6,-5,-4,-3,-2$
2. (a) 1
5. $a<a^2$
7. $a<a^2$

Exercise 6.11c

4. No. The rational numbers are dense
6. Yes.
8. $1/2^{n+1}$
10. No
12. -6
14. $\frac{7}{1}$
17. (a) 10 (c) 0

18.

(a)

A: −20 −15 −10 −5 0 5 10 15 20 25

A′: −20 −15 −10 −5 0 5 10 15 20 25

(c)

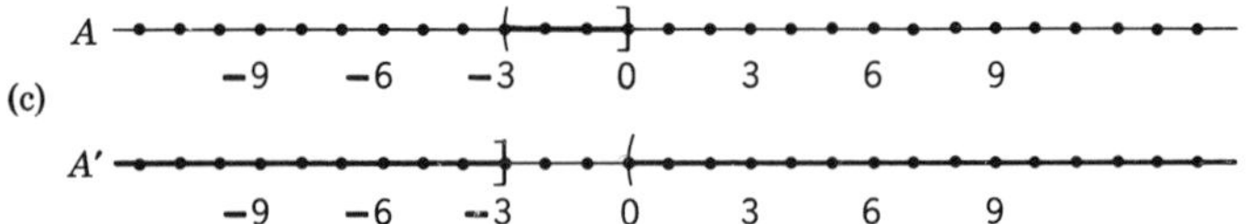

Exercise 6.12b

1. (a) $\frac{8}{9}$ (c) $\frac{10}{3}$
2. (a) $\frac{10}{9}$ (c) $\frac{49}{16}$

3. (a) $\frac{3}{2}$ (c) $\frac{5}{1}$ (e) $\frac{1}{31}$
4. (a) $-\frac{4}{5}$ (c) $-(\frac{3}{4}-\frac{2}{3})=-\frac{1}{12}$
6. (a) $\frac{2}{49}$ (c) $\frac{1}{1}$ (e) $\frac{1}{10}$
7. (a) $\frac{14}{9}$ (c) $\frac{2}{5}$

Exercise 6.12c

2. (a) $\frac{17}{12}$ or $1\frac{5}{12}$ (c) $\frac{87}{40}$ or $2\frac{7}{40}$
6. (a) $\frac{85}{161}$ (c) $\frac{34}{39}$ (e) $\frac{8}{23}$
7. $\frac{76}{63}$
8. (a) $\frac{1827}{50}$
9. (a) $\frac{5}{3}$ (c) $\frac{7}{2}$ (e) $\frac{29}{16}$
10. (a) .00000000006 (c) $2\frac{4}{7}$

Exercise 6.12d

4. 15
5. $2000
7. 40 ft.
9. 40 ft

Exercise 6.12e

1. (a) 32% (c) 75% (e) 16% (g) 24%
2. (a) $\frac{1}{4}$ (c) $\frac{2}{5}$ (e) $\frac{5}{4}$ (g) $\frac{4}{1}$
3. (a) 12 (c) 12.5% (e) 400

Exercise 6.13a

1.

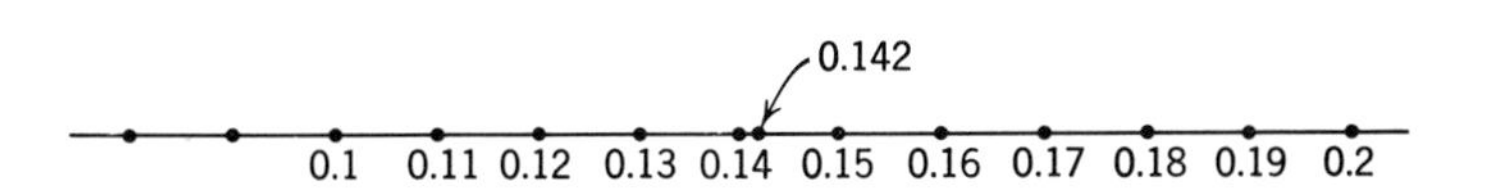

3. 1.41425; (table) 1.41421
5. $3.1416\overline{0}\ldots$; $3.141\overline{6}\ldots$; $3.141\overline{592}\ldots$; table value of π to 15 places: 3.141592653589793 . . .

Exercise 6.13e

1. (a) $\frac{32}{100}$ (c) $\frac{79}{1000}$
2. (a) 23 (c) 15
3. (a) $7\cdot10^2+0\cdot10^1+0\cdot10^0+1\cdot10^{-1}+2\cdot10^{-2}+5\cdot10^3$
(c) $1\cdot10^4+0\cdot10^3+0\cdot10^2+0\cdot10^1+0\cdot10^0+0\cdot10^{-1}$
$+0\cdot10^{-2}+0\cdot10^{-3}+1\cdot10^{-4}$
(e) $3\cdot10^0+1\cdot10^{-1}+4\cdot10^{-2}+1\cdot10^{-3}+6\cdot10^{-4}$
4. (a) 55.731984 (c) 0.1771 (e) 200,202.04 or 200,000 to one significant digit.
5. (a) $(2.99776)(1.673)(10^{-14})\cong(5.015)(10^{-14})$
(c) $(6.0228)(1.673)(10^{-1})\cong1.008$ (e) $(6.45\div2.4)(10^{-8})\cong(2.7)(10^{-8})$
6. π units

7. (*Hint*: $\widehat{AB} = \widehat{AB'}$, hence $\angle BOA = \angle B'O'A$. Since $OB \| C'O'$, $\angle BOA = \angle AO'C$. $\angle B'O'C' = 2(\angle AOB)$.)

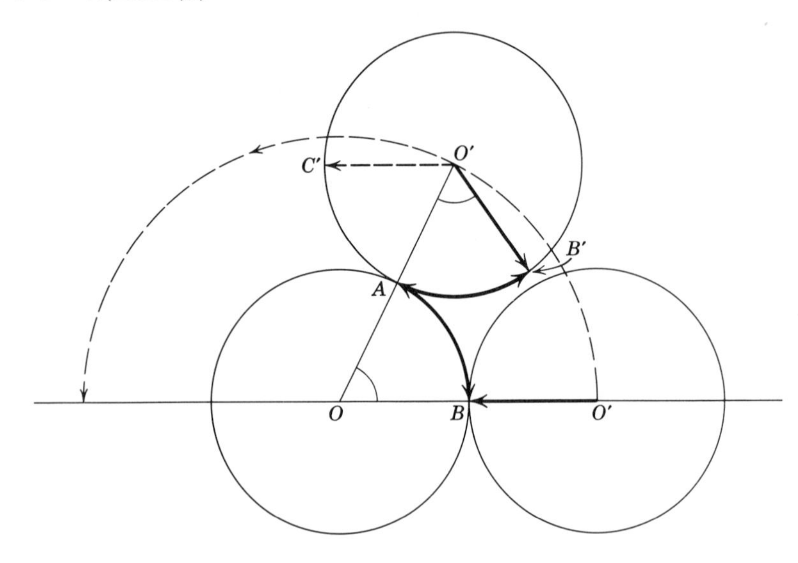

9. 7

Exercise 7.2

1. $\frac{6449}{4560} \cong 1.4142$; 0.0306; 0.000115
3. 1st approx. 17; 2nd approx. 17.32; 3rd approx. 17.320508
5. 1st approx. 1.4; 2nd approx. 1.414; 3rd approx. 1.4142136

Exercise 7.2a

1. Assume $5+\sqrt{2}=p/q$, where p/q is a rational number. Then $\sqrt{2}=p/q-5=(p-5q)/q$. But $(p-5q)/q$ is a rational number and $\sqrt{2}$ is not rational. Hence the assumption is false and $5+\sqrt{2}$ is irrational.
3. 2; -1
5. (a) for example, 17, 20, 25 (c) for example, 2, 10, $\sqrt{2}$
7. Circle with radius $\sqrt{2}$ or $\sqrt{5}$, etc.

Exercise 7.7

1. (a) $\{0, 1, 2, 3, 4, 5, 6, 7, 8, 9, 10\}$ (c) $0.\overline{18}\ldots$
2. (a) 12 (c) $0.1\overline{6}\ldots\ldots$
3. (a) 13 (c) $0.\overline{153846}\ldots$
4. (a) $\frac{1772}{9999}$ (c) $\frac{2906}{9900} = \frac{1453}{4950}$
5. (a) 0.1772 (c) 0.2935
6. (a) 5 (c) $\frac{9}{999}$ or $\frac{1}{111}$

Exercise 7.8

1. $\frac{1}{2}=0.5$; $\frac{1}{4}=0.25$; $\frac{1}{8}=0.125$; $\frac{1}{16}=0.0625$; $\frac{1}{5}=0.2$; $\frac{2}{5}=0.4$; $\frac{3}{5}=0.6$; $\frac{4}{5}=0.8$; $\frac{5}{5}=1.0$; $\frac{1}{10}=0.1$; etc.
3. (a) 1,006,764 (b) 1,007,578.06 (c) 814.06
5. Use more subdivisions.

Exercise 7.9

1. \$3.33
3. 6
5. 0.1428571429; 0.0909090909

Exercise 7.11

1. $3.14 = \pi \pm 0.01$

3. $\frac{1}{3 \cdot 10^7}$ or $0.0000000\bar{3}\ldots$

Exercise 7.12

1. Yes
3. sum: $2\sqrt{a}$; product: $a - b$; difference: $2\sqrt{b}$
5. 0.7071
7. 0.1165
9. 1.414; 3.162; 5.8284

Exercise 7.12a

1. 22.412
3. 31.6228
5. 0.7087

Exercise 7.12b

1. 1.7320508
3. 54.772256
5. 7.0710678
7. 1.843909

Exercise 8.3a

1. (a) −6 −5 −4 −3 −2 −1 0 1 2 3 4 5 6

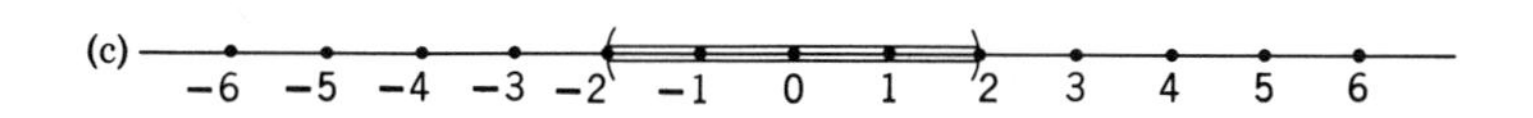

2. (a) open (c) open
3. (a) $7 \leqslant x \leqslant 10$ (c) $7 \leqslant x \leqslant 12$

Exercise 8.7

1. (a) Infinitely many (b) Only one
2. (a) Infinitely many (c) Only one
3. (a) Only one (b) Only one line but infinitely many points

4. 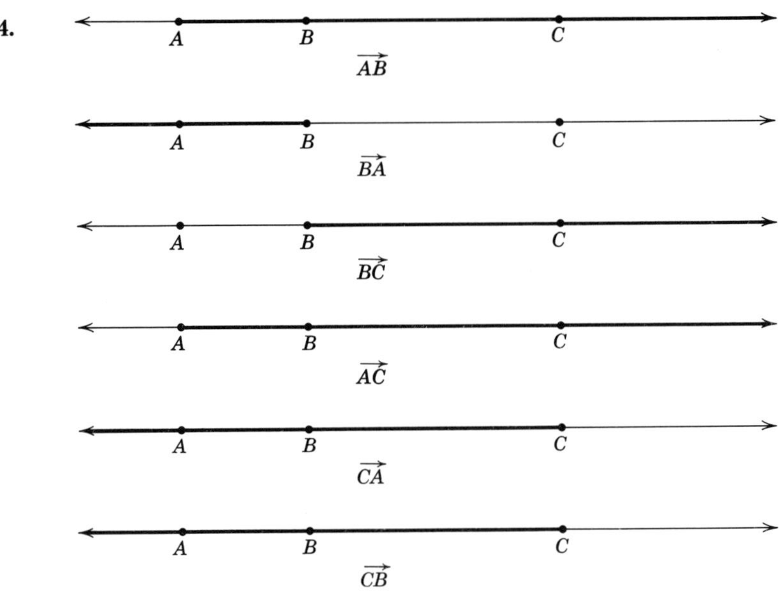

$\overrightarrow{AB} = \overrightarrow{AC}$ and $\overrightarrow{CA} = \overrightarrow{CB}$

5. 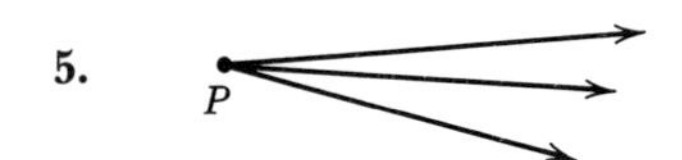

Infinitely many.

6. A B

Only one $\overset{\circ\!-\!\circ}{AB} \subset \overrightarrow{AB}$ Infinitely many.

7. 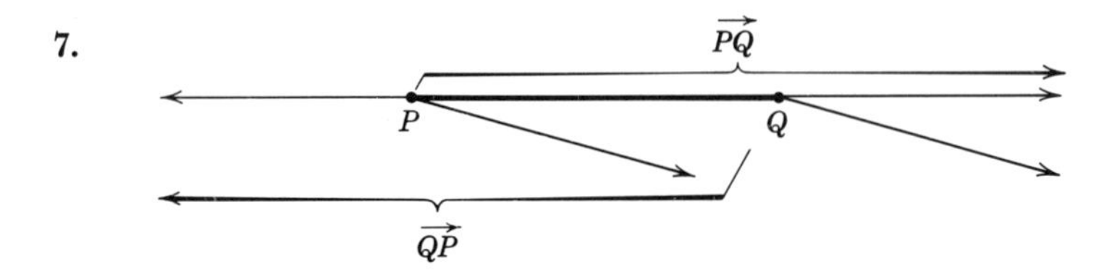

8. (a) None (c) 2
9. (a) 6
10. (a) $\overrightarrow{AC}$ (c) $\overset{\circ\!-\!\circ}{AD}$
11. (a) (c)
12. (a) $\angle BAC$, vertex A, sides $\overrightarrow{AB}$ and $\overrightarrow{AC}$
(c) $\angle XYZ$, vertex Y, sides $\overrightarrow{YX}$ and $\overrightarrow{YZ}$
13. (a) Yes
14. (a) No
15. (a) degrees (c) 37°

Exercise 8.8

1. $l_2 = m(\overline{AC}) + m(\overline{CB})$; $l_4 = m(\overline{AD}) + m(\overline{DC}) + m(\overline{CE}) + m(\overline{EB})$; $m(\overline{AC}) \leq m(\overline{AD}) + m(\overline{DC})$ and $m(\overline{CB}) \leq m(\overline{CE}) + m(\overline{EB})$ by the triangular inequality. Hence $m(\overline{AC}) + m(\overline{CB}) \leq m(\overline{AD}) + m(\overline{DC}) + m(\overline{CE}) + m(\overline{EB})$ and $l_2 \leq l_4$.
2. $l_2 = 2^n \cdot m(\overline{AC})$, where $m(\overline{AC})$ is the length of $\overline{AC}$, one of the equal line segments of the polygonal path of length l_{2^n}. Let B denote the midpoint of $\widehat{AC}$. Then $m(\overline{AC}) \leqslant m(\overline{AB}) + m(\overline{BC})$ by the triangular inequality. Hence $2^n \cdot m(AC) \leqslant 2^n(m(\overline{AB}) + m(\overline{BC}))$. But $l_{2^{n+1}} = 2^{n+1}(m(\overline{AB})) = 2^n[2(ab)] = 2^n(m(\overline{AB}) + m(\overline{BC}))$. Hence $l_{2^n} \leqslant l_{2^{n+1}}$.
3. 4.59 (Compare with arc length of approx. 4.71, using $\frac{22}{7}$ as an approximation to π.)
6. $\dfrac{3\pi}{2}$

Exercise 8.9

1. (a) 10 sq in. (c) 28 sq yd (e) 65 sq ft
2. (a) 12 sq ft (c) 80 sq cm (e) 105 sq cm
3. (a) 15,000 sq ft (c) 11,250 sq ft
5. (a) 20 sq in. (c) 30 sq ft
6. $(1.414)(1.732) \cong 2.449$

Exercise 8.9e

1. 100π
3. 20π
5. 40π
7. same

Exercise 8.10

1. (a) 240 cu in.; 256 sq in. (c) 22.5 cu ft; 54 sq ft
2. (a) 114 sq in.; 84 sq in.; 72 cu in. (c) 96 sq ft; 60 sq ft; 48 cu ft
   (e) 156 sq units; 126 sq units; 84 cu units
3. (a) $\frac{792}{7}$ sq in; $\frac{792}{7}$ cu in. (c) 2464 sq in.; $\frac{34496}{3}$ cu in.

Exercise 8.11

1. 37.5 ft
3. $166\frac{2}{3}$ yds
5. 2
7. ab

Exercise 8.12

1. (a) 2 sq in.
2. (a) $150\sqrt{3}$ sq in.
3. (a) $75\sqrt{3}$ sq in.
4. (a) 160 sq units
5. (a) 162 sq units

Exercise 8.13

3. $b \cong 0.195$; $\pi \cong 16(0.195) = 3.12$
5. Third approximation to π: 3.1056

Exercise 8.14a

1. $J \times J = \{(m, n) | m \text{ and } n \text{ are integers}\}$.

3. (a)

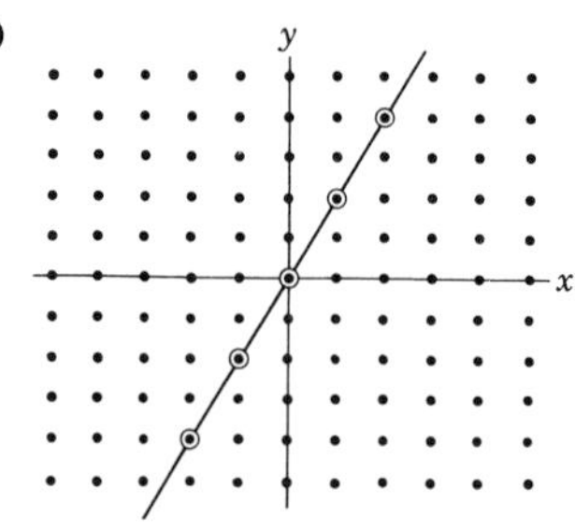

4. (a)

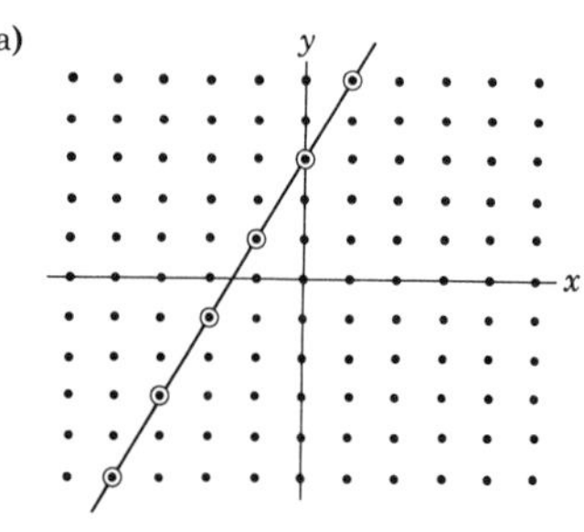

4. (c)

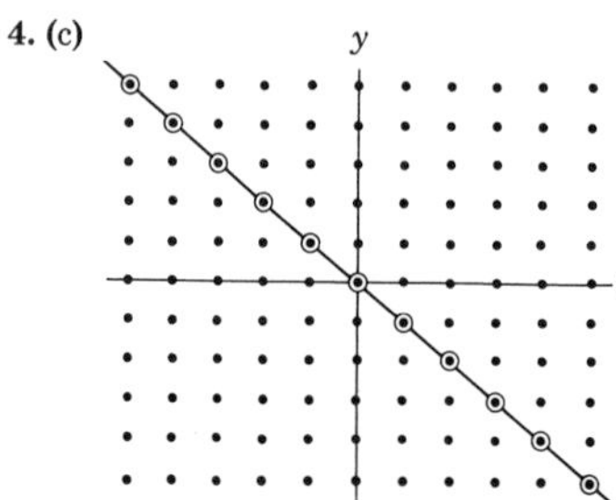

5. (a)

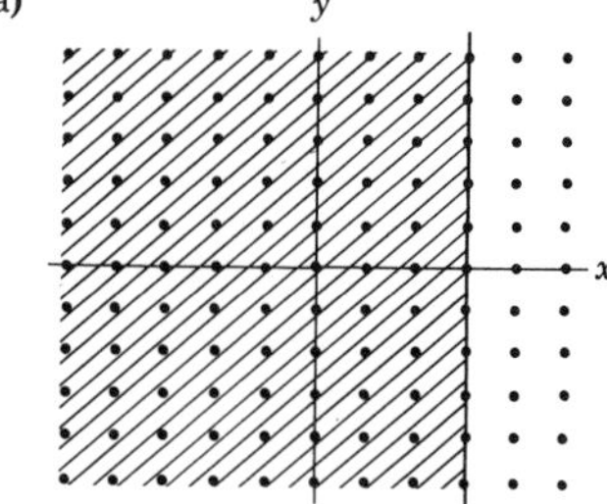

5. (c)

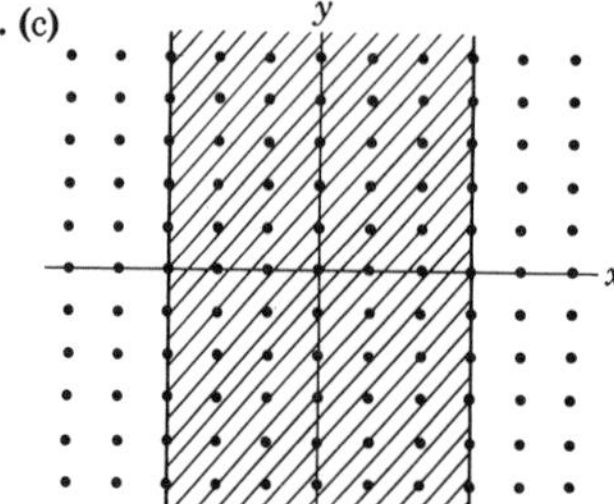

(e)

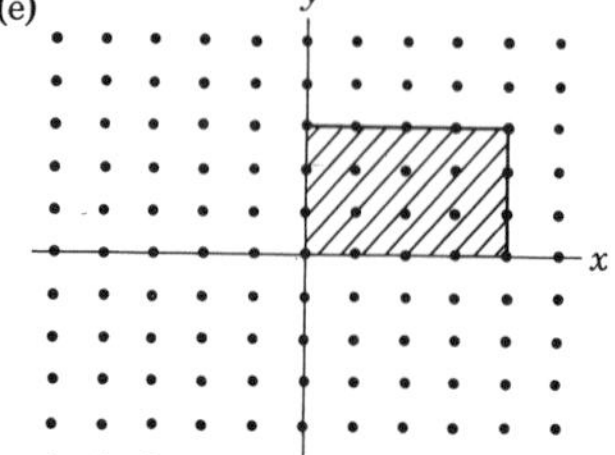

Exercise 8.14b

1. (a) 13 (b) 13 (d) $x^2 + y^2 = 13^2$
2. (a) $x^2 + y^2 = r^2$
3. (a) 7 (b) $|x| + |y|$

Index